# Clinical Chemistry
# in
# Diagnosis and Treatment

## JOAN F. ZILVA

M.D., B.Sc., F.R.C.P., F.R.C.Path., D.C.C. (Biochem.)

*Reader in Chemical Pathology, Westminster
Medical School, London; Honorary Consultant in Chemical
Pathology, Westminster Hospital, London*

## P. R. PANNALL

M.B., B.Ch. (Witwatersrand), F.F.Path. (S.A.), M.R.C.Path.

*Senior Lecturer in Chemical Pathology,
University of the Orange Free State Medical School;
Principal Specialist in Chemical Pathology, Academic Hospitals,
Bloemfontein; Formerly Registrar in Chemical Pathology,
Westminster Hospital, London*

*Second Edition*

LLOYD-LUKE (MEDICAL BOOKS) LTD
49 NEWMAN STREET
LONDON
1975

| | | | | | | |
|---|---|---|---|---|---|---|
| FIRST EDITION | . | . | . | . | . | 1971 |
| Reprinted . | . | . | . | . | . | 1972 |
| Reprinted . | . | . | . | . | . | 1973 |
| *Spanish translation* | | . | . | . | . | 1973 |
| SECOND EDITION | | . | . | . | . | 1975 |

PRINTED AND BOUND IN ENGLAND BY
THE WHITEFRIARS PRESS LTD
LONDON AND TONBRIDGE

ISBN 0 85324 118 X

# CLINICAL CHEMISTRY
## IN
# DIAGNOSIS AND TREATMENT

# FOREWORD

THIS book aims at giving, within a single cover, all the relevant bio-chemical and pathological facts and theories necessary to the intelligent interpretation of the analyses usually performed in departments of Clinical Chemistry or Chemical Pathology. The approach is firmly based on general principles and the authors have gone to great trouble to ensure that the development of ideas is logical and easy to follow. A basic knowledge of medicine and of elementary biochemistry is assumed but, given this, the reader should be able to understand even the more involved metabolic inter-relationships without undue difficulty and should thereafter be in a strong position to apply this knowledge in medical practice.

I have appreciated very much the opportunity of reading this book during its preparation and feel that it represents a distinctly novel approach to the interpretation of biochemical data. In my opinion it could be read with advantage by all the categories of reader mentioned in the preface, and moreover I believe that they will enjoy the experience.

N. F. MACLAGAN

*January, 1971*

# PREFACE TO THE SECOND EDITION

IN the four years since the first edition of this book appeared Chemical Pathology has continued to advance rapidly. This expansion has been greatest in the field of endocrinology and in our knowledge of protein fractions: chapters dealing with these topics have largely been rewritten. SI units are becoming generally accepted, and a short section is included on "Units in Chemical Pathology". Other chapters have been modified because of useful comments from students: for instance, the section on generalised renal disease has been radically modified, and the chapter on electrolytes has been split into two.

Reception has been generally favourable. We have carefully considered all the criticisms, and have tried to answer them while keeping a precarious balance between entering into too much detail for a book of this type, and making misleading statements. We are most grateful to all who have pointed out errors, and hope that these have been corrected, but take full responsibility for all remaining errors.

Two criticisms have been fairly general—that the "further reading lists" were inadequate, and that we omitted reference to "normal" values. We hope that we have rectified the first omission. We have mixed feelings about quoting so-called "normals", because these may vary from laboratory to laboratory: in the past we have suffered because students (and their seniors) have compared results issued by our laboratory with "normal ranges" quoted in other books. We have compromised by including a table listing the mean "normal" values for our laboratories, and leaving a column for students to fill in their own ranges. By this means we hope to give some idea of the values to be expected, without implying that ranges are rigid.

Many people have praised the inclusion of the chapters on "Chemical Pathology and the Clinician" and on "Interpretation of Results". We have expanded the latter to include a discussion of indications for (and for not) investigating. In each chapter we have tried to distinguish between findings which are useful clinically, and those which could be called "incidental": we have stressed that selective investigation, which is in the interest of the patient, requires the application of intelligence and a sound background of physiological knowledge. We feel strongly that one of the biggest dangers to the subject in the future is uncritical investigation, coupled with a poor understanding of factors that can affect results.

Once again we are indebted to more people than we have space to mention. We would particularly like to thank Dr. Philip Nicholson, Dr. Paul Carter, Dr. Peter Frost, Dr. Judith Scurr and Mr. Donald McLauchlan not only for the time and care they have taken in commenting on our typescripts, but also for continuing lively discussion and questioning of preconceived ideas. Professors J. R. Hobbs, G. M. Potgieter, D. M. Matthews, T. H. Bothwell, L. Eales and F. O. Müller have made helpful comments on individual chapters. Miss Valerie Waters, helped by Mrs. Pat Bester, Miss Helen Stuart and Miss Joanne Perry have typed our illegible script with speed, efficiency and accuracy, and Mr. David Gibbons of the Department of Medical Photography and Illustration of the Westminster Medical School, has again prepared the illustrations. We thank the publishers for continuing co-operation.

*January, 1975*

J.F.Z.
P.R.P.

# PREFACE TO THE FIRST EDITION

THIS book is intended primarily for medical students and junior hospital staff. It is based on many years practical experience of undergraduate and postgraduate teaching, and of the problems of a routine chemical pathology department. It is written for those who learn best if they understand what they are learning. Wherever possible, explanations of the facts are given: if the explanation is a working hypothesis (like, for instance, that for the ectopic production of hormones) this is stressed, and where no explanation is known, this is stated. Our experience of teaching and our discussions with students have led us to believe that many of them are willing to read a slightly longer book if it gives them a better understanding than a shorter one. Electrolyte and acid-base balance have been discussed in some detail because, in our experience, these are the most common problems of chemical pathology met with by junior clinicians and ones in which there are often dangerous misunderstandings. Some subjects, such as the porphyrias and conditions of iron overload, are discussed in greater detail than is necessary for undergraduate examinations: however, the incidence of these is high in some areas of the world, and an elementary source of reference seemed to be needed.

We have tried to stress the clinical importance of an understanding of the subject: by including in the chapter appendices some details of treatment, difficult to find together in other books, we hope that this one may appeal to clinicians as well as to students. Two chapters are included on the best use of a laboratory, including precautions which should be taken in collecting specimens and interpreting results. As we have stressed, pathologists and clinicians should work as a team, and full consultation between the two should be the rule. The diagnostic tests suggested are those which we have found, by experience, to be the most valuable. For instance, in the differential diagnosis of hypercalcaemia we find the steroid suppression test very helpful, phosphate excretion indices fallible and tedious to perform and estimation of urinary calcium useless.

Our own junior staff have found the drafts helpful in preparing for the Part 1 examination for Membership of the Royal College of Physicians, as well as for the primary examination for Membership of the Royal College of Pathologists, and our Senior Technicians are using it to study for the Special Examination for Fellowship in Chemical Pathology of the Institute of Medical Laboratory Technology. We feel that those studying for the primary examination for the Fellowship of the Royal College of Surgeons might also make use of it. It could provide a groundwork for study for the final examination for the Membership of the Royal College of Pathologists in Chemical Pathology, and for the Mastership in Clinical Biochemistry.

Initially each chapter should be worked through from beginning to end. For revision purposes there are lists and tables, and summaries of the contents of each chapter. The short sub-indices in the Table of Contents should facilitate the use of the book for reference.

Appendix A lists some analogous facts which we hope may help understanding and learning. The list must be far from complete, and the student should seek other examples for himself.

We wish to thank Professor N. F. Maclagan for his unfailing encouragement and his helpful advice and criticisms. We are also indebted to a great many other people, foremost amongst whom we should mention Dr. J. P. Nicholson who has read and commented on the whole book, and Professor M. D. Milne, Professor D. M. Matthews, Mr. K. B. Cooke and Dr. B. W. Gilliver for helpful advice and criticism on individual chapters. Many registrars and senior house officers in the Department, in particular Dr. Krystyna Rowland, Dr. Elizabeth Small, Dr. Nalini Naik and Dr. Noel Walmsley, have been closely involved in the preparation of the book and have provided invaluable suggestions and criticisms. Students, too, have read individual chapters for comprehensibility, and we would particularly like to thank Mr. B. P. Heather, Mr. J. Muir and Miss H. M. Merriman for helpful comments. Mr. C. P. Butler of the Westminster Hospital Pharmacy was most helpful during the preparation of the sections on therapy. Mrs. Valerie Moorsom and Mrs. Marie-Lise Pannall, with the help of Mrs. Brenda Sarasin and Miss Barbara Bridges, have borne with us during the typing of the drafts and the final transcript. The illustrations were prepared by Mr. David Gibbons of the Department of Medical Photography and Illustration of the Westminster Medical School.

Finally, we would like to thank the publishers for their co-operation and understanding during the preparation of this book.

*May, 1971*

J.F.Z.
P.R.P.

# CONTENTS

| | | |
|---|---|---|
| | FOREWORD | v |
| | PREFACE | vii |
| | UNITS IN CHEMICAL PATHOLOGY | xiii |
| | TABLE OF "NORMAL" VALUES | xvi |
| | ABBREVIATIONS USED IN THIS BOOK | xix |
| I | THE KIDNEYS: RENAL CALCULI | 1 |
| II | SODIUM AND WATER METABOLISM | 30 |
| III | POTASSIUM METABOLISM: DIURETIC THERAPY | 62 |
| IV | HYDROGEN ION HOMEOSTASIS: BLOOD GAS LEVELS | 76 |
| V | PITUITARY AND GONADAL HORMONES | 111 |
| VI | ADRENAL CORTEX: ACTH | 134 |
| VII | THYROID FUNCTION: TSH | 160 |
| VIII | CARBOHYDRATE METABOLISM AND ITS INTERRELATIONSHIPS | 179 |
| IX | PLASMA LIPIDS AND LIPOPROTEINS | 218 |
| X | CALCIUM, PHOSPHATE AND MAGNESIUM METABOLISM | 234 |
| XI | INTESTINAL ABSORPTION: PANCREATIC AND GASTRIC FUNCTION | 263 |
| XII | LIVER DISEASE AND GALL STONES | 289 |
| XIII | PLASMA PROTEINS AND IMMUNOGLOBULINS: PROTEINURIA | 314 |
| XIV | PLASMA ENZYMES IN DIAGNOSIS | 343 |
| XV | INBORN ERRORS OF METABOLISM | 358 |
| XVI | PURINE AND URIC ACID METABOLISM | 379 |
| XVII | IRON METABOLISM | 388 |
| XVIII | THE PORPHYRIAS | 405 |
| XIX | VITAMINS | 416 |

XX    PREGNANCY AND ORAL CONTRACEPTIVE THERAPY    .    431
XXI    BIOCHEMICAL EFFECTS OF TUMOURS    .    .    .    .    438
XXII    THE CEREBROSPINAL FLUID    .    .    .    .    .    452
XXIII    CHEMICAL PATHOLOGY AND THE CLINICIAN    .    .    457
XXIV    REQUESTING TESTS AND INTERPRETING RESULTS    .    469
INDEX    .    .    .    .    .    .    .    .    .    479

# UNITS IN CHEMICAL PATHOLOGY

RESULTS in chemical pathology have been expressed in a variety of units. For instance, electrolytes were usually quoted in mEq/l, protein in g/100 ml and cholesterol in mg/100 ml. The units used might vary from laboratory to laboratory: calcium might be expressed as mg/100 ml or mEq/l; in Britain, urea results were expressed as mg/100 ml of *urea*, whereas in the United States it is usual to report mg/100 ml of *urea nitrogen*. This situation can be confusing and with patients moving, not only from one hospital to another, but from one country to another, dangerous misunderstandings could arise.

## SYSTÈME INTERNATIONAL D'UNITÉS (SI UNITS)

International standardisation is obviously desirable; such standardisation has already been introduced into many branches of science and technology.
The main recommendations for chemical pathology are as follows:
1. Where the molecular weight (MW) of the substance being measured is known, the unit of quantity should be the *mole* or submultiple of a mole.

$$\text{Number of moles (mol)} = \frac{\text{wt. in g}}{\text{MW}}$$

In chemical pathology millimoles (mmol), micromoles ($\mu$mol) and nanomoles (nmol) will be the most common units.
2. The unit of volume should be the *litre*. Units of concentration will therefore be mmol/l, $\mu$mol/l or nmol/l.

### Examples
1. *Results previously expressed as mEq/l*

$$\text{Number of equivalents (Eq)} = \frac{\text{wt. in g}}{\text{Equivalent wt.}}$$

$$= \frac{\text{wt. in g} \times \text{valency}}{\text{MW}}$$

(a) In the case of univalent ions, such as sodium and potassium, the units will be numerically the same. A sodium of 140 mEq/l becomes 140 mmol/l.

(b) For polyvalent ions, such as calcium or magnesium (both divalent), the old units are numerically divided by the valency. For instance, a magnesium of 2·0 mEq/l becomes 1·0 mmol/l.

2. *Results previously expressed as mg/100 ml*
If results were previously expressed in mg/100 ml the method of conversion to mmol/l is to divide by the molecular weight (to convert from mg to mmol),

and to multiply by 10 (to convert from 100 ml to a litre). Thus effectively the previous units are divided by 1/10 of the molecular weight. For instance, the molecular weight of urea is 60, and of glucose 180. A urea value of 60 mg/100 ml and a glucose value of 180 mg/100 ml are both equivalent to 10 mmol/l.

The factor of 10 is, of course, only used for concentrations. The total amount of urea excreted in 24 hours in mg is numerically 60 times that in mmol.

### Exceptions

1. *Units of pressure* (e.g. mm Hg) are expressed as pascals (or kilopascals— kPa). 1 kPa = 7·5 mm Hg, so that a $Po_2$ of 75 mm Hg is 10 kPa. Pascals are SI units.

2. *Proteins.* Body fluids contain a complex mixture of proteins of varying molecular weights. It is therefore recommended that the gram (g) be retained, but that the unit of volume be the litre. Thus a total protein of 7·0 g/100 ml becomes 70 g/l.

3. The expression 100 ml is to be expressed as *decilitre* (dl).

4. *Enzyme units* are not to be changed yet. Note that *the definition of international units for enzymes does not state the temperature of the reaction* (p. 345).

5. Some constituents, such as IgE and some hormones, are still expressed in "international" or other special units.

At the moment different laboratories are at different stages of implementation. In this book we have adopted the following policy.

1. Where the old and new units are numerically the same, we have given only the new units (e.g. for sodium and potassium).

2. For proteins we give only g/l.

3. Where it is generally accepted that the new units be adopted we have given these, with the equivalent old units in brackets.

A conversion table for some of the commoner results is given opposite.

Note that:

$$1 \text{ mol} = 1\,000 \text{ mmol}$$
$$1 \text{ mmol} (10^{-3} \text{ mol}) = 1\,000 \text{ } \mu\text{mol}$$
$$1 \text{ } \mu\text{mol} (10^{-6} \text{ mol}) = 1\,000 \text{ nmol (nanomoles)}$$
$$1 \text{ nmol} (10^{-9} \text{ mol}) = 1\,000 \text{ pmol (picomoles)}.$$

## FURTHER READING

BARON, D. N., BROUGHTON, P. M. G., COHEN, M., LANSLEY, T. S., LEWIS, S. M., and SHINTON, N. K. (1974). *J. clin. Path.*, **27**, 590.

BARON, D. N. (1974). *Brit. med. J.*, **4**, 509.

BOLD, A. M., and WILDING, P. (1975). *Clinical Chemistry Conversion Scales for SI units with Adult Normal (Reference) Values.* Oxford: Blackwell Scientific Publications.

## SOME APPROXIMATE CONVERSION FACTORS FOR SI UNITS

| | From SI Units | To SI Units |
|---|---|---|
| Bilirubin | $\mu$mol/l $\times$ 0·058 = mg/dl | mg/dl $\div$ 0·058 = $\mu$mol/l |
| Calcium | | |
|   Plasma | mmol/l $\times$ 4 = mg/dl | mg/dl $\div$ 4 = mmol/l |
|   Urine | mmol/24 h $\times$ 40 = mg/24 h | mg/24 h $\div$ 40 = mmol/24 h |
| Cholesterol | mmol/l $\times$ 39 = mg/dl | mg/dl $\div$ 39 = mmol/l |
| Cortisol | | |
|   Plasma | nmol/l $\times$ 0·036 = $\mu$g/dl | $\mu$g/dl $\div$ 0·036 = nmol/l |
|   Urine | nmol/24 h $\times$ 0·36 = $\mu$g/24 h | $\mu$g/24 h $\div$ 0·36 = nmol/24 h |
| Creatinine | | |
|   Plasma | $\mu$mol/l $\times$ 0·011 = mg/dl | mg/dl $\div$ 0·011 = $\mu$mol/l |
|   Urine | $\mu$mol/24 h $\times$ 0·11 = mg/24 h | mg/24 h $\div$ 0·11 = $\mu$mol/24 h |
| Gases | | |
|   Po$_2$   Pco$_2$ | kPa $\times$ 7·5 = mm Hg | mm Hg $\div$ 7·5 = kPa |
| Glucose | mmol/l $\times$ 18 = mg/dl | mg/dl $\div$ 18 = mmol/l |
| Iron   TIBC | $\mu$mol/l $\times$ 5·6 = $\mu$g/dl | $\mu$g/dl $\div$ 5·6 = $\mu$mol/l |
| Phosphorus | mmol/l $\times$ 3 = mg/dl | mg/dl $\div$ 3 = mmol/l |
| Proteins | | |
|   Serum | | |
|     Total     Albumin     Immuno-      globulins | g/l $\div$ 10 = g/dl | g/dl $\times$ 10 = g/l |
|   Urine | | |
|     Concentration | g/l $\times$ 100 = mg/dl | mg/dl $\div$ 100 = g/l |
|     Daily Output | g/24 h | No change |
| Urate | mmol/l $\times$ 17 = mg/dl | mg/dl $\div$ 17 = mmol/l |
| Urea | mmol/l $\times$ 6 = mg/dl | mg/dl $\div$ 6 = mmol/l |
| 5-HIAA   HMMA | $\mu$mol/24 h $\times$ 0·2 = mg/24 h | mg/24 h $\div$ 0·2 = $\mu$mol/24 h |
| Oestriol   17 Oxosteroids   Total 17-Oxogenic     steroids | $\mu$mol/24 h $\times$ 0·3 = mg/24 h | mg/24 h $\div$ 0·3 = $\mu$mol/24 h |
| Faecal " Fat " | mmol/24 h $\times$ 0·3 = g/24 h | g/24 h $\div$ 0·3 = mmol/24 h |

## "NORMAL" VALUES

The authors feel strongly that each laboratory should issue its own list of "normal" values, and that clinicians should consult that list, rather than a textbook, when interpreting results. However, so that students should have some idea of the order of magnitude of "normal" values, this list gives mean normals for the authors' laboratories. The student should fill in the blank column with the values of his own laboratory. Investigations marked with an asterisk indicate those most likely to have different values in different laboratories: this especially applies to enzyme values.

| Investigation PLASMA, SERUM or BLOOD (Plasma unless otherwise stated) | Approximate mean adult normal values for authors' laboratories | | "Normal Range" for student's laboratory (fill in) |
|---|---|---|---|
| | "Old" Units | SI or other "New" Units | |
| *Amylase | Less than 200 Somogyi units | — | |
| Bilirubin | Less than 1 mg/dl | Less than 17 $\mu$mol/l | |
| Calcium | 9 mg/dl | 2·25 mmol/l | |
| Cholesterol | 200 mg/dl | 5·2 mmol/l | |
| *Cholinesterase | 350 mU/dl | — | |
| "Cortisol"  9 a.m. | 16 $\mu$g/dl | 440 nmol/l | |
| Midnight | 4 $\mu$g/dl | 110 nmol/l | |
| *Creatine Kinase | Up to 50 U/l | — | |
| Creatinine | 1·0 mg/dl | 88 $\mu$mol/l | |
| Electrolytes  Sodium | 140 mEq/l | 140 mmol/l | |
| Potassium | 4 mEq/l | 4 mmol/l | |
| Bicarbonate (standard and actual) | 24 mEq/l | 24 mmol/l | |
| Chloride | 100 mEq/l | 100 mmol/l | |
| Free Fatty Acids (FFA; NEFA) | 0·5 mEq/l | 500 $\mu$mol/l | |
| Gases and pH (whole blood)  $Po_2$ | 95 mmHg | 12·6 kPa | |
| pH | 7·4 | — | |
| $Pco_2$ | 40 mmHg | 5·3 kPa | |
| Glucose (whole blood) Fasting | 95 mg/dl | 5·3 mmol/l | |
| *HBD | 125 U/l at 35°C | — | |
| | 90 U/l at 25°C | — | |

| | | | |
|---|---|---|---|
| Iron | male | 120 µg/dl | 21·5 µmol/l |
| | female | 80 µg/dl | 14·3 µmol/l |
| Iron Binding Capacity (Total) | | 300 µg/dl | 54 µmol/l |
| Lipids | | | |
| Cholesterol (Total) | | 200 mg/dl | 5·2 mmol/l |
| *Phospholipids | | 8 mg/dl (as P) | 2·5 mmol/l |
| *Triglycerides (fasting) | | 88 mg/dl (as triolein) | 1·0 mmol/l |
| Magnesium | | 1·6 mEq/l | 0·8 mmol/l |
| *5'-Nucleotidase | | 10 U/l | — |
| *Phosphatases Acid (Tartrate labile) | Up to | 0·9 KA Units / 1·6 U/l at 37°C | — |
| Alkaline | Up to | 10 KA Units / 71 U/l at 37°C | — |
| *Phosphate | | 4 mg/dl (as P) | 1·3 mmol/l |
| *Proteins (Serum) | | | |
| Total | | 6·0 g/dl (about 0·3 g/dl higher for plasma) | 60 g/l (about 3 g/l higher for plasma) |
| Albumin | | 4·0 g/dl | 40 g/l |
| Protein bound iodine | | 6·0 µg/dl | 472 nmol/l |
| Thyroxine (T₄) and indices | | | |
| $T_4$ | | 8 µg/dl | 103 nmol/l |
| Resin uptake ratio | | 0·95 | — |
| Free thyroxine index | | 8 µg/dl | 103 nmol/l |
| *Transaminases ALT (SGPT) | | 12 U/l at 35 °C | — |
| AST (SGOT) | | 9 U/l at 25°C / 15 U/l at 35°C / 12 U/l at 25°C | — |
| Urate | male | 5·5 mg/dl | 0·33 mmol/l |
| | female | 4·5 mg/dl | 0·27 mmol/l |
| Urea | | 20 mg/dl | 3·3 mmol/l |

| Investigation URINE or FAECES | | Approximate mean adult normal values for authors' laboratories | | "Normal Range" for student's laboratory (fill in) |
|---|---|---|---|---|
| | | "Old" Units | SI or other "New" Units | |
| **URINE** | | | | |
| 5-HIAA | | 6 mg/24 h | 31 µmol/24 h | |
| HMMA (VMA) | | 4 mg/24 h | 20 µmol/24 h | |
| Steroids | | | | |
| 17 Oxosteroids | } male | 13 mg/24 h | 45 µmol/24 h | |
| Total 17 Oxogenic steroids | female | 10 mg/24 h | 35 µmol/24 h | |
| "Cortisol" | male | 250 µg/24 h | 690 nmol/24 h | |
| | female | 200 µg/24 h | 550 nmol/24 h | |
| **FAECES** | | | | |
| Fat (collected over 5-day period) | | Up to 5 g/24 h (as stearic acid) | Up to 18 mmol/24 h (as fatty acid) | |

| | |
|---|---|
| ACP | Acid Phosphatase |
| ACTH | Adrenocorticotrophic Hormone |
| ADH | Antidiuretic Hormone ("Pitressin": Vasopressin) |
| ALA | $\delta$ Aminolaevulinic Acid |
| ALP | Alkaline Phosphatase |
| ALS | Aldolase |
| ALT | Alanine Transaminase (= SGPT) |
| AMS | $\alpha$-Amylase |
| AST | Aspartate Transaminase (= SGOT) |
| BEI | Butanol Extractable Iodine |
| BJP | Bence Jones Protein |
| BMR | Basal Metabolic Rate |
| BSP | Bromsulphthalein |
| BUN | Blood Urea Nitrogen $\left(\dfrac{28}{60} \times \text{Blood urea in mg/dl}\right)$ |
| CBG | Cortisol Binding Globulin (Transcortin) |
| CC | Cholecalciferol |
| CK | Creatine Kinase (= CPK) |
| CoA | Coenzyme A |
| CPK | Creatine Phosphokinase (= CK) |
| CRF | Corticotrophin Releasing Factor |
| CSF | Cerebrospinal Fluid |
| 1:25-DHCC | 1:25-Dihydroxycholecalciferol |
| DIT | Di-iodotyrosine |
| DNA | Deoxyribonucleic Acid |
| DOC | Deoxycorticosterone |
| DOPA | Dihydroxyphenylalanine |
| DOPamine | Dihydroxyphenylethylamine |
| DPN | Diphosphopyridine Nucleotide (= NAD) |
| ECF | Extracellular Fluid |
| EDTA | Ethylene Diamine Tetra-acetate (Sequestrene) |
| EM Pathway | Embden-Meyerhof Pathway (Glycolytic Pathway) |
| ESR | Erythrocyte Sedimentation Rate |
| FAD | Flavin Adenine Dinucleotide |
| FBS | Fasting Blood Sugar |
| FFA | Free Fatty Acids (= NEFA) |
| FMN | Flavin Mononucleotide |
| FSH | Follicle Stimulating Hormone |
| FTI | Free Thyroxine Index |
| GFR | Glomerular Filtration Rate |
| GGT | $\gamma$ Glutamyltransferase ($\gamma$ Glutamyltranspeptidase) |
| GH | Growth Hormone |

| | |
|---|---|
| GMD | Glutamate Dehydrogenase |
| GOT | Glutamate Oxaloacetate Transaminase (= AST) |
| G-6-P | Glucose-6-Phosphate |
| GPD | Glucose-6-Phosphate Dehydrogenase |
| GPT | Glutamate Pyruvate Transaminase (= ALT) |
| GTT | Glucose Tolerance Test |
| HAA | Hepatitis Associated Antigen (= Australia Antigen) |
| HBD | Hydroxybutyrate Dehydrogenase |
| 25-HCC | 25-Hydroxycholecalciferol |
| HCG | Human Chorionic Gonadotrophin |
| HCS | Human Chorionic Somatomammotrophin (= HPL) |
| HDL | High Density Lipoprotein |
| HGH | Human Growth Hormone |
| HGPRT | Hypoxanthine Guanine Phosphoribosyl Transferase |
| 5HIAA | 5-Hydroxyindole Acetic Acid |
| HMMA | 4-Hydroxy-3-Methoxymandelic Acid (= VMA) |
| HPL | Human Placental Lactogen (= HCS) |
| 5HT | 5-Hydroxytryptamine (= Serotonin) |
| 5HTP | 5-Hydroxytryptophan |
| ICD | Isocitrate Dehydrogenase |
| ICF | Intracellular Fluid |
| ICSH | Interstitial Cell Stimulating Hormone (= LH) |
| Ig | Immunoglobulin |
| LATS | Long Acting Thyroid Stimulator |
| LCAT | Lecithin Cholesterol Acyl Transferase |
| LD | Lactate Dehydrogenase (= LDH) |
| LDH | Lactate Dehydrogenase (= LD) |
| LDL | Low Density Lipoprotein |
| LH | Luteinising Hormone (= ICSH) |
| LH-RH | LH Releasing Hormone |
| MIT | Mono-iodotyrosine |
| MSH | Melanocyte Stimulating Hormone |
| NAD | Nicotinamide Adenine Dinucleotide (= DPN) |
| NADP | Nicotinamide Adenine Dinucleotide Phosphate (= TPN) |
| NEFA | Non-Esterified Fatty Acids (= FFA) |
| 5'NT | 5' Nucleotidase (= NTP) |
| NTP | 5' Nucleotidase (= 5'NT) |
| 17-OGS | 17-Oxogenic Steroids (T-17 OGS) |
| 11-OHCS | 11-Hydroxycorticosteroids ("Cortisol") |
| 17-OHCS | 17-Hydroxycorticosteroids (17 oxogenic steroids) |
| OP | Osmotic Pressure |
| PBG | Porphobilinogen |
| PBI | Protein Bound Iodine |
| PHLA | Post-Heparin Lipolytic Activity |
| PIF | Prolactin-Release Inhibiting Factor |
| PP factor | Pellagra Preventive Factor (nicotinamide: niacin) |
| PRPP | Phosphoribosyl Pyrophosphate |
| PTH | Parathyroid Hormone (Parathormone) |

| | |
|---|---|
| RF | Releasing Factor (= RH) |
| RH | Releasing Hormone (= RF) |
| RNA | Ribonucleic Acid |
| RU | Resin Uptake (of $T_3$ or $T_4$) |
| SG | Specific Gravity |
| SGOT | Serum Glutamate Oxaloacetate Transaminase (= AST) |
| SGPT | Serum Glutamate Pyruvate Transaminase (= ALT) |
| SHBD | Serum Hydroxybutyrate Dehydrogenase (= HBD) |
| $T_3$ | Tri-iodothyronine |
| $T_4$ | Thyroxine (Tetra-iodothyronine) |
| TBG | Thyroxine Binding Globulin |
| TBPA | Thyroxine Binding Prealbumin |
| TBW | Total Body Water |
| TCA cycle | Tricarboxylic Acid Cycle (= Krebs' Cycle or Citric Acid Cycle) |
| TIBC | Total Iron Binding Capacity (usually measure of transferrin (siderophilin)) |
| T-17-OGS | Total 17-Oxogenic Steroids (17-OGS) |
| TP | Total Protein |
| TPN | Triphosphopyridine Nucleotide (= NADP) |
| TPP | Thiamine Pyrophosphate |
| TRF | Thyrotrophin Releasing Factor |
| TRH | Thyrotrophin Releasing Hormone |
| TSH | Thyroid Stimulating Hormone (Thyrotrophin) |
| UDP | Uridine Diphosphate |
| UTP | Uridine Triphosphate |
| VLDL | Very Low Density Lipoprotein |
| VMA | Vanillyl Mandelic Acid (= HMMA) |
| Z–E syndrome | Zollinger-Ellison Syndrome |

# Chapter I

# THE KIDNEYS : RENAL CALCULI

## THE KIDNEYS

THE kidneys excrete waste products of metabolism and are the most important organs in the maintenance of normal body homeostasis. The renal tubules reabsorb some metabolically important substances, such as glucose, from the glomerular filtrate, secrete other substances into it, and exchange ions across the cell wall (for instance potassium and hydrogen in exchange for sodium ions). It should be remembered that, although the renal tubular cells are of the greatest importance from the point of view of homeostasis, many of these processes occur in other cells of the body, including those of the intestinal mucosa. The normal functioning of these cells in the kidney depends on:

1. An adequate volume of glomerular filtrate with which the exchanges can occur (and therefore on normal glomerular function).

2. Concentrations of ions in the tubular cells representative of those in the body as a whole (for example, potassium and hydrogen ions).

3. The presence of such hormones as antidiuretic hormone (ADH) and aldosterone.

4. The integrity of the feed-back mechanisms controlling the hormones.

5. Functionally intact renal cells.

If all these factors are normal the kidney retains just as much of each constituent as the body requires.

Other functions of the kidney, which will not be dealt with further in this chapter, are:

1. The production of erythropoietin—a hormone stimulating erythropoiesis. The student is referred to textbooks of haematology for further details.

2. The production of renin (p. 37).

3. The conversion of 25-hydroxycholecalciferol to the active 1:25 dihydroxycholecalciferol (p. 237).

### NORMAL RENAL FUNCTION

**Glomerular Filtration**

Glomerular filtration is a passive process. Protein and protein-bound plasma constituents are filtered in negligible amounts by the normal glomerulus and most of the small amount of protein that is filtered is probably reabsorbed. Diffusible plasma constituents such as sodium, potassium and *ionised* calcium are present at concentrations almost identical with those in extracellular fluid. Changes in the blood supply to the glomerulus, or reduction of the permeability of the membrane, affect the *volume, but not the composition,* of the filtrate. Increased permeability of the glomerulus may lead to proteinuria.

## Tubular Function

The glomerular filtrate is modified in two ways. Many substances are dealt with actively by the tubular cells, while others are reabsorbed passively.

*Passive transport* can be explained by the presence of physicochemical gradients. For instance, a hydrostatic pressure gradient accounts for glomerular filtration and in the tubules water will move passively from an area of relatively low to relatively high osmotic pressure, or solute will move in the opposite direction: ions can also move along an electrochemical gradient produced by reabsorption of charged ions (for instance, active reabsorption of cation may be accompanied by passive reabsorption of anion). For such movements to take place the cell wall must be permeable to the substances concerned, and selective permeability may explain preferential absorption of one ion or another. Passive reabsorption requires no metabolic energy. Cell death may affect passive reabsorption by alteration of permeability of the cell walls.

*Active transport* can occur against physicochemical gradients and this process requires energy supplied by oxygen and ATP, and is directly affected by cell death and enzyme poisons. Such types of exchange occur in all cells in the body (for instance the "sodium pump" p. 64): in absorptive cells, such as those of the renal tubule and the intestinal mucosa, the process is one of passing substances through the cell from the lumen into the blood stream; in other cells transport is in the same direction on all sides of the cell, and substances pass in or out of the cell rather than through it.

Many substances, such as urea and hydrogen ion, can reach concentrations in the urine well above those in the blood. Whether this differential concentration is the result of reabsorption of water without the relevant solute (urea) or of secretion by the tubular cell (hydrogen ion), its maintenance depends on the relative impermeability of that part of the tubule distal to the site of concentration. If this breaks down high concentrations may not be reached.

We will now consider some of the more important urinary constituents dealt with by the renal tubular cells.

**Glucose and amino acids** are normally almost completely reabsorbed by the proximal renal tubule, and fluid entering the descending limb of

the loop of Henle is usually free of these constituents, which can then be
reutilised in body metabolism. Glucose is a "threshold" substance: the
tubular cells have a limited capacity to reabsorb it, and at concentrations
in the filtrate of above approximately 10 mmol/1 (180 mg/dl) glucose is
usually present in the urine (but see p. 189). The maximum amount
which the tubules can completely reabsorb (Tm) may be reduced due
to tubular factors (see also p. 14).

Phosphate is also reabsorbed by an active process in the proximal
tubule, but reabsorption is rarely complete. This reabsorption is
inhibited by parathyroid hormone (PTH), and this action accounts for
some of the changes in the plasma when PTH is circulating in excess
(p. 236). "Vitamin D" also has some inhibitory effect on phosphate
reabsorption. The presence of phosphate in the urine provides some of
its buffering power (p. 87).

Calcium and magnesium.—The reabsorption of calcium and magnesium is poorly understood. It is probably increased by parathyroid
hormone, and these ions may compete with sodium.

Urate is probably normally completely reabsorbed in the proximal
tubule. Normal urinary urate is derived from active tubular secretion.

Sodium.—Sodium can be reabsorbed by three mechanisms:

(a) *Isosmotic reabsorption.*—About 70 per cent of the sodium in the
glomerular filtrate is reabsorbed by an active process in the proximal
tubule. This reabsorption is limited by the availability of chloride (see
below).

(b) *Exchange with hydrogen ion.*—Sodium reabsorption in exchange
for hydrogen ion is linked with bicarbonate reabsorption, and is dependent on the enzyme carbonic anhydrase, present in cells throughout
the renal tubule (see p. 85 for more detailed discussion).

(c) *Exchange with potassium ion.*—Sodium is reabsorbed in exchange
for potassium ion in the distal tubule. This exchange is stimulated by
aldosterone.

Chloride.—Although an active mechanism for chloride reabsorption
has been suggested, probably most of the chloride in the glomerular
filtrate is reabsorbed passively in the proximal tubule along the electrochemical gradient created by sodium reabsorption. Sodium cannot be
reabsorbed isosmotically (mechanism (a)) without an accompanying
anion, because the resulting electrochemical gradient would soon halt
the process. Reabsorption of sodium is limited by the availability of
chloride, the most abundant anion in the glomerular filtrate (p. 90).

Potassium.—Urinary potassium can be affected in two ways:

(a) *Active reabsorption.*—Potassium is almost completely reabsorbed
by an active process in the proximal tubule, and fluid entering the distal
tubule contains little potassium.

(b) *Exchange with sodium ion.*—In the distal tubule potassium is secreted in exchange for sodium, and this process is stimulated by aldosterone. Hydrogen and potassium ions compete for this exchange.

**Hydrogen ion.**—Hydrogen ion is secreted throughout the tubule in exchange for sodium. Final adjustment takes place in the distal tubule. As potassium competes for this exchange, disturbances of hydrogen ion homeostasis can be initiated by abnormalities of potassium metabolism (p. 64).

**Bicarbonate.**—Bicarbonate is recovered from the glomerular filtrate although the tubular cells are impermeable to it. It is converted to carbon dioxide when hydrogen ion is secreted into the urine, and the carbon dioxide diffuses passively into the tubular cell where it is reconverted to bicarbonate in the presence of carbonic anhydrase and returned to the blood stream (p. 85).

**Urea.**—Urea diffuses passively into the blood from the proximal tubule as water reabsorption increases its concentration in the filtrate. However, this diffusion is limited, especially in the distal tubule. Urinary urea can reach concentrations well above those in blood.

**Creatinine.**—Creatinine, in man, is not reabsorbed but is secreted by the renal tubule in small amounts.

**Water.**—Water is always reabsorbed *passively* along an osmotic gradient. However, *active* solute transport is necessary to produce this gradient. There are two main processes involved in water reabsorption:

(a) Isosmotic reabsorption of water in the proximal tubule.

(b) Differential reabsorption of water and solute in the loop of Henle, distal tubule and collecting ducts.

(a) *Isosmotic reabsorption of water in the proximal tubule.*—Approximately 200 litres of water are filtered daily by the glomeruli, and yet only about 2 litres of urine enter the bladder. Therefore the nephron as a whole reabsorbs 99 per cent of the filtered water, about 70 to 80 per cent (that is 140 to 160 litres a day) being returned to the body by the proximal tubules.

The proximal tubules pass through the renal cortex and their walls are freely permeable to water. Active reabsorption of solutes such as sodium and glucose from the glomerular filtrate is accompanied by passive reabsorption of an osmotically equivalent amount of water. Blood flow is brisk in this area and solute and water are removed rapidly. As water and solute reabsorption are almost concurrent, fluid entering the loop of Henle, though much reduced in volume, is still almost isosmotic and this process cannot adjust extracellular osmolality; it merely reclaims the bulk of filtered water and solute.

(b) *Differential reabsorption of water and solute in the loop of Henle, distal tubule and collecting duct.*—Normally between 40 and 60 litres of water a day enter the loops of Henle. Not only is this volume further reduced to about 2 litres, but, if changes in extracellular osmolality are to be corrected, the proportion of water reabsorbed must be varied according to the body's needs. With extremes of water intake urinary osmolality can vary from about 40 to about 1400 mosmol/kg. (These figures should be compared with the normal for plasma, and therefore for glomerular filtrate, of about 290 mosmol/kg.) It will be seen that the proportion of solute to water reabsorption must be capable of varying by a factor of approximately 35. As the proximal tubule cannot dissociate water and solute reabsorption, this must occur between the end of the proximal tubule and the end of the collecting duct.

It is generally agreed that two mechanisms are involved, although it must be stressed that there are differences of opinion as to the exact details.

i. *Countercurrent multiplication,* an *active* process occurring probably in the *loop of Henle,* whereby high medullary osmolality is created, and urinary osmolality is reduced. This acts by itself in the absence of ADH, and a dilute urine is produced.

ii. *Countercurrent exchange,* a *passive* process, only occurring in the *presence of ADH,* whereby water without solute is reabsorbed from the *distal tubules and collecting ducts* into the *ascending vasa recta* along the osmotic gradient created by multiplication; by this means the urine is concentrated and the plasma diluted.

*Countercurrent multiplication.*—The most generally held theory considers that this occurs in the loops of Henle, sodium being actively pumped from the ascending to the descending limb while fluid is flowing through the loop.

Fluid entering the descending limb from the proximal tubule is almost isosmotic—that is, is of the same osmolality as that in the general circulation. This is normally a little under 300 mosmol/kg, and for ease of discussion we will use the figure 300 mosmol/kg.

Suppose that the loop has been filled, no pumping has taken place, and the fluid in the loop is stationary. Osmolality throughout the loop and the adjacent medullary tissue will be at about 300 mosmol/kg.

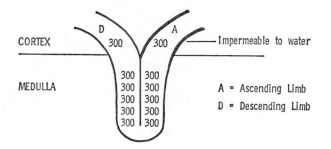

Suppose 1 mosmol of solute per kg is pumped from the ascending limb (A) into the descending limb (D), the fluid column remaining stationary.

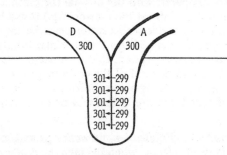

If this pumping were continued and there were no flow limb D would become very hyperosmolal and limb A equally hypo-osmolal.

Let us now suppose that the fluid flows so that each figure "moves two places".

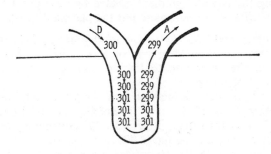

As this happens more solute is pumped from limb A to limb D.

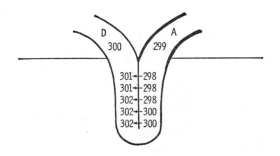

If the fluid again flows "two places", then the situation will be:

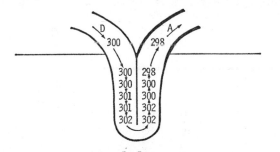

If these steps occur simultaneously and continuously, the wall of the *distal part* of limb A being *impermeable to water*, the consequences, if blood flow is not rapid would be:

(i) Increasing osmolality in the tips of the loops. As the *walls of the loops are permeable* to water and solute, osmotic equilibrium would be reached with all the surrounding tissues and the deeper layers of the medulla including the blood in the vasa recta, which will also be of increasing osmolality.

(ii) Hypo-osmolal fluid leaving the ascending limb.

The final result might be:

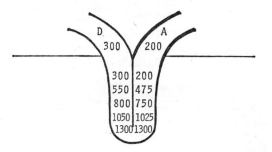

In the absence of ADH the walls of the distal tubules and collecting ducts are impermeable to water, no further change in osmolality occurs, and hypo-osmolal urine would be passed.

*Countercurrent exchange* is essential, *together with multiplication*, for *concentration of urine*. It can only occur in the presence of ADH, and depends on the "random" apposition of collecting ducts and ascending vasa recta, a result of the close anatomical relations of *all* medullary constituents (Fig. 1) (apposition to descending vasa recta will also occur, but this will have little effect on urinary osmolality). The action

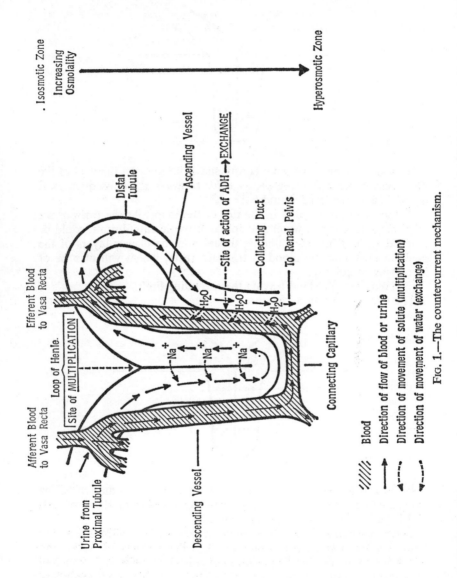

Fig. 1.—The countercurrent mechanism.

of ADH makes the walls of the distal part of the tubule and the collecting ducts permeable to water, which then passes along the osmotic gradient created by multiplication; urine is thus concentrated as the collecting ducts pass into the increasingly hyperosmolal medulla. The increasing concentration of the fluid as it passes down the ducts would reduce the osmotic gradient if it did not meet even more concentrated blood flowing in the opposite (countercurrent) direction. The gradient is thus maintained, and water can continue to be reabsorbed until urine reaches the osmolality of the deepest layers (four or five times that of plasma). The diluted blood is carried towards the cortex and soon enters the general circulation, thus tending to reduce plasma osmolality.

Whether this is the exact mechanism or not, both concentration and dilution of urine depend on active processes, which may be deranged if tubules are damaged.

Let us now look at the process in more detail as it works at extremes of water intake, bearing in mind that the tendency to changes in *plasma* osmolality is normally rapidly corrected.

*Water load.*—A high water intake dilutes the extracellular fluid and the fall in osmolality *cuts off ADH* production (p. 39). As the walls of the collecting ducts are then impermeable to water, *countercurrent multiplication* is acting *alone*, and as we have seen, a dilute urine will be produced. However, plasma osmolality would not be corrected unless the hyperosmolality created by multiplication could be carried into the general circulation.

It has been found that during a maximal water diuresis the osmolality at the tips of the papillae may only reach about 600 mosmol/kg, rather than the maximum of about 1400 mosmol/kg. Since increasing the circulating volume increases renal blood flow, the *more rapid flow in the vasa recta* may tend to "wash out" medullary hyperosmolality, returning some of the solute, without extra water, into the circulation. Thus, not only is more water than normal lost in the urine, but more solute is also "reclaimed" into the general circulation.

*Water restriction.*—Water restriction, by increasing plasma osmolality, *leads to ADH production* and allows countercurrent exchange. Reduced circulating volume results in *sluggish flow in the vasa recta*, allowing build up of the medullary hyperosmolality produced by multiplication and therefore increasing this exchange.

*Osmotic diuresis.*—Under physiological circumstances by far the largest contribution to the osmolality of the glomerular filtrate comes from sodium salts, and active removal of sodium in the proximal tubule is followed by passive reabsorption of water. Suppose that another osmotically active substance is circulating in significant amounts, that this substance is freely filterable at the glomerulus, but that it cannot be reabsorbed either actively or passively to any significant extent in the proximal tubule. *Mannitol,* which cannot diffuse significantly through cell walls, is such a substance. As water reabsorption takes place with that of sodium the mannitol will become more concentrated, and its osmotic effect will inhibit further water diffusion. Since less water is reabsorbed a larger volume than usual will enter the loop of Henle and flow through the rest of the tubular system. Water reabsorption in the distal tubule and collecting duct has a limited capacity, and water diuresis will result. Moreover, as some sodium reabsorption occurs without water, the sodium concentration falls in the fluid in the distal part of the proximal tubule: back diffusion of sodium from blood along this concentration gradient opposes the action of the sodium pump. Sodium reabsorption is therefore impaired, although to a lesser extent than that of water. Note that urine leaving the proximal tubule is still isosmotic with blood, but that the contribution to the osmotic pressure of the tubular fluid from sodium is less than that in blood; the difference is made up by mannitol (in the absence of osmotic diuretics the sodium concentrations are the same in the proximal tubular lumen and in the blood). *Urea* can diffuse back to a limited extent in the proximal tubule, and *glucose* can be actively reabsorbed up to a "threshold" value. However, if these substances are filtered at very high concentrations the reabsorptive capacity is exceeded and they then act as osmotic diuretics (see p. 39).

Summarising the functions of the kidney:

1. Excretion of most substances depends initially on normal permeability of the *glomerulus* (cortex), and on a hydrostatic pressure gradient between the blood flowing through it and the lumen of the glomerulus. If either of these factors is significantly reduced the tubules may not receive adequate amounts of material for "fine adjustment" to the body's needs.

2. The *proximal tubule* (cortex) reabsorbs many substances which can be reutilised by the body almost completely (e.g. glucose and amino acids). About 70 per cent of the sodium and water filtered at the glomerulus is reabsorbed at this site, the active process being reabsorption of sodium.

3. The *loop of Henle* (medulla) by means of countercurrent multiplication, creates the osmotic gradient which, in the presence of a normally functioning ADH feed-back mechanism, enables water reabsorption to be increased or reduced according to the body's need, with formation of urine of higher or lower osmolality than that of the extracellular fluid.

4. The *distal tubule* (cortex) makes the final adjustment of sodium, potassium and hydrogen ions by means of exchange mechanisms.

5. Final adjustment of water excretion takes place in the *distal tubule* and the *collecting ducts* (medulla) under the influence of ADH, and is dependent on normal functioning of the loop of Henle.

## CHEMICAL PATHOLOGY OF KIDNEY DISEASE

Although some diseases affect primarily the glomeruli and others the renal tubules, it is rare that disturbances of each of these are completely isolated from one another: the parts of the nephron are so closely associated anatomically, as well as being dependent on a common blood supply, that the end result of almost all renal disease is disturbance of the nephron as a whole. In most cases not all nephrons are equally affected, and while some may be completely non-functional, others may be quite normal, and yet others may have some functions disturbed to a greater extent than others. Some of the effects of chronic renal failure can be explained by this patchy distribution of disease (compare the effects of patchy pulmonary disease, p. 105).

However, in the initial stages some diseases affect either tubules or glomeruli predominantly and can occasionally be arrested or even reversed at this stage. It is probably easier to understand renal failure if these two types of lesion are isolated and discussed first.

### Glomerular Dysfunction

**Consequences of a reduced glomerular filtration rate.**—If the GFR is reduced an abnormally small volume of filtrate is produced: its composition remains that of a plasma ultrafiltrate. Abnormally small amounts of all constituents of the filtrate are presented to the tubular cells, and as the flow is relatively sluggish, stay in contact with them for longer than normal.

1. The reduced volume of filtrate may, of itself, result in *oliguria* (less than 400 ml per day): this will occur only if the GFR is reduced to less than about 30 per cent of its normal value (i.e. to less than about 30 ml/minute). If the low GFR is due to factors that also cause maximal ADH secretion, water reabsorption in the collecting ducts will also be high, and *urine of high osmolality and specific gravity* will be passed. These findings are due to the presence of substances not normally reabsorbed significantly by the tubular cells: the concentration of urinary urea and creatinine, for instance, will be high.

2. The total daily amount of *urea and creatinine* lost in the urine depends mainly on the amount filtered at the glomerulus during the same period, since tubular action on them is quantitatively insignificant in this context. The amount filtered at the glomerulus depends on plasma concentration and the GFR. If the rate of excretion falls below the rate of production the plasma levels will rise: as a higher plasma level

results in a higher rate of excretion the concentration may equilibrate temporarily at this raised concentration. If, however, the rise in concentration cannot compensate for the fall in filtered volume, levels continue to rise.

*Phosphate and urate*, like urea, are products of catabolism. These can be reabsorbed in the proximal tubule, and reabsorption may be complete if the amount filtered is low. Here again, if production is not balanced by excretion (and in this case by anabolism), plasma levels will rise.

3. Isosmotic sodium reabsorption is almost total in the proximal tubule, and less than usual is available for exchange with hydrogen ions throughout the tubule and with potassium ions in the distal tubule. This has two important results:

(*a*) Reduction of hydrogen ion secretion throughout the nephron. There is resultant systemic *acidosis* with *a low plasma bicarbonate concentration.*

(*b*) Reduction of potassium secretion in the distal tubule with *potassium retention.* The tendency to hyperkalaemia is aggravated by movement of potassium out of cells due to the acidosis (p. 64).

If the low GFR is due to low renal blood flow aldosterone secretion will be maximal (p. 37), and any sodium reaching the distal tubule will be almost completely reabsorbed in exchange for $H^+$ and $K^+$. *Urinary sodium* concentration will then be *low*. Thus the laboratory findings in advanced cases will be as follows:

*Effect on plasma*

    Uraemia, high plasma creatinine concentration
    Acidosis with a low plasma bicarbonate
    A tendency to hyperkalaemia
    Hyperuricaemia, hyperphosphataemia, hypocalcaemia (p. 245).

*Effects on urine*

    Oliguria
    Urinary osmolality (and specific gravity), urea and sodium concentrations appropriate to the clinical state. If the patient is *hypotensive or dehydrated the osmolality, specific gravity and urea concentration should be high and the sodium concentration low* (less than 30 mmol/1).

N.B. These urinary findings may be reversed if the dehydration is due to diabetes insipidus or to an osmotic diuresis (p. 47). Additional findings will depend on the cause of the low GFR.

**Causes of reduced glomerular filtration rate.**—As stated earlier, the relatively uncomplicated picture is usually seen only in the early stages of the disease. Later the nephron as a whole is affected.

1. Reduction in differential hydrostatic pressure in the glomerulus.

   (*a*) Reduced glomerular blood flow ("pre-renal uraemia"). Renal circulatory insufficiency due to:

   Low systemic blood pressure (haemorrhage, dehydration, "shock")
   Congestive cardiac failure
   Bilateral renal artery stenosis
   Acute oliguric renal failure

   (*b*) Increased intraluminal pressure
   ? Acute oliguric renal failure (p. 18)
   Obstruction of the ureters or urethra ("post-renal uraemia").

2. Disease of the glomerulus
   Acute glomerulonephritis
   Chronic glomerulonephritis.

1. (a) *Renal circulatory insufficiency* is probably the commonest cause of a low GFR with normal tubular function, and is most commonly due to haemorrhage or dehydration. If blood pressure (or, more accurately, renal blood flow) is restored within a few hours the condition is reversible: if the condition persists for longer periods of time the danger of ischaemic damage to the tubules increases, and when glomerular function is restored the picture of acute oliguric renal failure manifests itself within a few days. If the low GFR is due to this cause the patient will be hypotensive and possibly clinically dehydrated and, in addition to the laboratory findings listed above, there may be haemoconcentration (p. 41). Uraemia due to renal dysfunction is aggravated if there is increased protein breakdown, due either to tissue damage or to the presence of blood in the gastro-intestinal tract: the liberated amino acids are converted to urea in the liver. Increased tissue breakdown also aggravates hyperkalaemia and acidosis.

In congestive cardiac failure circulation to the kidney may be sufficiently impaired to cause a mild uraemia: the clinical findings are those of the primary condition, and haemodilution is often present. In renal artery stenosis hypertension of renal origin is usually present.

In acute oliguric renal failure it has been shown that renal blood flow, especially to the cortex, is reduced, and may account for some of the oliguria in this condition.

(b) *Increased intraluminal pressure* reduces the hydrostatic pressure gradient essential for normal filtration. "Post-renal uraemia" may be due to obstruction by calculi, polyps or other neoplasms, strictures or prostatic hypertrophy. Long continued back pressure may lead to renal damage, with uraemia persisting after the relief of the obstruction.

2. *Glomerulonephritis* reduces the permeability of the glomerular membrane; in chronic glomerulonephritis the tubules also are usually involved in scarring and the biochemical findings are therefore those of generalised renal dysfunction (p. 16). In acute glomerulonephritis fluid and sodium retention occur, usually in equivalent amounts so that plasma sodium concentrations remain normal. The overloading may be aggravated by injudicious fluid administration: in such cases there is haemodilution, and if the administered fluid is hypotonic, a fall in plasma sodium concentration. There is often a history of sore throat, and protein, casts and red cells in small amounts are found in the urine. Proteinuria and occasional erythrocytes may also be found in the chronic condition.

*Increased glomerular permeability* occurs in the *nephrotic syndrome*. Most plasma proteins can then pass the glomerulus and proteinuria of several grams a day occurs. The main effects are on *plasma proteins* and the subject is discussed more fully in Chapter XIII. Uraemia occurs only in the late stages of the disease when many glomeruli cease to function.

### Tubular Dysfunction

**Consequences of generalised tubular dysfunction.**—Even if the GFR is normal, damage to the tubular cells impairs the adjustment of the composition and volume of the urine.

1. The countercurrent mechanism may be impaired. Water reabsorption is reduced and large volumes of dilute urine are passed.

2. The tubules cannot secrete hydrogen ion and therefore cannot reabsorb bicarbonate normally and cannot acidify the urine.

3. Reabsorption of sodium and exchange mechanisms involving sodium are impaired and the urine contains an inappropriately high concentration of sodium relative to the state of hydration (p. 19). Failure of sodium reabsorption in the proximal tubule contributes to impairment of water reabsorption at this site.

4. Potassium reabsorption in the proximal tubule is impaired and potassium depletion may develop.

5. Reabsorption of glucose, phosphate, urate and amino acids is impaired and there is often glycosuria, phosphaturia and generalised aminoaciduria (acquired Fanconi syndrome). The plasma phosphate and urate levels may be low.

Uraemia occurs only if fluid and electrolyte depletion causes renal circulatory insufficiency.

Thus the laboratory findings in advanced cases may be:

*Effects on plasma*

Acidosis with low plasma bicarbonate

Hypokalaemia*

Hypophosphataemia and hypouricaemia*

Haemoconcentration (if the patient is dehydrated) (p. 41)

A normal blood urea* (unless the patient is dehydrated)

*Effects on urine*

Polyuria*

Inappropriately low osmolality (and specific gravity) and inappropriately low urea concentration*

An inappropriately high concentration of sodium* (more than about 30 mmol/l even if the patient is dehydrated or hypotensive).

Contrast those findings marked * with those of glomerular damage·

*It must be stressed that the finding of inappropriate urinary concentrations depends on the patient being hypotensive or dehydrated at the time of the investigation.* If the patient is normotensive and well hydrated measurement of urinary concentrations adds nothing to diagnostic precision (Chapter II).

In most cases there is mild proteinuria, and casts are present in the urine.

**Causes of predominantly tubular damage.**—In all these cases the glomeruli are involved in the later stages of the disease. A few examples only are given here.

1. Recovery phase of acute oliguric renal failure (see next page) due to:

Prolonged renal circulatory insufficiency

Proximal tubular necrosis due to various poisons (e.g. carbon tetrachloride)

2. Progressive damage to the tubules due to:

Hypercalcaemia

Hypokalaemia

Hyperuricaemia

Wilson's disease (copper, p. 372).

Bence Jones protein in myeloma (p. 332)

Galactosaemia (p. 371)
Various poisons, especially heavy metals
Early pyelonephritis

3. Inborn errors of specific tubular functions. These are discussed in the relevant chapters and summarised in Chapter XV.

*Progressive damage to the tubular cells* is frequently due to precipitation of substances such as calcium around the renal tubules; prolonged potassium depletion causes vacuolation of the tubular cells. Many of the other substances listed under this heading can cause acute oliguric renal failure if present in high enough concentration. In the early stages of these chronic conditions the picture may be that described under predominant tubular lesions: for instance, cases of chronic hypercalcaemia can present complaining of polyuria, and are often found to have hypokalaemia. In the later stages the whole nephron is involved in scarring with impairment of glomerular function.

Pyelonephritis, while affecting the kidneys as a whole, has a predilection for the medulla. The ability to form a concentrated urine, and other tubular functions, are therefore lost early in the disease.

### Generalised Renal Failure

Probably neither isolated glomerular nor isolated tubular failure occurs. However, in the conditions described in the previous pages, relatively crude methods of investigation do not detect the minor degree of dysfunction of the other part of the nephron. In most cases of "pure" tubular dysfunction, however, reduced urea or creatinine clearances may indicate minimal glomerular dysfunction in the presence of "normal" plasma urea and creatinine levels.

Between these two extremes there is a spectrum of conditions in which the proportions of tubular and glomerular dysfunction vary. The findings will depend on the relative contributions from these two factors, and an attempt has been made to indicate this in Table I. The dotted line indicates that the proportions are variable.

1. When the GFR falls below about 30 per cent of normal, substances almost unaffected by tubular action such as *urea and creatinine* will be retained with a consequent rise in their plasma concentration.

2. The degree of retention of *potassium, urate and phosphate* will depend on the balance between the degree of glomerular retention and the degree of loss due to failure of proximal tubular reabsorption: at the glomerular end of the spectrum so little is filtered that, despite failure of reabsorption, blood levels rise; at the tubular end of the spectrum glomerular retention is more than balanced by failure to reabsorb any filtered potassium, urate and phosphate. Similarly, the

TABLE I

FINDINGS IN PLASMA AND URINE IN RENAL DYSFUNCTION
(see text)

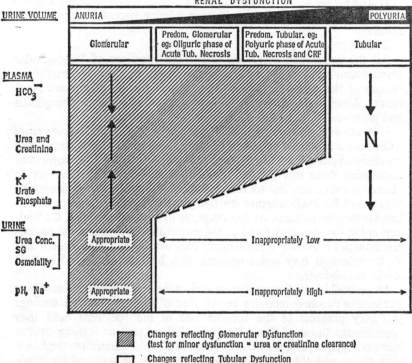

RENAL DYSFUNCTION

| URINE VOLUME | ANURIA | | | POLYURIA |
|---|---|---|---|---|
| | Glomerular | Predom. Glomerular eg: Oliguric phase of Acute Tub. Necrosis | Predom. Tubular. eg: Polyuric phase of Acute Tub. Necrosis and CRF | Tubular |

PLASMA
$HCO_3^-$

Urea and Creatinine

$K^+$
Urate
Phosphate

N

URINE
Urea Conc.
SG
Osmolality — Appropriate — Inappropriately Low

$pH, Na^+$ — Appropriate — Inappropriately High

▨ Changes reflecting Glomerular Dysfunction
(test for minor dysfunction = urea or creatinine clearance)

☐ Changes reflecting Tubular Dysfunction
(test for minor dysfunction = water concentration test)

urine volume depends on the balance between the volume filtered, and the proportion reabsorbed by the tubules (remember that normally 99 per cent of filtered water is reabsorbed). At the glomerular end of the spectrum nothing is filtered and the patient is anuric: at the tubular end, although filtration is reduced, tubular reabsorption is so impaired that the patient suffers from polyuria.

3. While *plasma levels* of urea and creatinine depend largely on glomerular function, *urinary concentrations* depend almost entirely on tubular function. However little is filtered at the glomerulus, the composition of the filtrate is that of a plasma ultrafiltrate. Any variation from this in urine passed is due to tubular activity. The fewer tubules working, the nearer urine concentrations will be to those of plasma.

Urine concentrations inappropriate to the state of hydration suggest tubular damage, whatever the degree of glomerular damage.

**Causes of generalised renal dysfunction.**—These can be acute or chronic.

1. *Acute*

Acute oliguric renal failure ("acute tubular necrosis"; acute renal failure).

2. *Chronic*

All the conditions described under the headings of "Glomerular Dysfunction" and "Tubular Dysfunction" can progress to generalised disease of the nephron, giving the common picture of chronic renal failure. Many other diseases (for instance chronic glomerulonephritis and polycystic disease) affect the kidney as a whole.

1. *Acute oliguric renal failure: "acute tubular necrosis"; acute renal failure* often follows a period of reduced GFR due to renal circulatory insufficiency, and from the therapeutic point of view it is important to distinguish these two phases. The oliguria is probably due not to glomerular damage, but to reduced cortical blood flow: this may be aggravated by back-pressure on the glomeruli due to obstruction of tubular flow by oedema. At this stage, as indicated in Table I, the findings are at the glomerular end of the spectrum; the fact that the tubules are damaged is evident if the urinary concentrations are inappropriate. Water retention may cause oedema with haemodilution unless fluid intake is restricted.

As cortical blood flow increases and as the tubular oedema resolves, glomerular function recovers before that of the tubules. The findings gradually progress to the tubular end of the spectrum until they approximate those of "pure" tubular lesions and as urinary output increases the polyuria may cause water depletion. Electrolyte depletion may occur, and the initial hyperkalaemia may be replaced by hypokalaemia. Mild acidosis (common to both glomerular and tubular lesions) persists until late. Finally, recovery of the tubules restores renal function to normal.

2. *In chronic renal failure* individual nephrons will be affected to different degrees. As more glomeruli are involved the rate of urea excretion falls and cannot balance the rate of production by protein catabolism: as a consequence the plasma urea level rises and the urea concentration in the filtrate of the normal nephrons rises. This may cause an osmotic diuresis in these nephrons: in other nephrons, tubules may be damaged out of proportion to the glomeruli. Both the tubular dysfunction in nephrons with normal glomeruli and the osmotic diuresis through intact nephrons contribute to the polyuria which may occur in chronic renal failure. If water loss is replaced by a high fluid intake, urea excretion can continue through healthy glomeruli at a high

rate: the increased excretion through normal glomeruli may eventually balance the reduced permeability of others, and a new steady state is reached at a higher level of blood urea. At this stage the findings are near the tubular end of the spectrum in Table I, the glomerular dysfunction being indicated by the high plasma urea level. If these subjects are kept well hydrated they may remain in a stable condition with a moderately raised plasma urea for years. Potassium levels are variable, but tend to be raised. If nephron destruction continues the condition approximates more and more to the glomerular end of the spectrum, until oliguria precipitates a steep rise in plasma urea and potassium concentrations. This stage is often terminal. Before assuming that the latter is the case, care should be taken to ensure that the sudden rise is not due to electrolyte and water depletion.

In prolonged chronic renal failure osteomalacia may occur and may occasionally require cautious treatment (p. 256).

### Differential Diagnosis of Oliguria with Uraemia

Oliguria with uraemia may be due to renal circulatory insufficiency, to acute glomerulonephritis, or to generalised renal damage. The diagnosis of acute glomerulonephritis can usually be made on clinical grounds, and on the finding of microscopic haematuria. However, the differentiation between renal circulatory insufficiency and generalised tubular damage may be difficult, and the treatment is radically different in the two cases. In most situations the clinical history and examination will differentiate between the two. If doubt remains and *if the patient is dehydrated or hypotensive* (i.e. has a low renal blood flow), measurement of urinary sodium and urea concentration, and urinary osmolality or specific gravity may help. Under these conditions a concentrated urine, containing less than 30 mmol/l of sodium, should be passed if the tubules are functioning. Failure to concentrate the urine implies true renal failure.

### INVESTIGATION OF RENAL FUNCTION

Plasma urea and creatinine levels depend on the balance between their production and excretion.

*Urea* is the product of amino acid breakdown in the liver and is therefore derived from protein, either in the diet or in body tissues: the rate of production is accelerated by a high protein diet, or by increased endogenous catabolism due to starvation or tissue damage. The capacity of the normal kidney to excrete urea is high, and in the presence of normal renal function only extremely high protein diets can cause rises of plasma urea to levels above the "normal" range. In patients with a gross increase in tissue catabolism due to severe tissue damage or to acute starvation (in chronic starvation plasma levels tend to be low due to protein depletion) urea levels may rise above "normal", especially as renal function is often mildly impaired in such cases due to renal circulatory factors. In spite of these reservations, a significantly high plasma urea concentration almost certainly indicates impaired renal function (this is almost certainly so at levels above 13 mmol/l [urea above 80 mg/dl: BUN above 37 mg/dl]). The probability is increased if a cause for renal dysfunction is present, or if there are protein, casts or cells in the urine. In the few cases when doubt remains, clearance studies (see below) or measurement of plasma creatinine may resolve it.

Plasma *creatinine* is not affected by diet, because it is produced endogenously by creatine breakdown. The circulating creatinine is derived from metabolism of tissue creatine: although plasma levels would be expected to increase with increased tissue breakdown, they do so less than those of urea. In theory, therefore, creatinine would seem preferable to urea estimation as an index of renal function, and for this reason is used by many laboratories for this purpose. However, in spite of an improvement in the precision and speed of creatinine estimation with the advent of automated methods, the accuracy of the urea method is still significantly higher, especially at lower levels. Which of these parameters is used is largely a matter of local choice.

If urea or creatinine levels are significantly raised, and especially if they are rising, if there is oliguria or a history suggestive of renal disease, or if there is proteinuria, and the urine contains protein, casts, cells or bacteria in significant amounts, they can usually be safely assumed to be due to renal impairment. For routine clinical purposes progress can be followed merely by measuring these concentrations in the plasma, because they reflect changes in renal clearance.

### Detection of Minor Degrees of Renal Damage

More than 60 per cent of the kidney must be destroyed before either plasma urea or creatinine levels are significantly raised (this is only valid for urea if the patient is taking a normal or low protein diet, and if there is no excessive protein catabolism, p. 19). Damage may be suspected for other reasons, such as evidence of renal infection, a previous episode of uraemia, or the finding of protein in the urine. If plasma urea or creatinine concentrations are normal more refined tests are useful: by the time they are abnormal presence of disease is not in doubt. This is a general principle which also applies, for example, to the detection of diabetes mellitus (p. 194) and liver damage (p. 298).

Renal concentrating ability.—This is a simple test and can be carried out on the ward. The ability to form a concentrated urine in response to water deprivation depends on normal tubular function (countercurrent multiplication), and on the presence of ADH. Failure of this ability is usually due to renal disease, but if there is any doubt the test can be repeated after administration of ADH (pitressin) (Appendix, p. 28).

**Clearance tests of glomerular function.**—Clearance tests measure the amount of blood which could theoretically be completely cleared of a substance per minute. This figure can be calculated if the amount excreted per minute is divided by its concentration in the blood: the amount excreted per minute can be calculated on a timed urine collection by multiplying the concentration of the substance in the specimen by the volume passed in the time, and dividing by the time in minutes. In the case of urea therefore:

Urea Clearance (ml/minute) =

$$\frac{\text{Urinary [urea]} \times \text{Urine volume (ml)}}{\text{Plasma [urea]} \times \text{Time of collection (minutes)}}$$

Care should be taken that plasma and urinary urea concentrations are expressed in the same units (e.g. mmol/l).

If a true estimate of GFR is to be made a substance should be chosen which is excreted solely by glomerular filtration and is not reabsorbed or secreted by the tubules. *Inulin* is thought to be such a substance: inulin is not produced in the mammalian body and measurement of its clearance requires administration either by constant infusion to maintain its level in the blood steady during the period of the test, or by a single injection followed by serial blood sampling so that its concentration at the mid-point of urine collection can be calculated. Such exogenous clearances are not very practicable for routine use, and this objection applies to clearances of radioactive *vitamin $B_{12}$* and chromated ethylenediamine tetracetic acid (*EDTA*), which have also been claimed to be cleared by glomerular action only.

Substances produced in the body are usually present in a fairly steady circulating concentration for the period of the test, and blood need only be taken at the mid-point of the urine collection. Urea and creatinine clearances are commonly used for clinical purposes. Neither of these substances fulfils the criterion that they are not reabsorbed from or added to the glomerular filtrate by the tubular cells. *Urea* diffuses back into the blood stream from the proximal tubule, and urea clearance values are lower than those of inulin: *creatinine* is secreted in small amounts by the tubule and gives clearance values higher than those of inulin. Either clearance gives an approximation to the GFR adequate for clinical purposes, and although the estimation of urea is more accurate than that of creatinine, there is little to choose between them. It should be noted that the rate of protein breakdown, while it may affect levels of plasma urea, does *not* affect its rate of clearance by the kidney.

If urea clearance is chosen, care should be taken to inhibit the urea splitting activity of organisms such as *Proteus vulgaris*, which may be present in the urine, by the use of mercury salts as a preservative. If such organisms are present, short periods of collection (for example, three collections of 1 hour) are preferable to a long one (for example, 24 hours).

*The biggest error of any clearance method is in the timed urine collection.*

It should be noted that clearance tests will give equally low values with renal circulatory insufficiency, or with "post-renal" causes, as with true renal damage, and cannot distinguish between them.

### Incidental Abnormal Findings in Renal Failure

In assessing the severity and progress of renal failure it is usual to estimate plasma urea or creatinine, electrolytes (especially potassium) and bicarbonate concentrations. Other abnormalities occur which, although not useful in diagnosis or assessment of renal dysfunction, may be misinterpreted if the cause is not recognised. Plasma *urate* levels rise in parallel with plasma urea, and a high level does not necessarily indicate primary hyperuricaemia (p. 386). Plasma *phosphate* levels also rise, and those of *calcium* fall (p. 245). In chronic uraemia secondary hyperparathyroidism may lead to bone disease with a high level of *alkaline phosphatase*. Hypocalcaemia should only be treated if there is evidence of bone disease (p. 256). *Anaemia* is commonly present, and is normocytic and normochromic in type: it does not respond to iron therapy.

BIOCHEMICAL PRINCIPLES OF TREATMENT OF<br>RENAL DYSFUNCTION

*Oliguric renal failure*—The oliguria of dehydration or haemorrhage, which is due to a reduction of glomerular filtration rate only, should be treated with the appropriate fluid (p. 59).

In oliguric renal failure due to parenchymal damage the aims are:

1. To restrict fluid and sodium, giving only enough fluid to replace that lost in the previous 24 hours (p. 41).

2. To provide adequate non-protein calories to prevent aggravation of uraemia and hyperkalaemia by increased endogenous catabolism.

3. To prevent dangerous hyperkalaemia (p. 73).

Diuretics tend to increase renal blood flow. In some cases of acute oliguric renal failure recovery may be hastened by administration of an osmotic diuretic, such as mannitol, or other diuretics such as frusemide (furosemide) or ethacrynic acid in large doses.

*In chronic renal failure with polyuria* the aim is to replace fluid and electrolytes lost. Sodium and water depletion may aggravate the uraemia.

*Haemodialysis or peritoneal dialysis* removes urea and toxic substances from the blood stream, and corrects electrolyte balance, by dialysing the patient's blood against fluid containing no urea, and normal plasma concentrations of electrolytes, *ionised* calcium and other plasma constituents. The blood is either passed over a dialysing membrane before being returned to the body, or the folds of the peritoneum are used as a dialysing membrane with their capillaries on one side, and suitable fluid injected into the peritoneal cavity on the other. A relatively slow reduction of urea concentration is preferable to a rapid one because of the danger of cellular overhydration when extracellular osmolality falls abruptly (p. 34). Dialysis is used in cases of potentially recoverable acute oliguric renal failure, to tide the patient over a crisis, or as a regularly repeated procedure in suitable cases of chronic renal failure. It may also be used to prepare patients for renal transplantation, and to maintain them until the transplant functions adequately.

## RENAL CALCULI

Renal stones are usually composed of normal products of metabolism which are present in the normal glomerular filtrate, often at concentrations near their maximum solubility: quite minor changes in urinary composition may cause precipitation of such constituents, whether in the substance of the kidney (see section on tubular damage, p. 15), as crystals, or as calculi. Although this discussion concerns stone formation it should be remembered that crystalluria and parenchymal damage can occur under the same circumstances, and that the treatment of all such conditions is the same.

### CONDITIONS FAVOURING CALCULUS FORMATION

1. A **high urinary concentration** of one or more constituents of the glomerular filtrate.

(a) *A low urinary volume*, with normal renal function, due to restricted fluid intake or excessive fluid loss over long periods of time (this is particularly common in the tropics). This condition favours formation of most types of stone, especially if one of the other conditions listed below is also present.

(b) An abnormally *high rate of excretion* of the metabolic product forming the stone, due either to an increased level in the glomerular filtrate (secondary to high plasma concentrations) or to a failure of normal tubular reabsorption from the filtrate.

2. **Changes in pH** of the urine, sometimes due to bacterial infection, which favour precipitation of different salts at different hydrogen ion concentrations.

3. **Urinary stagnation** due to obstruction to urine outflow.

4. **Lack of normal inhibitors.**—It has been suggested that normal urine contains an inhibitor, or inhibitors, of calcium oxalate crystal growth that are absent in the urine of some patients with a liability to recurrent calcium stone formation.

### COMPOSITION OF URINARY CALCULI

1. Calcium containing stones
    (a) Calcium oxalate ⎱ with or without magnesium
    (b) Calcium phosphate ⎰ ammonium phosphate.
2. Uric acid containing stones.
3. Cystine containing stones.
4. Xanthine containing stones.

#### Calculi Composed of Calcium Salts

These account for between 70 and 90 per cent of all renal stones. Precipitation of calcium is favoured by hypercalcuria, and the type of salt depends on urinary pH and on the availability of oxalate or phosphate. All patients presenting with renal calculi should have a plasma calcium estimation performed, and if this is normal it should be repeated at regular intervals.

*Hypercalcaemia* causes hypercalcuria if renal function is normal, and estimation of urinary calcium in such cases does not help in the diagnosis. The causes and differential diagnosis of hypercalcaemia are discussed on p. 249.

In many subjects with calcium containing renal calculi the plasma calcium level is normal. It is in such cases that the estimation of the daily excretion of urinary calcium may be useful. The commonest cause of *hypercalcuria with normocalcaemia* is the so-called *idiopathic hypercalcuria,* a name which reflects our ignorance of the aetiology of the condition: because some of these cases may represent an early stage of primary hyperparathyroidism, plasma calcium estimations should be carried out at regular intervals, especially if the plasma phosphate level is low. Any *increased release of calcium from bone,* as in actively progressing osteoporosis (in which loss of matrix causes secondary decalcification) or in prolonged acidosis (in which ionisation of calcium salts is increased) causes hypercalcuria, but rarely hypercalcaemia. One type of renal tubular acidosis (p. 95) not only increases the renal load of calcium but, because of the relative alkalinity of the urine, favours its precipitation in the kidney and renal tract: this is a *rare* cause.

An increased excretion of *oxalate* favours the formation of the very insoluble calcium oxalate, even if calcium excretion is normal. The source of the increased oxalate may be the diet: the very rare inborn errors of oxalate metabolism, primary hyperoxaluria, should be considered if renal calculi occur in childhood.

It has already been mentioned that *alkaline conditions* favour calcium precipitation, and whereas calcium oxalate stones form at any urinary pH, a high pH favours formation of calcium phosphate: this type of stone is particularly common in chronic renal infection with urease containing (urea splitting) organisms (for example, *Proteus vulgaris*), which convert urea to ammonia.

A significant proportion of cases remains in which there is no apparent cause for the calcium precipitation.

Calcium containing calculi are usually *hard and white*. They are *radio-opaque*. Calcium phosphate stones are particularly prone to form "staghorn" calculi in the renal pelvis, easily visualised by straight x-ray.

**Treatment of calcium containing calculi.**—This depends on the cause. Excretion of calcium should be reduced

(*a*) By treating the primary condition (especially hypercalcaemia);

(*b*) If this is not possible, by reducing intake of calcium in the diet, and possibly by decreasing calcium absorption by administration of oral phosphate (p. 261);

(*c*) By reducing the concentration of urinary calcium by maintaining a high fluid intake (unless renal failure is present). It is the concentration rather than the total 24-hour output which determines the tendency to precipitation.

## Uric Acid Stones

These account for about 10 per cent of all renal calculi, and are sometimes associated with *hyperuricaemia* (with or without clinical gout). Precipitation is favoured in an *acid urine*. In a large proportion of cases no predisposing cause can be found.

Uric acid stones are usually *small, friable* and *yellowish-brown* in colour, but can be large enough to form "staghorn" calculi. They are *radiotranslucent*, but may be visualised on an intravenous pyelogram.

**Treatment** of hyperuricaemia is discussed on p. 384. If the plasma urate concentration is normal, fluid intake should be kept high and the urine alkalinised. A low purine diet (p. 379) may help to reduce urate production and therefore excretion.

## Cystine Stones

These are rare. In normal subjects the concentration of urinary cystine is well within its solubility. In severe cases of the inborn error cystinuria (p. 365) the solubility may be exceeded and the patient may present with *radio-opaque* renal calculi. Like uric acid, cystine is more soluble in alkaline than acid urine and the principles of treatment are the same as for uric acid stones. Penicillamine can also be used in therapy (p. 366).

## Xanthine Stones

These are very uncommon and may be the result of the rare inborn error, xanthinuria (p. 386). Xanthine stones following the use of xanthine oxidase inhibitors, such as allopurinol, have not been reported.

## SUMMARY

### THE KIDNEYS

1. Normal renal function depends on a normal glomerular filtration rate (GFR) and normal tubular function.

2. In most cases of renal disease glomerular and tubular dysfunction coexist.

3. A low GFR leads to:
   Uraemia and retention of other nitrogenous end-products, including urate, and of phosphate.
   Acidosis.
   Hyperkalaemia.
   Oliguria.

4. A low GFR with normal tubular function is most commonly due to renal circulatory insufficiency. It can also be the result of obstruction to urinary outflow, or of glomerular damage.

5. In the nephrotic syndrome the glomerulus is more permeable than normal to proteins.

6. Tubular damage leads to:
   Acidosis.
   Hypokalaemia.
   Hypophosphataemia and hypouricaemia.
   Polyuria. The urine is inappropriately dilute and contains an inappropriately high sodium concentration in relation to the patient's state of hydration.

7. In acute oliguric renal failure there is a reduced GFR and the findings are usually those associated with this.

8. The differentiation between the oliguria of a primary glomerular lesion with normal tubular function and of acute tubular necrosis is best made on clinical grounds and, if necessary, by examining the urine.

9. In most cases plasma urea or creatinine levels reflect changes in renal clearance and are adequate for diagnosing and following up cases of renal disease. Tubular function may be tested by assessing the ability of the kidney to concentrate urine.

10. Clearance tests are valuable if a minor degree of renal damage is suspected in spite of a plasma urea or creatinine level within the "normal" range.

## RENAL CALCULI

1. The formation of renal calculi is favoured by:
   (a) A high urinary concentration of the constituents of the calculi. This may be due to oliguria, or a high rate of excretion of the relevant substances.
   (b) A pH of the urine which favours precipitation of the constituents of the calculi.
   (c) Urinary stagnation.

2. Calcium containing calculi account for 70 to 90 per cent of all renal stones. They are most commonly idiopathic in origin but hypercalcaemia, especially that of primary hyperparathyroidism, should be excluded as a cause.

3. Uric acid stones account for a further 10 per cent of renal calculi. Rare causes are cystinuria and xanthinuria.

## FURTHER READING

DE WARDENER, H. E. (1973). *The Kidney,* 4th edit. London: J. & A. Churchill.

KASSIRER, J. P. (1971). Clinical evaluation of kidney function—Tubular function. *New Engl. J. Med.,* **285,** 499.

WILLS, M. R. (1971). *Biochemical Consequences of Chronic Renal Failure.* Aylesbury: Harvey, Miller and Medcalf.

SMITH, L. H. (1972). Nephrolithiasis. Current concepts in pathogenesis and management. *Postgrad. Med.,* **52,** 165.

# APPENDIX

## URINE CONCENTRATION TEST

In the normal subject,restriction of water intake for a period of hours results in maximal stimulation of ADH secretion (p. 39). ADH acts on the collecting ducts, water is reabsorbed and a concentrated urine is passed. If the counter-current multiplication [mechanism is impaired (p. 5) maximal water re-absorption cannot take place, and if ADH levels are low the effect is similar.

If the feed-back mechanism is intact ADH levels are sufficient to produce maximal effect. Under these circumstances administration of exogenous ADH will not improve renal concentrating power. If, however, the primary disease is diabetes insipidus with normal tubular function, administration of ADH will convert this to normal.

*This test should not be performed if the patient is already dehydrated.* In such cases the demonstration of a low urine to plasma osmolality ratio (see below) is diagnostic without resort to fluid restriction. *The patient should be kept under observation during the test,* which should be terminated, after collecting blood and urine, if he becomes distressed.

### Procedure

The patient is allowed no food or water after 6 p.m. on the night before the test.

On the day of the test:—

7 a.m.—The bladder is emptied, and the *specimen discarded.*

8 a.m.—The bladder is emptied. The osmolality (or specific gravity) of this specimen is measured. If this reading is above 850 mosmol/kg (specific gravity 1·022) the test may be terminated: if it is below this figure the osmolality (or specific gravity) of urine passed at 9 a.m. should be measured.

Blood may also be taken for determination of plasma osmolality.

### Interpretation

A maximum osmolality of less than 850 mosmol/kg, or a specific gravity less than 1·022 indicates impaired renal concentrating power, either due to tubular disease or to diabetes insipidus: the urine to plasma osmolality ratio should be above 3·0 in normal subjects. In most cases it is clear which of these possibilities is implicated. Where this is still in doubt the pitressin test may be carried out.

## PITRESSIN TEST

The above procedure is followed, but 5 units of the oily suspension of vasopressin (pitressin) tannate is injected intramuscularly at 7 p.m. on the evening before the test. If the failure to concentrate is due to tubular disease this will not be improved by the pitressin: if it is due to diabetes insipidus the urine will now be concentrated normally.

**Caution.**—1. Hydrometers are frequently inaccurate. The specific gravity of distilled water should read 1·000 (the hydrometer is usually calibrated at room temperature): if it does not, a suitable correction should be made.

2. The hydrometer should be floating freely when the reading is taken. If it touches the wall of the container false values will be obtained. The procedure should be carried out by someone experienced with it.

Urinary urea concentration may be a more valuable estimation. Values above 330 mmol/l (2 g/dl) usually indicate good renal function. If an osmometer is available, measurement of urinary osmolality gives more precise results.

3. Sugar, protein and contrast media used in intravenous pyelography contribute to urinary specific gravity and in their presence high readings are not necessarily indicative of normal renal function. Protein, because of its high molecular weight, contributes very little to osmolality. Glucose does, however, significantly affect it: 150 mmol/l (2·7 g/dl) of glucose in the urine adds 0·001 to the specific gravity and 150 mosmol/kg to osmolality. In the presence of glycosuria, estimation of urinary urea concentration may be valuable.

### INVESTIGATION OF THE PATIENT WITH RENAL CALCULI

1. If the stone is available, send it to the laboratory for analysis.

2. Exclude *hypercalcaemia* and *hyperuricaemia*. If either is present it should be treated.

3. If the *plasma calcium is normal*, collect a 24-hour specimen of urine (in a container containing acid to keep calcium in solution) for *urinary calcium estimation*. If hypercalcuria is present it should be treated.

4. *If all these tests are negative*, screen the urine for *cystine*. This is especially important if there is a family history. If the qualitative test is positive, cystine should be estimated quantitatively on a 24-hour specimen of urine (p. 366).

5. If the patient is acidotic and the urine is alkaline, perform an ammonium chloride load test (p. 109).

6. If renal calculi occur in *childhood*, a low *plasma urate* and high *urinary xanthine* suggest xanthinuria. If these are normal the 24-hour excretion of *oxalate* should be estimated to exclude primary hyperoxaluria as a cause.

## Chapter II

# SODIUM AND WATER METABOLISM

THE control of sodium and water balance are so closely linked that an understanding of one is impossible without an understanding of the other.

Sodium is widely distributed in the body. It is predominantly outside cells (unlike potassium, which is almost all intracellular); quantitatively significant, but relatively metabolically inert, sodium is present in bone salts.

As sodium is normally the most abundant extracellular cation it accounts, together with its associated anions, for most of the osmotic activity of the extracellular fluid (ECF). Osmotic activity depends on *concentration*, and is determined by the relative amounts of water and sodium, rather than absolute amounts of either. Hypo- or hypernatraemia is caused by imbalance between the two, and the clinical picture is due to the consequent osmotic changes.

If sodium and water are lost in equivalent amounts the osmolal concentration is unaffected: symptoms are then those of inadequate circulatory volume. Of course, osmotic and volume changes may, and often do, occur together.

Transport of sodium, of potassium and of hydrogen ions across cell membranes are interdependent, and disturbances of potassium and hydrogen ion homeostasis often accompany those of sodium: changes in associated anions such as chloride and bicarbonate often occur at the same time. The separation of the contents of this chapter from those on potassium and hydrogen ion homeostasis is arbitrary, and for ease of discussion only. A clinical situation should be assessed with all these factors in mind.

## WATER AND SODIUM BALANCE

A 70 kg man contains approximately 45 litres of water and 3000 mmol of metabolically active sodium. Maintenance of this total amount depends on the balance between intake and loss. Water and electrolytes are taken in food and drink, and are lost in urine, faeces and sweat: in addition, about 500 ml of water is lost daily in expired air.

## Loss Through the Kidneys and Intestinal Tract

*Renal loss* of sodium and water depends on the balance between that filtered at the glomerulus and that reabsorbed during passage through the tubules, and therefore on normal glomerular and tubular function. *Faecal loss* is mainly dependent on the balance between the sodium and water secreted in bile, saliva and gastric, pancreatic and intestinal secretions, and reabsorbed during passage through the intestine: the contribution from oral intake is small. This loss therefore depends on the integrity of the intestinal epithelial cells, and on the time for which intestinal contents are in contact with them. Intestinal hurry or loss of secretions from the upper gastro-intestinal tract by vomiting or through fistulae are important causes of sodium and water depletion.

It is a sobering thought that approximately 200 litres of water and 30,000 mmol of sodium are filtered through the glomerulus, and a further 10 litres of water and 1500 mmol of sodium are secreted into the intestinal tract each day. In the absence of homeostatic mechanisms for reabsorption, the whole of the extracellular water and sodium could be lost in about 2 hours. It is, therefore, not surprising that failure of these mechanisms causes such extreme disturbances of water and sodium balance. Normally 99 per cent of this initial loss is reabsorbed, and net daily losses amount to about 1·5 to 2 litres of water and 100 mmol of sodium in the urine and 100 ml and 15 mmol in faeces.

## Loss in Sweat and Expired Air

Normal daily water loss in sweat and expired air amounts to about 900 ml: about 30 mmol of sodium is lost a day in sweat. Although anti-diuretic hormone (ADH) and aldosterone have some effect on the composition of sweat, this is relatively unimportant, and sweat loss is primarily controlled by body temperature. Respiratory water loss depends on respiratory rate, and control of this bears no relation to the body requirements for water. Normally loss by sweat and respiration is rapidly corrected by changes in renal and intestinal loss. However, as neither can be controlled to meet sodium and water requirements, they may contribute considerably to abnormal balance when homeostatic mechanisms fail, or in the presence of gross depletion, whether due to poor intake or to excessive loss by other routes.

# DISTRIBUTION OF WATER AND SODIUM IN THE BODY

In mild disturbances of water and electrolyte metabolism the total amount of these in the body is of less importance than their distribution within it.

## DISTRIBUTION OF ELECTROLYTES

The body has two main fluid compartments of very different electrolyte composition. The compartments are:

1. **The intracellular compartment,** in which *potassium* is the predominant cation.

2. **The extracellular compartment,** in which *sodium* is the predominant cation.

The extracellular fluid can be subdivided into:

(*a*) The interstitial fluid which is of very low protein concentration.

(*b*) The intravascular fluid (plasma) which contains protein in high concentration.

### Distribution of Electrolytes Between Cells and Extracellular Fluid

The intracellular concentration of sodium is less than a tenth of that in the extracellular fluid (ECF), while that of potassium is about thirty times as much. In absolute amounts about 95 per cent of the metabolically active sodium in the body is outside cells, and about the same proportion of potassium is intracellular.

Other ions tend to move across cell walls with sodium and potassium. The hydrogen ion has already been mentioned. For example, magnesium and phosphate are predominantly intracellular, and chloride extracellular ions. The distribution of all these, and of bicarbonate, will be affected by the same conditions as those mentioned.

### Distribution of Electrolytes Between Plasma and Interstitial Fluid

The vascular wall, unlike that of the cell, is freely permeable to small ions. Plasma contains protein in significant concentration and interstitial fluid does not. Electrolyte concentrations are therefore very slightly higher in the latter to balance the osmotic effect of the protein concentration inside vessels. The difference is small and clinically insignificant, and for practical purposes one can assume that plasma electrolytes are representative of those in the extracellular fluid as a whole.

## DISTRIBUTION OF WATER IN THE BODY

A little over half the body water is inside cells. Of the extracellular water about 15–20 per cent is in the plasma. The remainder makes up the extravascular, extracellular interstitial fluid.

The distribution of water across cell walls depends on the *in vivo* effective osmotic difference between intra- and extracellular fluid: that across blood vessel walls is determined by the balance between the *in vivo* effective osmotic pressure of plasma and the net outward hydrostatic pressure. Correct interpretation of plasma electrolyte results depends on a clear understanding of these factors.

## Osmotic Pressure

Net movement of water across a membrane permeable only to water depends on the concentration difference of dissolved particles on the two sides. For any given weight/volume concentration, the larger the particle (the higher the molecular weight) the fewer there are in unit volume, and the less osmotic effect they will exert. However, if the membrane is freely permeable to smaller particles, as well as to water, these exert no osmotic effect, and the larger molecules become more important in affecting movement of water. To explain water distribution in the body it is important to appreciate these two factors:

1. Number of particles per unit volume;
2. Particle size related to membrane permeability.

### Plasma Osmotic Pressure: Distribution of Water across Cell Walls

For the definitions of osmolarity and osmolality see p. 36.

Measured plasma osmolality.—Plasma osmotic pressure is usually determined by measuring the depression of its freezing point below that of pure water. The result is a measure of its *total* osmotic pressure—the osmotic effect which would be exerted by the sum of all the dissolved molecules and ions across a membrane permeable only to water. Table II shows that by far the largest contribution to this (90 per cent

TABLE II

APPROXIMATE CONTRIBUTIONS OF PLASMA CONSTITUENTS TO PLASMA OSMOLALITY

| | Concentration (mmol/l unless otherwise stated) | mosmol/kg |
|---|---|---|
| Sodium | 135 | 270 |
| Associated anions | 135 | |
| Potassium | 3·5 | 7 |
| Associated anions | 3·5 | |
| Calcium | 2·5 (10 mg/dl) | 1·5 |
| | Assuming 40% protein bound (p. 235) ionised calcium = 1·5 | |
| Associated anions | | 1·5 or 3 |
| Magnesium | 1 | 1 |
| Associated anions | | 1 or 2 |
| Urea | 5 (30 mg/dl) | 5 |
| Glucose | 5 (90 mg/dl) | 5 |
| Protein | 70 g/l | Approximately 1 |
| | | Approximately 295 |

or more) comes from *sodium and its associated anions*, the effect of protein being negligible. The only major difference between extra-vascular fluid and plasma is in their protein content: thus *total plasma osmolality is almost identical with the osmolality of the fluid bathing cells*.

Hydrostatic pressure differences across cell walls are negligible, and cell hydration depends on the osmotic difference between intra- and extracellular fluid. The cell membrane is freely permeable to water, but although different substances diffuse, or are actively transported, across it at different rates, it is a general rule that solute enters cells more slowly than does water. Normally the total intracellular osmolality, due predominantly to potassium and associated anions, equals that of the ECF, mostly due to sodium salts, and there is no *net* movement of water in or out of cells. In pathological circumstances rapid changes of extracellular solute concentration affect cell hydration: slower changes, by allowing time for redistribution of solute, have less effect. The speed of change is even more important than the magnitude, and over-zealous treatment of abnormal extracellular osmolality may produce more dangerous alteration in cell hydration than the initial abnormality.

Because *sodium* and its associated anions account for about 90 per cent of plasma osmolality in the normal subject (Table II), rapid changes of sodium concentration affect cell hydration, a rise causing cellular dehydration and a fall cellular overhydration. Normal levels of *urea* and *glucose* contribute very little to measured plasma osmolality. However, concentrations 15-fold or more above normal can occur in severe uraemia and hyperglycaemia, and then these solutes contribute to it very significantly. Urea can diffuse into cells slowly, and during prolonged uraemia its osmotic *effect* is reduced by this fact: acute changes can alter cell hydration. Glucose is actively transported into cells, but once there is rapidly metabolised: intracellular levels remain low, and severe hyperglycaemia has a marked influence on cell hydra-tion. Although uraemia and hyperglycaemia can cause cellular dehydra-tion, the contribution of normal concentrations of urea and glucose to plasma osmolality is so small that low levels of these solutes, unlike those of sodium, do not cause cellular overhydration.

Because the speed of change is even more important than the absolute level, over-zealous treatment of established hypernatraemia (with hypo-tonic fluids), uraemia (by haemodialysis), or hyperglycaemia (with large doses of insulin) may produce dangerous cerebral cellular overhydra-tion.

Concentrations of other solutes of low osmolal concentration, such as those of calcium, potassium or magnesium, rarely vary by a factor of more than 3, even in pathological states; they do not cause significant osmolality changes.

Substances not transported into cells, such as mannitol, can be infused to reduce cerebral oedema. They can also be used as osmotic diuretics. (Hypertonic glucose or urea can also be used in this way). They are included in total plasma osmolality measurements.

▶ **Calculated plasma osmolarity.**—Under most circumstances plasma osmolarity can be calculated sufficiently accurately for clinical purposes if plasma sodium, potassium, urea and glucose concentrations are known. If all measurements are in mmol/l, the *approximate* total osmolarity in mosmol/l will be:

$$2[Na^+] + 2[K^+] + [urea] + [glucose]$$

The factor of 2 applied to sodium and potassium concentrations allows for associated anions, and assumes complete ionisation.

This calculation is *not* valid:

1. If an unmeasured osmotically active substance, such as mannitol, has been infused.

2. If there is gross hyperproteinaemia or lipaemia (see section on "Osmolarity and Osmolality").

## Plasma Colloid Osmotic Pressure: Distribution of Water across Blood Vessel Walls

The distribution of water across capillary walls is unaffected by electrolyte concentration, but affected by the osmotic effect of *plasma proteins*.

The maintenance of blood pressure depends on the retention of intravascular water at a pressure higher than that of interstitial fluid. Thus hydrostatic pressure is tending to force fluid into the extravascular space. In the absence of any effective osmotic pressure across vascular walls water would be lost rapidly from the intravascular compartment.

Unlike the cell membrane, the blood vessel wall is freely and rapidly permeable to small molecules; sodium exerts almost no osmotic effect at this site. The smallest molecule present intravascularly at significant concentration, and which is virtually absent extravascularly, is albumin (MW about 70,000). The vascular wall is almost impermeable to it. Albumin concentration is therefore the most important factor opposing the net outward hydrostatic pressure: the higher molecular weight proteins although together present in much the same concentration as albumin, contribute much less to this effect because of their larger size. This effective osmotic pressure across blood vessel walls is the *colloid osmotic* or *oncotic pressure*.

*Because proteins contribute negligibly to measured plasma osmolality* (Table II), *measurements of the latter cannot be used to assess osmotic effects across blood vessel walls.*

### Units of Measurement of Osmotic Pressure: Osmolarity and Osmolality

Concentrations of molecules can be expressed in two ways:

1. In molarity, the number of moles (or mmol) *per litre of solution*.
2. In molality, the number of moles (or mmol) *per kg of solvent*.

If the molecules are dissolved in pure water at concentrations such as are found in biological fluids, these two figures will differ very little. Plasma, however, is a complex solution. It contains, amongst other things, dissolved proteins, and the total volume of solution consists of that of water plus that of protein. The small molecules are dissolved only in the water, and at a protein concentration of 70 g/l the volume of water is about 6 per cent less than that of total solution (that is, the molality will be about 6 per cent greater than the molarity). Methods of measuring individual ions such as sodium, measure them in molarity (mmol/l).

Osmotic pressure can also be expressed in two ways. (In the present context the number of mosmol is almost numerically equal to the number of mmol):

1. In *osmolarity* expressed as *mosmol/l of solution*.
2. In *osmolality* expressed as *mosmol/kg of solvent*.

The usual method of measuring plasma osmotic pressure by freezing point depression depends on *osmolality*, whereas calculated osmotic pressure from dissolved ions and molecules is in osmolarity. The osmometer therefore measures osmotic concentration in *body water*. It is this *osmolal*, rather than osmolar, concentration which exerts its effect across cell walls and which is concerned in homeostatic mechanisms.

Under normal circumstances, although measured osmolality should be higher than calculated osmolarity (because of the protein content of plasma), little significant difference is found between the two figures. This is because incomplete dissociation of molecules to ions reduces the osmotic effect by about the same amount as the volume occupied by protein raises it. Calculated osmolarity is then adequate for clinical purposes. *This is not true in hyperproteinaemic or hyperlipaemic states*, when measurement of plasma osmolality is advantageous. Moreover, as urine does not normally contain either protein or lipids, *direct comparison between urine and blood can only be made by measurement of osmolality of both*. Calculation of urinary osmolarity is not feasible because of the considerable variation in concentration of different solutes, and osmolality should always be measured.

## CONTROL OF SODIUM AND WATER METABOLISM

The following is a simplified account of a complex situation and the student should bear this in mind.

## CONTROL OF SODIUM

Intake of sodium is probably not actively controlled. The most important factor controlling loss is the mineralocorticoid hormone, aldosterone.

### Aldosterone

Aldosterone is secreted by the zona glomerulosa of the adrenal cortex (p. 134). It affects sodium-potassium (and possibly sodium-hydrogen ion) exchange across *all* cell membranes. We shall concentrate on its effect on renal tubular cells, but we should bear in mind that it also affects faecal sodium loss, and the distribution of electrolytes in the body.

In the distal renal tubule aldosterone increases sodium reabsorption from the glomerular filtrate in exchange for potassium or hydrogen ion. The net result in the body is retention of sodium and loss of potassium. In the presence of high levels of circulating aldosterone *urinary sodium concentrations are low*.

Many factors have been implicated in the feed-back control of aldosterone secretion. Such factors as local electrolyte concentration in the adrenal and the kidney are probably of little clinical or physiological importance compared with the effect on the adrenal of the renin-angiotensin system, although this has been questioned by some authorities.

### The Renin-Angiotensin System

*Renin* is a proteolytic enzyme secreted by a cellular complex situated near the renal glomeruli (and therefore called the juxtaglomerular apparatus). In the blood stream it acts on a renin substrate (an $\alpha_2$ globulin) to form *angiotensin I*. This decapeptide is further split by a peptidase, located predominantly in the lungs, to *angiotensin II*. This peptide hormone has two actions:

1. It acts directly on blood vessel walls, causing vasoconstriction. It therefore helps to maintain blood pressure.

2. It stimulates the cells of the zona glomerulosa to secrete aldosterone.

The most important stimulus to renin production seems to be reduced renal blood flow (possibly changes in the mean blood pressure in renal vessels are the actual stimuli). Poor renal blood flow is often associated with an inadequate systemic blood pressure and the two effects of angiotensin II ensure that this is corrected:

1. Vasoconstriction raises the blood pressure before the circulating volume can be restored.

2. Sodium retention occurs due to the action of aldosterone: as we shall see later, this will usually be accompanied by water retention with restoration of the circulating volume.

We may make an oversimplified statement based on the effect and control of aldosterone secretion. Aldosterone causes sodium retention. The stimulus to its secretion is renin, which is controlled, effectively, by circulating blood volume, Therefore, *blood volume controls sodium retention.*

## Non-Aldosterone Factors Affecting Renal Sodium Excretion

Aldosterone is probably the most important factor affecting sodium excretion. However, proximal tubular factors also affect it. These are:

1. Changes in *glomerular filtration rate.*
A very *low GFR decreases sodium excretion.*
A very *high GFR causes natriuresis.*

Changes in GFR, by altering the amount of sodium filtered in unit time, vary the amount presented to the proximal tubule. If this is much reduced sodium will be almost completely reabsorbed in the proximal tubule; if much increased, proximal tubular mechanisms are swamped.

2. Changes in *renal blood flow.*
A *low tubular blood flow* has been shown to increase proximal tubular reabsorption and *decrease sodium excretion.*
An *increased tubular blood flow* has been shown to decrease proximal tubular reabsorption and *increase sodium excretion.*

3. Changes in *oncotic pressure of plasma in tubular blood vessels.*
*Haemoconcentration* increases proximal tubular reabsorption and *decreases sodium excretion.*
*Haemodilution* decreases proximal tubular reabsorption and *increases sodium excretion.*

Note that, in general, these effects potentiate each other and that of aldosterone. Thus, expansion of the plasma volume increases renal blood flow, cutting off aldosterone secretion, increasing the GFR, and increasing the tubular flow: if the expansion is with protein-free fluid there will be haemodilution. All these factors cause natriuresis.

A further hormonal factor controlling sodium excretion has been postulated: this is the *natriuretic hormone* or "third factor" (the GFR is the "first factor" and aldosterone the "second factor"). This hormone, which is said to be secreted in response to an expansion of plasma volume, may inhibit sodium reabsorption in the proximal tubule. It has been suggested that it fails to be secreted in oedematous states. It is probably more acceptable to think of "third factor" as being non-hormonal, and to be the net effect of the non-aldosterone renal factors mentioned above.

In certain inborn errors of renal tubular function the kidneys cannot respond to aldosterone; this may also be the case in the recovery phase of acute tubular necrosis, and with tubular damage of any kind: for instance, prolonged hypokalaemia in primary aldosteronism may prevent maximal tubular response to aldosterone.

In the later discussion we will assume the renin-aldosterone mechanism to be of overriding importance in the control of sodium excretion.

## CONTROL OF WATER

**Intake** of water is controlled by thirst. The hypothalamic thirst centre responds to increased osmolality in the blood circulating through it (or, more accurately, to the osmotic difference between this and the cells of the centre), and therefore, usually to changes in sodium concentration. This occurs whether or not the fluid volume is normal.

**Loss** of water is also controlled via the hypothalamus in response to osmolality changes. Increased plasma osmolality (usually due to increased sodium concentration) stimulates secretion of antidiuretic hormone (ADH). The mechanism is very sensitive and a rise of osmotic pressure of only 2 per cent quadruples ADH output, while a similar fall cuts if off completely. Such changes (of about 3 mmol/l of sodium) may *not* be obvious using relatively crude laboratory methods.

**Antidiuretic hormone** is a peptide produced by the hypothalamus and secreted by the posterior pituitary gland. By increasing the permeability of distal renal tubular cells it increases passive water reabsorption along the osmotic gradient produced by the countercurrent mechanism (p. 7). Under its influence a *concentrated urine of high osmolality and specific gravity is passed.*

Water intake and loss are both controlled by plasma osmolality. Plasma osmolality normally depends largely on its sodium concentration. We have already pointed out that fluid volume controls sodium retention. We can now make the further *simplified* statement that *sodium concentration controls the amount of water in the body.* (It must be stressed that in *uraemia or hyperglycaemia*, or *after infusion of substances such as mannitol*, this is not true.)

### Non-ADH Factors Affecting Water Excretion

The distal tubule has a limited capacity for reabsorption of water, even in the presence of a maximal amount of ADH: thus if a large volume reaches this site water diuresis will occur.

The glomerular filtrate may contain high concentrations of osmotically active substances which are not completely reabsorbed during passage through the proximal tubules and loop of Henle: these inhibit water and, to a lesser extent, sodium reabsorption in the proximal tubule (p. 10) and a larger volume than normal reaches the distal tubule. This is the basis of action of osmotic diuretics such as mannitol and, to a lesser extent, urea and hypertonic glucose. Patients being fed intravenously, or who, because of tissue damage, are breaking down larger than usual quantities of protein and producing excessive amounts of urea from the released amino acids, may become dehydrated even in the presence of adequate amounts of ADH (p. 47).

As in the case of sodium and aldosterone, such non-ADH effects are usually relatively unimportant.

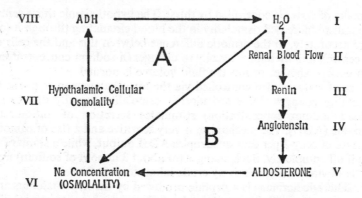

FIG. 2.—Simplified cycle of sodium and water homeostasis. (Note that in uraemia, hyperglycaemia and after infusion of hyperosmolal solutions such as mannitol, factors other than sodium cause significant changes in osmolality.)

## INTERRELATIONSHIP BETWEEN SODIUM AND WATER HOMEOSTASIS

Let us now look more closely at the simplified statements made above concerning the control of ADH and aldosterone secretion. We said that, effectively, sodium controls ADH secretion and water controls aldosterone secretion. But ADH controls water loss and aldosterone controls sodium loss. The homeostasis of sodium and water is interdependent and this simplified scheme is shown in Fig. 2. The thirst mechanism is not shown here but it should be remembered that an increase of osmolality not only reduces water loss but increases thirst and therefore intake. The diagonal line which divides the rectangle into two triangles, A and B, indicates that changes in hydration can also directly alter sodium concentration.

We shall refer to this scheme repeatedly while explaining changes occurring in pathological states. The reader should bear in mind that it is a simplified scheme and does not allow for the effect of fluid shifts across cell walls on extracellular concentrations and volume: nor does it allow for changes in other osmotically active solute than sodium.

## ASSESSMENT OF SODIUM AND WATER BALANCE

Although it is easy to measure the intake of water and sodium of patients receiving liquid feeds (either oral or intravenous), it is less easy to do so when a solid diet is being taken. Fortunately, accurate measurement is rarely necessary in such cases.

Renal loss is also easy to measure but that in formed faeces, sweat and expired air ("insensible loss") is more difficult to assess and may be important when homeostatic mechanisms have failed, in the presence of abnormal losses by extrarenal routes, in unconscious patients and in infants (see p. 47). The aim should be to ensure that such subjects are normally hydrated, and then to keep them "in balance". The medical attendant has the difficult task of imitating normal homeostasis without the aid of the sensitive and interlinked mechanisms of the body.

### ASSESSMENT OF THE STATE OF HYDRATION

Assessment of the state of hydration of a patient depends on observation of his clinical state, and on laboratory evidence of haemoconcentration or haemodilution. It must be stressed that both these methods are crude, and that quite severe disturbances of water balance can occur before they are obvious clinically or by laboratory methods.

In extracellular dehydration (other than that due to haemorrhage), water and often electrolytes are lost from the vascular compartment, and the concentrations of large molecules and of cells rise: there is therefore a rise in all plasma protein fractions, in haemoglobin level, and in the haematocrit reading (*haemoconcentration*). Conversely, in overhydration these concentrations fall (*haemodilution*). These findings, of course, may also be affected by pre-existing abnormalities of protein or red cell concentrations.

The problem of assessing hydration can be a difficult one, but can usually be resolved if the history and clinical and laboratory findings are all taken into account.

### ASSESSMENT OF FLUID BALANCE

By far the most important measurement in assessing changes in day-to-day fluid balance is that of intake and output. "Insensible loss" is usually assumed to be about 900 ml per day, but this must be balanced against "insensible" production of about 500 ml water daily by metabolism. The *net* "insensible loss" is therefore the difference between these two—about 400 ml a day. A normally hydrated patient, unable to control his own balance, should be given this basic volume of fluid daily with, in addition, the volume of measured losses (urine, vomitus, etc.) during the preceding 24 hours. If he is thought to be abnormally hydrated fluid intake should be adjusted until hydration is restored. Inevitably the intake for any day must be calculated from the output in the preceding 24 hours. This is adequate when the patient starts in a normal state of hydration.

It should be remembered that a pyrexial patient may lose a litre or more of fluid in sweat, and that if he is also overbreathing, or on a respirator, respiratory water loss can be considerable. In such cases the allowance of 400 ml daily for insensible loss may not be enough.

Many very ill patients are incontinent of urine, and measurement of even this volume may be impossible. Changes in body fluid may be assessed by daily weighing, since 1 litre of water weighs 1 kg. Over short periods of time changes in solid body weight will be small, and alteration in weight can be assumed to be due to changes in fluid balance. Unfortunately, very ill patients may be unable to sit in weighing chairs. Some hospitals have weighing beds, in which the patient is weighed with the bed and the bedclothes. In this situation, care must be taken to keep the weight of the bedclothes constant.

Assessment of fluid balance in severely ill patients may present grave problems. Every attempt must be made to make *accurate* measurements of fluid intake and loss. In most circumstances carefully kept fluid charts and daily weighing are of more importance than frequent plasma electrolyte estimations. Inaccurate charting is useless, and may be dangerous.

## ASSESSMENT OF SODIUM BALANCE

If fluid balance is accurately controlled *in a normally hydrated patient* plasma sodium concentrations are the best means of assessing sodium balance. As we shall see later, the proviso of normal hydration is important, and plasma electrolyte values *alone* should never be used to assess the need for therapy. They must be correlated with the history and the present state of hydration.

## CLINICAL FEATURES OF WATER AND SODIUM DISTURBANCES

We are now in a position to explain the immediate clinical consequences of water and sodium disturbances. These depend on changes in extracellular osmolality and hence in cellular hydration (sodium) and changes in circulating volume (water)

## CLINICAL FEATURES OF DISTURBANCES OF SODIUM CONCENTRATION

If sodium balance is disturbed without a corresponding change in that of water, most of the immediate clinical features are due to a change in *extracellular sodium concentration* and hence to a change in plasma osmolality.

Gradual changes, by allowing time for electrolyte exchange across cells walls, may produce little clinical effect until they are extreme.

**Hyponatraemia** causes cellular overhydration. The clinical effect of overhydration of cerebral cells is to cause *headache, confusion* and, later, *fits*. Passage of water into cells reduces the apparent change in extracellular sodium concentration and aggravates any extracellular water depletion.

In the presence of uraemia and hyperglycaemia, however, hyponatraemia may exist with normal, or rarely, even high plasma osmolality. Moreover, because sodium concentration is measured on a molar rather a molal basis, in gross hyperproteinaemia or hyperlipaemia sodium concentration in plasma water (the physiologically important concentration) may be normal in spite of measured hyponatraemia (see p. 36). Under these circumstances the hyponatraemia may be "appropriate".

**Hypernatraemia** causes cellular dehydration and this causes thirst. Again, the effect of dehydration of cerebral cells is to cause mental confusion and, later, coma. Passage of water from cells reduces the change in extracellular sodium concentration by dilution. This effect may be aggravated by uraemia or hyperglycaemia.

Rapid changes in the volume of cerebral cells may cause tearing of small blood vessels, with cerebral haemorrhages ranging from petechiae to major cerebral catastrophes. It is most important to bear this in mind during treatment.

## CLINICAL FEATURES OF DISTURBANCES OF FLUID VOLUME

If the extracellular osmolality is unchanged the clinical features of disturbances of water metabolism are due to changes in circulating volume.

**Water deficiency** causes hypovolaemia. If cellular hydration is not altered by changes in effective osmolality across cell membranes clinical signs of "dehydration" and thirst may be absent. *Hypotension and collapse* follow, and death may result from circulatory insufficiency. Laboratory findings are those of *haemoconcentration* with a raised plasma protein concentration and haematocrit. Because renal perfusion is poor, glomerular filtration rate may fall and there may be *uraemia* and other signs of nitrogen retention.

**Water excess**, in the absence of changes in effective OP across cell walls, causes *hypertension* and overloading of the heart with *cardiac failure*. *Oedema* occurs when albumin levels are low, because the reduced oncotic pressure together with increased intravascular hydrostatic pressure causes passage of water into interstitial fluid. Laboratory findings are characteristic of *haemodilution* and, unless there is renal failure, the plasma urea level tends to be low.

## DISTURBANCES OF SODIUM AND WATER METABOLISM

Disturbances of sodium and water metabolism are most commonly due to excessive losses from the body. Sometimes inadequate intake contributes to the deficiency. In the presence of normal homeostatic mechanisms excessive intake is rarely of clinical importance.

Disturbances can also be due to abnormal amounts of the homeostatic hormones, aldosterone and ADH, or to failure of the end organs, particularly the kidney and intestine, to respond to them.

### WATER AND SODIUM DEFICIENCY

Pure water and pure sodium deficiency are rare. However, conditions in which either is lost in excess of the other are relatively common. Predominant sodium depletion is almost always accompanied by some water depletion and *vice versa*.

TABLE III

APPROXIMATE SODIUM CONCENTRATIONS IN BODY FLUIDS (mmol/l)

| Plasma | Gastric | Biliary and pancreatic | Small intestinal | Ileal | Ileostomy (new) | Diarrhoea | Sweat |
|--------|---------|------------------------|------------------|-------|------------------|-----------|-------|
| 140 | 60 | 140 | 110 | 120 | 130 | 60 | 60 |

**Predominant Water Depletion**

This syndrome is due to loss of water in excess of loss of sodium. It is usually the result of the loss of fluid containing a lower concentration of sodium than that of plasma, or of deficient water intake. Sweat and gastric juice contain concentrations of sodium much lower than those in plasma (see Table III); in diabetes insipidus, during an osmotic diuresis, or in the rare inborn error associated with failure of the renal tubules to respond to ADH, urine of low sodium concentration is passed. *As hyperosmolality due to predominant water loss causes thirst, effects are only seen if water is not available or cannot be taken in adequate quantities.*

The clinical situations associated with predominant water loss are:

1. Water deficiency in the presence of normal homeostatic mechanisms.

    (i) Excessive fluid loss—(*a*) Loss of excessive amounts of sweat.

                                   (*b*) Loss of gastric juice.

                                   (*c*) Loss of fluid stools of low sodium content (usually in infantile gastro-enteritis).

                                   (*d*) Excessive respiratory loss, especially on a respirator.

                                   (*e*) Loss of fluid from extensive burns.

    (ii) Deficiency of fluid intake—Inadequate water supply, or mechanical obstruction to its intake.

2. Failure of homeostatic mechanisms for water retention.

    (i)   Inadequate response to thirst mechanism (for example in comatose patients and infants). It is doubtful if significant water depletion can develop if thirst mechanisms are normal and water is available.

    (ii)  Deficiency of ADH (diabetes insipidus).

    (iii) Overriding of ADH action by osmotic diuresis.

    (iv) Failure of renal tubular cells to respond to high levels of ADH (nephrogenic diabetes insipidus).

In most clinical states associated with predominant water depletion more than one of these factors is responsible.

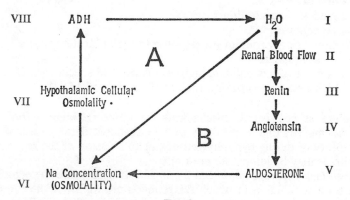

Fig. 2

**1. Predominant water depletion with normal homeostatic mechanisms.—** The reader should refer to Fig. 2 during the following explanation.

Triangle B—Immediate Effects:

(a) Loss of water in excess of sodium increases the plasma sodium concentration and osmolality (diagonal line I–VI).

(b) Loss of water reduces renal blood flow and stimulates aldosterone production (I–V). Sodium is retained and the rise in plasma sodium is aggravated (V–VI).

Triangle A—Compensatory Effects:

Increased plasma osmolality stimulates:

(a) Thirst, increasing water intake if water is available and if the patient can respond to it (not shown on diagram).

(b) ADH secretion (VI–VIII). Urinary volume falls (VIII–I) and water loss is minimised.

If adequate amounts of water are available depletion is rapidly corrected. In the absence of intake, or in the presence of continued loss by extrarenal routes, the mechanisms in Triangle A cannot function and homeostasis breaks down. Hypernatraemia is therefore an early finding in water depletion: it is important to realise that it may be present before clinical signs of dehydration are obvious.

The clinical signs are those of:
1. Hypernatraemia (p. 43).
2. Later, water deficiency (p. 43).
3. ADH secretion leading to oliguria.

The findings are:
1. Haemoconcentration.
2. Hypernatraemia.
3. Low urinary sodium (in response to high aldosterone levels).
4. Mild uraemia.
5. A urine of low volume, high osmolality and specific gravity and high urea concentration (due to the action of ADH).

**2. Failure of homeostatic mechanisms for water retention.—**_Diabetes insipidus_ may be due to pituitary or hypothalamic damage caused by head injury or during hypophysectomy, or to invasion of the region by tumour. It may be idiopathic in origin.

_Hereditary nephrogenic diabetes insipidus_ is a rare inborn error of renal tubular function in which ADH levels are high but the tubules cannot respond to it (compare pseudohypoparathyroidism, p. 245): the newborn infant passes large volumes of urine and rapidly becomes dehydrated. Urinary loss is difficult to assess at this age, and the cries of thirst may be misinterpreted. During the *recovery phase of acute oliguric renal failure*, or in other conditions with predominant tubular damage, there may also be failure to respond to ADH.

The reader should refer again to Fig. 2 and to the immediately preceding section.

The first stage is identical with that described in the previous section. Because Triangle A is not functioning (*b*) cannot occur, and the point is soon reached at which intake cannot adequately replace loss.

The clinical signs and findings are those described in the previous section, with the exception that there is polyuria, not oliguria, and that the urine is of low osmolality, specific gravity and urea concentration. These findings are the result of ADH deficiency. If any doubt remains as to the diagnosis, water may be deliberately withheld (see Appendix to Chapter I). In the absence of ADH, or when the tubules fail to respond to it, concentration of urine fails to occur. In the case of true ADH deficiency a concentrated urine will be passed if exogenous ADH (pitressin) is given. In nephrogenic diabetes insipidus maximal amounts of ADH are already circulating, and administration of it will not influence water reabsorption.

**The unconscious patient.**—The syndrome of predominant water depletion with hypernatraemia is seen most commonly in the unconscious or confused patient or in the infant with gastro-enteritis or pneumonia. In such subjects there is usually more than one cause of water depletion.

1. They are often pyrexial. Loss of hypotonic *sweat* is increased.

2. They may be *overbreathing* because of pneumonia, acidosis or brain stem damage. Water loss is increased.

3. Humidifiers on *respirators* are rarely adequate and artificially respired patients, who tend to be hyperventilated, may become water depleted.

4. They may be given hypertonic intravenous infusions, either to provide nutrient (dextrose or amino acids) or to produce osmotic diuresis in cases of poisoning (thus increasing the urinary loss of poison). There may be tissue damage and hence breakdown of protein to urea. The coma may be due to diabetes mellitus with glycosuria. All these factors will cause an *osmotic diuresis* overriding the effect of ADH.

5. There may be true diabetes insipidus due to head injury.

6. *The subject cannot respond to hyperosmolality by drinking.*

In the presence of factors 4 and 5 a high urine volume contributes to dehydration (and is *not* an indication of "good" hydration) and the extent of loss may not be realised if the subject is incontinent. Homeostatic mechanisms may not be capable of responding to hyperosmolality (5 and 6) and hypernatraemia is an early finding. It may occur before clinical signs of dehydration are evident and is dangerous because of the resulting cellular dehydration.

Such a syndrome is also seen in unconscious or confused patients with extensive *burns*. The water in the exuded ECF evaporates, while some of the electrolyte is reabsorbed into the circulation. Tissue breakdown and intravenous feeding contribute to an osmotic diuresis.

### Predominant Sodium Depletion

No normal body fluids contain significantly higher concentrations of sodium than those in plasma (see Table III, p. 44). In the presence of normal homeostatic mechanisms the commonest cause of predominant sodium depletion is loss of sodium and water followed by replacement with fluids of low sodium concentration. This is most commonly iatrogenic but may occur if patients have lost sodium in diarrhoea, vomiting or sweat, and then drink a large volume of water.

The clinical conditions accompanied by predominant sodium deficiency are:

1. Sodium deficiency in the presence of normal homeostatic mechanisms.

    (*a*) Vomiting  
    (*b*) Diarrhoea  
    (*c*) Loss through fistulae     } Followed by replacement with fluids low in sodium.  
    (*d*) Excessive sweating  

2. Failure of homeostatic mechanisms for sodium retention.

    (*a*) Addison's disease (absence of aldosterone)  
    (*b*) "Pseudo" Addison's disease (failure of the renal tubules to respond to aldosterone). This is very rare.

**1. Predominant sodium depletion with normal homeostatic mechanisms.—** The reader should refer to Fig. 2.

Triangle A. Immediate effects and emergency compensation:

    (*a*) Sodium concentration tends to fall, cutting off ADH secretion (VI–VIII).  
    (*b*) Water is lost in the urine (VIII–I).  
    (*c*) Osmolality is restored to normal (diagonal line I–VI).

This is the emergency mechanism (Triangle A), preventing shift of water across cell membranes at the expense of water loss.

Triangle B. Compensatory Mechanisms:

    (*a*) Loss of water stimulates aldosterone secretion (I–V).  
    (*b*) Sodium is retained (V–VI) and the initial sequence of events in Triangle A is reversed. Sodium and water are restored to normal.

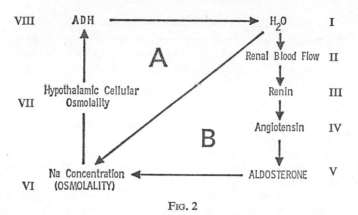

Fig. 2

In the presence of continuing sodium deficiency the homeostatic mechanisms of Triangle B cannot function effectively. Osmolality is maintained at the expense of continued water loss. Only when this is extreme will hyponatraemia occur.

Clinical signs are:
1. Those of water deficiency (p. 43).
2. *Late*, those of hyponatraemia (p. 43).

The findings are those of:
1. Haemoconcentration.
2. Renal circulatory insufficiency with mild uraemia.
3. *Late* hyponatraemia.
4. Low urinary sodium concentration (due to aldosterone secretion).

**2. Failure of homeostatic mechanisms for sodium retention.**—Addison's disease with hypoaldosteronism is by far the most important cause of this type of disturbance. During the recovery phase of acute oliguric renal failure large amounts of sodium can be lost in urine due to residual tubular dysfunction.

The reader should again refer to Fig. 2 and to the immediately preceding section.

The first stage is identical with that described in the preceding section. The homeostatic mechanisms of Triangle B are not functioning, so that the second stage cannot occur. Osmolality is maintained until late by water loss. The clinical and laboratory findings are those described in the preceding section except that, because of aldosterone deficiency, *urinary sodium concentration is high.*

## WATER AND SODIUM EXCESS

In the presence of normal homeostatic mechanisms an excess of water and electrolytes is of relatively little importance, because it is rapidly corrected. These syndromes are commonly associated with failure of homeostatic mechanisms.

### Predominant Excess of Water

Water overloading occurs in two circumstances in which normal homeostasis has failed.

1. In renal failure, where fluid of low sodium concentration has been replaced in excess of that lost. Fluid balance should be carefully controlled in such patients.

2. In the presence of "inappropriate" ADH secretion (the term "inappropriate" is used in this book to describe continued secretion of a hormone under conditions in which it should normally be cut off). ADH is one of the peptide hormones which can be manufactured by malignant tissue of non-endocrine origin (see Chapter XXI for a further discussion). "Inappropriate" secretion (possibly from the pituitary or hypothalamus itself) occurs in a variety of other conditions, including infections. Such ADH production is not under normal feed-back control and therefore continues to be produced in circumstances in which its secretion should be completely cut off (in the presence of low extracellular osmolality). This fact is evidenced by the production of a relatively concentrated urine in the presence of a dilute plasma.

If we refer again to Fig. 2 we will see how excessive intake is normally corrected.

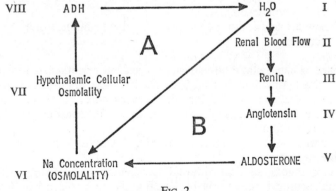

FIG. 2

Triangle B. Immediate effects:
(a) Excess water tends to lower plasma sodium (diagonal arrow), (I–VI).

(*b*) Increased renal blood flow cuts off aldosterone production, increasing urinary sodium loss and therefore further decreasing plasma sodium (V–VI).

Triangle A. Compensatory effects:
    ADH is cut off (VI–VIII) and large volumes of dilute urine are passed (VIII–I).

When renal glomerular function is poor, or in the presence of "inappropriate" ADH secretion (which is not subject to feed-back control), Triangle A is out of action. The second stage cannot take place.
    The clinical consequences are:
    1. Those of water excess.
    2. If overhydration is rapid, those of hyponatraemia.

Typically these patients may be remarkably well clinically with very low plasma sodium concentrations. This contrasts with the clinical state in true sodium depletion, when hypotension leads to collapse.
    The findings are:
    1. Haemodilution.
    2. Hyponatraemia.

If the cause is glomerular failure there will be uraemia. If it is due to "inappropriate" ADH secretion the plasma urea level will tend to be low.

#### Predominant Excess of Sodium

Predominant sodium excess is most commonly due to an "inappropriate" secretion of excess aldosterone or other corticoids in Conn's syndrome (primary aldosteronism), or in Cushing's syndrome, or to excessive stimulation of a normally functioning adrenal in secondary aldosteronism. Cushing's syndrome is described more fully on p. 144.
    **Primary aldosteronism** (Conn's syndrome).—Primary aldosteronism, or excess aldosterone secretion not subject to feed-back control, is most commonly due to a benign aldosterone secreting adenoma of the adrenal cortex.
    The reader should refer again to Fig. 2.
    1. Excess aldosterone causes urinary sodium retention (V–VI).
    2. The increased sodium concentration stimulates ADH secretion. (VI–VIII) and water retention (VIII–I).
    3. Water retention tends to return plasma sodium concentrations to normal (diagonal arrow, I–VI).
    4. Triangle B is out of action and aldosterone secretion cannot be

cut off. This tends to maintain plasma sodium levels at or near the upper end of the normal range.

5. Continued action of aldosterone causes sodium retention at the expense of potassium loss, and potassium depletion occurs.

The clinical features are:

1. Those of water excess (p. 43). Patients are hypertensive but rarely oedematous.
2. Those of hypokalaemia (p. 70).

Findings are:

1. Hypokalaemia (due to excess aldosterone).
2. A high plasma bicarbonate (for explanation see p. 64).
3. A plasma sodium in the high normal range or just above it.
4. A low urinary sodium in the early stages. However, sodium excretion may rise later, possibly because of hypokalaemic tubular damage (p. 15), and consequent unresponsiveness to aldosterone, or because of the effect of the expanded plasma volume on non-aldosterone factors which affect sodium excretion (p. 38).

Hypokalaemic alkalosis occurs in potassium depletion from any cause. However, the association of these findings in a patient without an obvious cause (such as administration of purgatives or diuretics) for potassium loss and with hypertension should suggest the diagnosis of primary aldosteronism. Further suggestive evidence would be the finding of high aldosterone with low renin levels (in all cases of *secondary* aldosteronism *both* are high). Although these estimations are unavailable in most routine laboratories, in the rare cases in which they are needed they can be performed in special centres.

**Secondary aldosteronism: Oedema.**—Any of the conditions already described in the sections on water and sodium depletion, in which aldosterone secretion is stimulated following reduction in renal blood flow could, strictly, be called secondary aldosteronism. The term is more commonly used to indicate the conditions in which, because the initial abnormality is not corrected, long-standing increased aldosterone secretion itself produces abnormalities.

Aldosterone is secreted following stimulation of the renin-angiotensin system by a low renal blood flow. This may occur either because of local abnormalities in renal vessels or because of a reduced circulating volume.

Secondary aldosteronism, in the usually accepted sense of the word, occurs in the following conditions:

1. Redistribution of extracellular fluid, leading to a reduction of plasma volume in the presence of normal or high total extracellular

fluid volume. These conditions are due to a reduced plasma oncotic pressure, and are therefore associated with low plasma albumin levels. Oedema is present. Such conditions are:

(a) Liver disease.
(b) Nephrotic syndrome.
(c) Protein malnutrition.

2. Damage to the renal vessels, reducing renal blood flow. These conditions are usually not associated with oedema:

(a) Essential hypertension.
(b) Malignant hypertension.
(c) Renal hypertension (e.g. renal artery stenosis).

3. Cardiac failure. In this case two factors may cause low renal blood flow. Firstly the cardiac output may be low, with poor renal perfusion pressure. Secondly, high intravascular hydrostatic pressure on the venous side of the circulation may cause redistribution of fluid and oedema.

The mechanisms in Fig. 2 are brought into play:

1. Reduced renal blood flow stimulates aldosterone secretion (I–V).
2. Sodium retention stimulates ADH secretion (V–VIII) and therefore water retention (VIII–I).

Intravascular volume tends to be restored, but in conditions with hypoalbuminaemia or in cardiac failure more fluid passes into the interstitial fluid and the cycle restarts. A vicious circle is set up in which circulating volume can only be maintained by water retention and oedema results. In non-oedematous states hypertension occurs.

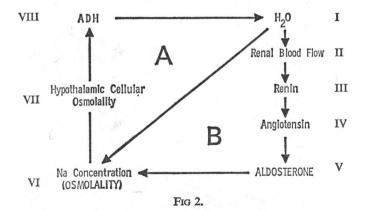

FIG 2.

This cycle should only stimulate water retention in parallel with sodium retention. However, many of these patients have hyponatraemia.

TABLE IV

HYPOTHETICAL NUMERICAL EXAMPLES TO ILLUSTRATE CONDITIONS ASSOCIATED WITH NORMAL AND ABNORMAL PLASMA SODIUM CONCENTRATIONS

| | Example | ECF (litres) | E.C. Na (mmol) | Plasma Na (mmol/l) | Haemoconc. Dilution | Clinical Features | Principle of Treatment |
|---|---|---|---|---|---|---|---|
| **Normonatraemia (Equivalent Na and H$_2$O changes)** | | | | | | | |
| Normal | — | 20 | 2,800 | 140 | None | — | — |
| Na + H$_2$O depletion | Early Addison's disease | 15 ↓ | 2,100 ↓ | 140 | Conc. (Urea ↑) | Water deficiency (p. 43) | Treat cause Isosmolal saline |
| Na + H$_2$O excess | Non-oedematous secondary aldosteronism | 30 ↑ | 4,200 ↑ | 140 | Dil. | Water excess (p. 43) | Treat cause Restrict Na + H$_2$O |
| **Hyponatraemia (Relative H$_2$O excess)** | | | | | | | |
| H$_2$O excess | Inappropriate ADH secretion | 25 ↑ | 2,800 | 112 ↓ | Dil. (Urea N or ↓) | Water excess (p. 43) Hypo-osmolality (p. 43) | Restrict H$_2$O |

| | | | | | | | |
|---|---|---|---|---|---|---|---|
| | Na depletion | Diarrhoea with H₂O replacement | 20 | 2,240 ↓ | 112 ↓ | None | Hypo-osmolality (p. 43) | Give hyperosmolal saline |
| | Na + H₂O excess | Oedematous secondary aldosteronism (commonest form) | 30 ↑ | 3,360 ↑ | 112 ↓ | Dil. | Water excess (p. 43) Hypo-osmolality (p. 43) | Restrict Na and H₂O Diuretics. |
| | Na + H₂O depletion | Late Addison's disease | 15 ↓ | 1,680 ↓ | 112 ↓ | Conc. (Urea ↑) | Water deficiency (p. 43) Hypo-osmolality (p. 43) | Give steroids + normosmolal or hyperosmolal saline |
| Hypernatraemia (Relative Na excess) | H₂O depletion | Unconscious patient | 18 ↓ | *2,800* | 155 ↑ | Mild conc. | Hyper-osmolality (p. 43) | Give hypo-osmolal fluid *slowly* |
| | Sodium excess | Excessive intake usually in infants (*Very rare*) | 20 | 3,100 ↑ | 155 ↑ | None | Hyper-osmolality (p. 43) | Remove Na by dialysis |

Values in first line taken as "normals" to which others are related. "Normal" values in italics throughout.
No account has been taken of shifts of fluid across cell walls which would slightly reduce changes in plasma sodium levels.
Note especially relatively slight reduction in volume associated with hypernatraemia.

The reasons for this are not clear, but from the therapeutic point of view it must be remembered that in the presence of oedema the total amount of sodium in the body is high, even if there is hyponatraemia (see Table IV).

Hypokalaemia may be present, but is a less common finding than in primary aldosteronism. Again the reason is not clear, but it may be due to a redistribution of potassium between cells and extracellular fluid. Hypokalaemia is, however, more readily precipitated by diuretic therapy in secondary hyperaldosteronism than in normal subjects.

The clinical features of these cases are those of the primary condition.

The findings are:

1. A normal or low plasma sodium concentration.
2. A low urinary sodium excretion.
3. Findings due to the primary abnormality, e.g. hypoalbuminaemia, uraemia, etc.

### THE DIAGNOSTIC VALUE OF URINARY SODIUM CONCENTRATION

It will be noticed that urinary sodium excretion is low when aldosterone secretion is high. In the absence of primary aldosteronism this finding indicates low renal blood flow. Similarly, except in Addison's disease, high urinary sodium excretion indicates high renal blood flow. The estimation gives no information about the state of body sodium stores. In most cases the state of hydration and sodium metabolism can be assessed by noting the clinical picture and plasma findings. Only in the rare cases in which doubt still remains or in the differential diagnosis of renal tubular and glomerular lesions (p. 19) is urinary sodium estimation indicated.

## THE CLINICAL SIGNIFICANCE OF PLASMA SODIUM CONCENTRATIONS

Table IV summarises the situations in which hypernatraemia and hyponatraemia may be found, and indicates some aspects of the clinical picture which may help to differentiate the causes. The two important points to remember are:

1. **Hypernatraemia** almost invariably indicates water depletion.
2. **Hyponatraemia** is much more commonly due to water excess than to sodium depletion: if this is so the body content of sodium may be normal, but is more often high. If it is due to the sodium depletion the patient will be dehydrated, both on clinical and laboratory findings, with mild to moderate uraemia.

Plasma sodium levels down to about 120 mmol/l occur in many ill patients, and, unless there is evidence of dehydration, require no treatment. In states in which plasma osmolality is contributed to significantly by other substances, as in severe uraemia and hyperglycaemia, mild hyponatraemia may be "appropriate".

(The student is urged to read the indications for electrolyte estimation on p. 75.)

## BIOCHEMICAL BASIS OF TREATMENT OF SODIUM AND WATER DISTURBANCES

### TREATMENT OF SODIUM AND WATER DISTURBANCES

It cannot be stressed too strongly that treatment should not be based on plasma sodium concentrations alone. Correct treatment of, for example, hyponatraemia depends on a knowledge of the state of hydration and renal function of the patient, and involves assessment of the history, clinical findings, laboratory findings indicating haemoconcentration or haemodilution, and the level of blood urea. In rare circumstances it is also useful to know the urinary sodium and urea concentrations and osmolality or specific gravity. The cause of the disturbance should be sought and treated, but mild hyponatraemia rarely requires treatment in itself. Hypernatraemia, on the other hand, should always be treated by *slow* infusion of hypotonic fluid.

Table IV uses hypothetical values to illustrate the various combinations of disturbances of sodium and water metabolism and their treatment. Solutions available for intravenous use are given in the Appendix. These rules can only be a guide to treatment in often complicated clinical situations. They may, however, help to avoid some of the more dangerous errors of electrolyte therapy.

When homeostatic mechanisms have failed, and especially in renal failure, normal hydration should be maintained according to the principles outlined on p. 41.

## SUMMARY

1. Homeostatic mechanisms for sodium and water are interlinked. Potassium and hydrogen ion often take part in exchange mechanisms with sodium.

2. Distribution of fluid between cells and extracellular fluid depends on osmotic differences between the intra- and extracellular fluid. Changes in this are usually due to changes in sodium concentrations.

3. Distribution of fluid between the intravascular and interstitial compartments depends on the balance between the hydrostatic pressure

and the effective plasma osmotic pressure: the latter depends largely on albumin concentration.

4. Aldosterone secretion is the most important factor affecting body sodium.

5. Aldosterone secretion is controlled by the renin-angiotensin mechanism which responds to changes in renal blood flow.

6. ADH secretion is the most important factor affecting body water.

7. ADH secretion is controlled by plasma osmolality. Plasma osmolality depends mainly on sodium concentration.

8. Clinical effects of disturbances of water and sodium metabolism are due to:

(a) Changes in extracellular osmolality, dependent largely on sodium concentration. In pathological conditions urea and glucose concentrations can be important.

(b) Changes in extracellular volume.

## FURTHER READING

CHRISTENSEN, H. N. (1964). *Body Fluids and the Acid-Base Balance.* Philadelphia: W. B. Saunders. (A programmed text).

MULROW, P. J., and FORMAN, B. H. (1972). Tissue effects of mineralocorticoids. *Amer. J. Med.*, 53, 561.

WILLIAMS, G. H., and DLUHY, R. G. (1972). Aldosterone biosynthesis. Interrelationship of regulatory factors. *Amer. J. Med.*, 53, 595.

# APPENDIX

## TABLE V

### Some Sodium Containing Fluids for Intravenous Administration

| | Electrolyte Content (mmol/l) | | | | Glucose (mmol/l with g/dl in brackets) | Ca (mmol/l) | Approximate osmolarity relative to plasma |
|---|---|---|---|---|---|---|---|
| | Na | K | Cl | HCO$_3$ | | | |
| **1. Saline** | | | | | | | |
| "Normal" (physiological) saline | 154 | — | 154 | — | — | — | × 1 |
| Twice "Normal" | 308 | — | 308 | — | — | — | × 2 |
| Half "Normal" | 77 | — | 77 | — | — | — | × ½ |
| Fifth "Normal" | 31 | — | 31 | — | — | — | × ⅕ |
| **2. "Dextrose" Saline** | | | | | | | |
| | 77 | — | 77 | — | 149 (2·69) | — | × 1 |
| | 31 | — | 31 | — | 240 (4·3) | — | × 1 |
| | 77 | — | 77 | — | 278 (5·0) | — | × 1½ |
| | 154 | — | 154 | — | 278 (5·0) | — | × 2 |
| | 154 | — | 154 | — | 556 (10·0) | — | × 3 |
| **3. Sodium Bicarbonate** | | | | | | | |
| 1·4% | 167 | — | — | 167 | — | — | × 1 |
| 2·8% | 334 | — | — | 334 | — | — | × 2 |
| 8·4% | 1000 | — | — | 1000 | — | — | × 6 |

TABLE V (*continued*)

| | Electrolyte Content (mmol/l) | | | | Glucose (mmol/l with g/dl in brackets) | Ca (mmol/l) | Approximate osmolarity relative to plasma |
|---|---|---|---|---|---|---|---|
| | Na | K | Cl | HCO₃ | | | |
| **4. Complex Solutions** | | | | | | | |
| Ringer's Solution | 147 | 4·2 | 156 | — | — | 2·2 | × 1 |
| Darrow's Solution | 122 | 35·8 | 104 | 53 (as lactate) | — | — | × 1 |
| Hartmann's Solution | 131 | 5·4 | 112 | 29 (as lactate) | — | 1·8 | × 1 |
| Sodium lactate 1/6 molar | 167 | — | — | 167 (as lactate) | — | — | × 1 |
| *Amino acid solutions* | | | | | | | |
| "Amigen" 5% | 35 | 18 | 22 | | 278 (5·0) | 2·5 | Hyperosmolar |
| "Aminosol 3·3%" | 53 | 0·15 | | | | | Hyperosmolar |
| "Aminosol 10%" | 160 | 0·5 | | | | | Hyperosmolar |
| "Trophysan 5" | 6 | 8 | | | | | Hyperosmolar |

TABLE VI

SODIUM FREE DEXTROSE SOLUTIONS FOR INTRAVENOUS ADMINISTRATION (USUALLY USED TO PROVIDE CALORIES)

| Dextrose | Glucose (mmol/l) | Osmolarity | Calories/l |
|---|---|---|---|
| 5% | 278 | Isosmolar | 205 |
| 10% | 556 | Hyperosmolar | 410 |
| 20% | 1112 | Hyperosmolar | 820 |
| 40% | 2224 | Hyperosmolar | 1640 |

## TABLE VII

### Other Hyperosmolar Solutions (Usually Used as Osmotic Diuretics)

| | | Concentration (mmol/l) | Calories/l |
|---|---|---|---|
| Mannitol | 10% | 550 | 0 |
| | 20% | 1100 | 0 |
| Urea | 4% | 667 | 0 (But may also contain fructose 10% giving 410 calories/l) |
| | 30% | 5000 | 0 |
| Sorbitol | 20% | 1100 | 800 |
| | 30% | 1650 | 1200 |
| Fructose | 10% | 556 | 410 |
| | 20% | 1112 | 820 |
| | 40% | 2224 | 1640 |

# Chapter III

# POTASSIUM METABOLISM:
# DIURETIC THERAPY

THE total amount of potassium in the body is about 3000 mmol.

Potassium is predominantly an *intracellular* ion, and only about 2 per cent of the body content is in the extracellular fluid. Plasma potassium levels are therefore a poor indication of total body amounts. This apparent disadvantage is more theoretical than real, because it is *plasma potassium concentrations* that are of immediate importance in therapy; either hyperkalaemia or hypokalaemia, if severe, is dangerous, and must be treated whatever the state of the intracellular potassium. However, some assessment of the overall situation should be made to anticipate therapeutic needs.

Like calcium and magnesium (p. 234) the very low extracellular concentration of potassium ions is important for normal neuromuscular activity and cardiac action.

## FACTORS AFFECTING PLASMA POTASSIUM CONCENTRATIONS

The predominantly intracellular location of potassium provides a reservoir of the ion, and plasma potassium concentration, unlike that of sodium, is relatively little affected by changes in water balance. The hyperkalaemia often associated with dehydration is due more to renal retention than directly to haemoconcentration.

Potassium enters and leaves the extracellular compartment by three main routes:

1. The intestinal mucosa
2. The kidney
   (*a*) The glomerulus
   (*b*) The tubular cells
3. The walls of *all* body cells

### The Intestine

Potassium leaves the extracellular compartment in all intestinal secretions. The approximate concentrations in these secretions are indicated in Table VIII. Although most of them are not very high, the daily volume secreted into the lumen of the gut is large (p. 31) and contains about 100 mmol/day. Most of this potassium is reabsorbed, together with the dietary intake, and very little potassium is present in formed faeces. As in the case of sodium, excessive intestinal loss of potassium in diarrhoea stools, in ileostomy fluid or via fistulae is mainly derived from intestinal secretions rather than from dietary intake. However, prolonged starvation can cause potassium deficiency and hypokalaemia.

TABLE VIII

APPROXIMATE POTASSIUM CONCENTRATIONS IN BODY FLUIDS (mmol/l)

| Plasma | Gastric | Biliary and pancreatic | Small intestinal | Ileal | Ileostomy (new) | Diarrhoea | Sweat |
|--------|---------|------------------------|------------------|-------|-----------------|-----------|-------|
| 4 | 10 | 5 | 5 | 5 | 15 | 40 | 10 |

### The Kidney

(a) **The glomerulus.**—The concentration of potassium in the glomerular filtrate is the same as that in plasma. Because of the very large filtered volume, daily loss by this route would be about a third of the body content (about 800 mmol) if there were no tubular function.

(b) **The tubules.**—Potassium is almost completely reabsorbed in the *proximal tubule* and renal tubular damage can cause potassium depletion.

Potassium is resecreted in the *distal tubule* and *collecting ducts* in exchange for *sodium*. Hydrogen ions compete with potassium for this exchange, which is stimulated by *aldosterone* (p. 37). If the proximal tubule is functioning, potassium loss in the urine depends on three factors:

(i) The amount of *sodium available for exchange*. This depends on the *GFR, sodium load* on the glomerulus, and *proximal tubular reabsorption*. The latter is inhibited by many diuretics (p. 69).

(ii) The relative *amounts of hydrogen and potassium ions* present in the cells of the distal tubule and collecting ducts, and the *ability to secrete $H^+$ in exchange for $Na^+$* (inhibited during therapy with carbonic anhydrase inhibitors and in some types of renal tubular acidosis).

(iii) The circulating *aldosterone* level. This is increased following water loss (which usually accompanies intestinal loss of potassium) (p. 46) and in almost all conditions requiring diuretic therapy.

## The Cell Wall

Potassium is the predominant intracellular cation and is continuously lost from the cell down a concentration gradient. This loss is opposed by the "sodium pump" situated at the cell surface. This pumps sodium out of the cell in exchange for potassium and hydrogen ions.

In most circumstances the shift of potassium across cell membranes is accompanied by a shift of sodium in the opposite direction, but the *percentage* change in extracellular sodium levels will be much less than that of potassium. A simplified example will demonstrate this. Let us assume extra- and intracellular volumes to be equal, the plasma sodium to be 140 mmol/l and potassium 4 mmol/l, with reversal of these concentrations inside cells. An exchange of 4 mmol/l of sodium for potassium across the cell wall would double the plasma potassium concentration (a very clinically significant change) while only reducing the plasma sodium concentration to 136 mmol/l. *Extracellular* levels are the physiologically important ones, and these shifts affect potassium more than sodium.

There is *net loss* of potassium from the cell:

(*a*) If potassium is lost from the ECF and replenished from the cell.

(*b*) If the sodium pump is inefficient, as in diabetic ketoacidosis and in anoxic states.

(*c*) In acidosis.

There is *net gain* of potassium by the cell:

(*a*) If the activity of the sodium pump is increased, as it is after the administration of glucose and insulin: this effect may be used to treat hyperkalaemia. It is the main cause of the change from hyperkalaemia to hypokalaemia during treatment of diabetic coma.

(*b*) In alkalosis. Induction of alkalosis can be used to treat hyperkalaemia.

### Interrelationship Between Hydrogen and Potassium Ions

Extracellular hydrogen ion concentration affects the entry of potassium into all cells: changing the relative proportions of $K^+$ and $H^+$ in distal tubular cells affects urinary loss of potassium. In acidosis decreased entry of potassium into the cells from the ECF, coupled with reduced urinary secretion of the ion, causes hyperkalaemia: in alkalosis hypokalaemia is due both to net increased entry of potassium into cells and to increased urinary loss.

As the relationship between $K^+$ and $H^+$ is reciprocal, changes in $K^+$ balance affect the hydrogen ion balance of the body. In potassium depletion loss from the ECF is followed by loss from cells. $H^+$ and $Na^+$ enter cells, and an *extracellular alkalosis* develops, but there is an *intracellular acidosis*. In the tubular cells the gain of $H^+$ results in more of this ion being available for exchange with intraluminal $Na^+$: the *acid urine* reflects the intracellular acidosis rather than the extracellular alkalosis.

Reabsorption and regeneration of $HCO_3^-$, both dependent on $H^+$ secretion (p. 85), are increased: chronic potassium depletion is therefore accompanied by a *high plasma* $[HCO_3^-]$. If other causes for a raised plasma bicarbonate (such as respiratory disease) are absent, and especially if factors known to cause potassium depletion (for instance, diuretic therapy) are present, this finding is a sensitive indication of potassium depletion, even in the absence of hypokalaemia. *The combination of hypokalaemia and a high plasma $[HCO_3^-]$ is more likely to be due to $K^+$ depletion, which is common, than primarily to metabolic alkalosis which is rare.* These are the changes of *chronic* potassium depletion.

In *acute* loss the slight lag in potassium release from cells may result in more severe hypokalaemia for the same degree of depletion than in the chronic state: bicarbonate levels are less likely to be raised, because significant bicarbonate retention by the kidney takes several days.

Theoretically, potassium excess could cause intracellular alkalosis and extracellular acidosis, with a low plasma bicarbonate. However, *the combination of hyperkalaemia and a low plasma $[HCO_3^-]$ is more likely to be due to metabolic acidosis, which is common, than primarily to potassium excess, which is extremely rare.* In *respiratory* acidosis the

TABLE IX

INTERRELATIONSHIPS OF PLASMA POTASSIUM AND BICARBONATE LEVELS

| Plasma $[K^+]$ | Plasma $[HCO_3^-]$ | Most likely cause | Examples of Clinical Conditions |
|---|---|---|---|
| Low N or $\downarrow$ | $\uparrow$ | Chronic $K^+$ depletion | Diuretic therapy *Chronic diarrhoea (chronic purgative-takers) |
| $\downarrow$ or $\downarrow\downarrow$ | N or $\downarrow$ | Acute $K^+$ depletion | Severe acute diarrhoea Fistulae etc. |
| $\downarrow$ | $\downarrow$ | Respiratory alkalosis | Overtreatment on respirator Hysterical overbreathing |
| $\uparrow$ | $\downarrow$ | Metabolic acidosis | *Renal failure Diabetic ketoacidosis |
| $\uparrow$ | $\uparrow$ | Respiratory acidosis | Bronchopneumonia |
| $\uparrow$ | N | Acute $K^+$ load | Excessive $K^+$ therapy |

* It should be stressed that this is *only a guide*. For instance, in severe diarrhoea bicarbonate loss may be so high that $[HCO_3^-]$ is reduced in spite of potassium depletion: similarly, renal tubular lesions cause depletion of both $HCO_3^-$ and $K^+$.

plasma bicarbonate is high (p. 98), but the plasma potassium will also tend to be high.

These various situations are summarised in Table IX.

## ABNORMALITIES OF PLASMA POTASSIUM LEVELS

In any clinical situation no single factor entirely accounts for the changes in plasma potassium concentration. For instance, in conditions associated with intestinal potassium loss, concomitant water loss causes secondary hyperaldosteronism; similarly, most conditions requiring diuretic therapy are associated with hyperaldosteronism. This hyperaldosteronism aggravates urinary loss and may also increase entry of potassium into body cells generally. However, if dehydration and sodium depletion are very severe, the reduced amount of filtered sodium means that less is available for exchange with potassium in the distal tubule and aldosterone cannot produce maximal effects: hypokalaemia will then be "unmasked" as the patient is rehydrated. This should be anticipated.

### HYPOKALAEMIA

This is usually the result of potassium depletion, although, if the rate of loss of potassium from cells equals or exceeds that from the body, potassium depletion may not cause hypokalaemia. It can occur without depletion if there is a shift into cells, as in alkalotic states and in the rare condition, familial periodic paralysis.

The causes of hypokalaemia may be classified as follows:

1. Predominantly due to **loss of potassium from the body.**

(a) Predominantly due to loss from the ECF in *intestinal secretions.*
  *Prolonged vomiting*
  *Diarrhoea*
  *Loss through intestinal fistulae*

The intestinal loss is usually aggravated by the secondary hyperaldosteronism consequent on water loss. This causes an inappropriately high urinary loss. The following important points should be noted:

(i) Fluid from a *recent ileostomy* and *diarrhoea stools* are particularly rich in potassium (Table VIII), but a prolonged drain of *any* intestinal secretion causes depletion.

(ii) *Habitual purgative-takers* may present with hypokalaemia and are often reluctant to admit to the habit.

(b) Predominantly due to loss from the ECF in *urine.*

(i) Increased activity of sodium : potassium exchange mechanisms in the distal nephron.

*Secondary hyperaldosteronism.*—This often aggravates other causes of potassium depletion.

*Cushing's syndrome and steroid therapy.*—Patients secreting excess of, or on prolonged therapy with, glucocorticoids tend to become hypokalaemic due to the mineralocorticoid effect on the distal tubule.

*Primary hyperaldosteronism* (p. 51).

*Synacthen or ACTH therapy and ectopic ACTH production* (p. 448).

(ii) Excess available sodium for exchange in the distal nephron.

*Diuretics inhibiting proximal sodium reabsorption* (p. 69). The increased sodium : potassium exchange is aggravated by secondary hyperaldosteronism.

(iii) Decreased renal sodium : hydrogen ion exchange, favouring sodium : potassium exchange.

*Carbonic anhydrase inhibitors* (p. 99).

*Renal tubular acidosis* (p. 95).

(iv) Reduced proximal tubular potassium reabsorption.

*Renal tubular failure* (e.g. polyuric phase of acute oliguric renal failure).

*"Fanconi syndrome"* (p. 14).

2. Predominantly due to **reduced potassium intake.**

*Chronic starvation.*—If water and salt intake are reduced, secondary hyperaldosteronism may aggravate the hypokalaemia.

3. Predominantly due to **redistribution** in the body. Loss from ECF into cells.

*Glucose and insulin therapy.*— This may be used to treat hyperkalaemia.

*Familial periodic paralysis* (very rare).—In this condition episodic paralysis occurs associated with spontaneous entry of potassium into cells.

4. Loss from ECF by **more than one route.**

(*a*) Into cells and urine
   *Alkalosis*

(*b*) Into cells, urine and intestine

   *Pyloric stenosis with alkalosis.*—The loss in urine and into cells are probably more important causes of hypokalaemia than the loss in gastric secretion.

## HYPERKALAEMIA

This occurs most commonly when the rate of potassium leaving cells is greater than its rate of excretion. The causes of hyperkalaemia may be classified as follows:

1. Predominantly due to **gain of potassium by the body.**
Gain by ECF from the *intestine or by an intravenous route.*
*Over-enthusiastic potassium therapy.*
*Failure to stop potassium therapy* when depletion has been corrected.

2. **Failure of renal secretion** of potassium.

(*a*) Decreased activity of sodium : potassium exchange mechanisms in distal nephron.
*Hypoaldosteronism* (as in Addison's disease)
(*b*) Too little sodium available for exchange in the distal nephron.
*Renal glomerular failure.*—Hyperkalaemia is usually aggravated by the concomitant acidosis and gain of potassium by the ECF from cells.
*Sodium depletion*
*Diuretics acting on the distal nephron* by antagonising aldosterone, or direct inhibition of "sodium pump" (p. 70).

3. Predominantly due to **redistribution** of potassium in the body.

Gain of potassium by ECF from cells.
*Severe tissue damage*
*Severe acute starvation* (as in anorexia nervosa). These cause cell damage and release of $K^+$ into the ECF.

4. **Gain by ECF by more than one route.**

Reduced renal excretion in spite of gain by ECF from cells.
*Acidosis*
*Anoxia*—Failure of "sodium pump" in all cells.
Failure in distal tubular cells causes potassium retention. If anoxia is very severe "lactic acidosis" aggravates the hyperkalaemia.

**Diabetic ketoacidosis.**—In early untreated diabetes potassium leaves the cells, and in spite of a high urinary loss and consequent body depletion hyperkalaemia is usual. This is due to partial failure of the "sodium pump" resulting from impaired glucose metabolism because of insulin lack. As the condition becomes more advanced two other factors contribute to hyperkalaemia.

(i) Dehydration with a low GFR.
(ii) Acidosis due to ketosis.

All these factors are reversed during insulin and fluid therapy.

As potassium enters cells extracellular levels fall and the depletion is revealed. *Plasma potassium levels should be monitored during therapy*, and potassium should be given as soon as concentrations start to fall.

## MEASUREMENT OF URINARY AND INTESTINAL LOSSES

Pure urinary or intestinal loss as a cause of hypokalaemia is very rare. Measurement of such losses with a view to quantitative replacement may lead to dangerous errors of therapy. Exchanges across cell walls cannot be measured. If gain of potassium by the ECF from cells is faster than loss from it into urine and intestine, replacement of measured loss could endanger the patient's life by aggravating hyperkalaemia; if loss from ECF into cells is predominant, therapy based on urinary excretion may be inadequate. Moreover, urinary losses may merely reflect plasma concentration; for instance, a high excretion is appropriate if trauma has damaged many cells, reducing cellular capacity and releasing the ion from cells.

*It is plasma potassium levels that are important*, and in rapidly changing states frequent estimation of these is the only safe way of assessing therapy. In chronic depletion the plasma bicarbonate level may help to indicate the state of cellular repletion.

## DIURETIC THERAPY

In oedematous states the fluid accumulation is accompanied by an excess of sodium in the body, *even if there is hyponatraemia* (p. 56). Diuretics act by inhibiting sodium reabsorption in the renal tubule, and secondarily causing water loss. All diuretics tend to affect potassium balance, and this effect should be anticipated.

Diuretics can be divided into two main groups.

1. *Those inhibiting proximal tubular reabsorption* of sodium (and/or, pumping of sodium in the loop of Henle) and therefore water reabsorption: the increased sodium load on the distal tubule and collecting ducts increases sodium : potassium exchange at this site (which is stimulated by the accompanying hyperaldosteronism). In our experience long-term diuretic therapy almost invariably causes significant *potassium depletion*, and sometimes symptomatic hypokalaemia, even if potassium supplements are given (although this has been questioned recently). *A high plasma bicarbonate* is common in long continued use of such diuretics. This is because loss of potassium from the ECF results in increased $Na^+ : H^+$ exchange in the distal nephron and at the "sodium pump" in all body cells. In this situation high plasma $HCO_3^-$ levels are

a more sensitive indication of $K^+$ depletion than plasma potassium levels.

Such diuretics are the *thiazide group*, *frusemide* (furosemide; "Lasix") and *ethacrynic acid* ("Edecrin"). It is claimed that frusemide and ethacrynic acid cause less potassium depletion than the thiazide group.

2. Those either directly *inhibiting aldosterone* or inhibiting the exchange mechanisms in the *distal tubule* and *collecting duct*. These cause *potassium retention* and may lead to hyperkalaemia: potassium supplements should *not* be used. Potassium retaining diuretics include:

*Spironolactone* ("Aldactone")—A competitive aldosterone antagonist.
*Amiloride* ("Midamor") ⎱ Inhibitors of the $Na^+ : K^+$ exchange
*Triamterene* ("Dytac") ⎰ mechanisms in the renal tubule.

This group of diuretics is often used, together with those causing potassium loss, when hypokalaemia cannot be controlled by potassium therapy.

In addition carbonic anhydrase inhibitors such as acetazolamide ("Diamox") (p. 99) inhibit sodium reabsorption and act as diuretics. They are rarely used for this purpose now because of the danger of acidosis.

### CLINICAL FEATURES OF DISTURBANCES OF POTASSIUM METABOLISM

The clinical features of disturbances of potassium metabolism (like those of, for example, sodium and calcium) are due to changes in extracellular concentration of the ion.

Hypokalaemia, by interfering with neuromuscular transmission, causes *muscular weakness*, *hypotonia* and *cardiac arrhythmias*. It may also aggravate paralytic ileus.

*Intracellular potassium depletion causes extracellular alkalosis* (see p. 65). This reduces ionisation of calcium salts (p. 235) and in long-standing potassium depletion of gradual onset the presenting symptom may be muscle *cramps* and *tetany*. This syndrome is accompanied by high plasma bicarbonate levels.

Prolonged potassium depletion causes lesions in renal tubular cells, and this may complicate the clinical picture.

Severe hyperkalaemia always carries the danger of cardiac arrest. Both hypokalaemia and hyperkalaemia cause characteristic changes in the electrocardiogram.

### TREATMENT OF POTASSIUM DISTURBANCES

Abnormalities of plasma potassium should be corrected whatever the state of the total body potassium. However, an attempt should be made to assess the latter so that sudden changes in plasma potassium (for instance, during treatment of diabetic coma) can be anticipated. Treatment should be controlled by frequent plasma potassium estimations.

**Hyperkalaemia.**—Treatment of hyperkalaemia is based on three principles. In severe hyperkalaemia the first two principles are used:

1. Very severe hyperkalaemia can cause cardiac arrest. Calcium and potassium have opposing actions on heart muscle, and the immediate danger can be minimised by infusion of calcium salts (usually as gluconate) (see Appendix). This allows time to institute measures to lower plasma potassium.

2. Plasma potassium can be reduced rapidly (within an hour) by increasing the rate of entry into cells. Glucose and insulin speed up glucose metabolism and the action of the "sodium pump". Induction of alkalosis by infusion of bicarbonate also increases the rate of entry into cells (see Appendix). For purely practical reasons this treatment (which involves intravenous infusion) cannot be continued indefinitely, but its use allows long-term treatment to be instituted.

3. In moderate hyperkalaemia a slower acting method can be used. Potassium can be removed from the body at a rate higher than, or equal to, that at which it is entering the extracellular fluid by using oral ion exchange resins. These are unabsorbed and exchange potassium for sodium or calcium ions. It will be seen that plasma potassium is lowered at the expense of body depletion. This potassium may have to be replaced later.

**Hypokalaemia.**—If hypokalaemia is *mild*, potassium supplements should be given *orally* until plasma potassium and bicarbonate levels return to normal. These levels should be monitored regularly. Undertreatment is more common than overtreatment during oral therapy. The normal subject loses about 60 mmol of potassium daily in the urine, much larger amounts being excreted during diuretic therapy. By the time hypokalaemic alkalosis is present the total deficit is probably several hundred mmol. A patient with hypokalaemia should be given *at least* 80 mmol a day: much more may be needed if plasma levels fail to rise.

In *severe* hypokalaemia, particularly if the patient is unable to take oral supplements, *intravenous potassium* should be given cautiously (see Appendix). Diarrhoea, if present, not only reduces absorption of oral potassium supplements, but may itself be aggravated by them. Such a situation may also be an indication for intravenous therapy.

### SUMMARY

1. Changes in plasma potassium levels are the net result of changes between ECF and cells, kidney and gut.

2. In any clinical situation many factors are involved, and monitoring of plasma potassium levels is the only safe guide to treatment.

3. As hydrogen and potassium ions compete for exchange with sodium across cell walls and in the renal tubule, disturbances of hydrogen ion homeostasis and potassium balance often coexist. A raised plasma bicarbonate may indicate intracellular potassium depletion.

4. Clinical manifestations of disturbances of potassium metabolism are due to its action on neuromuscular transmission and on the heart.

5. Diuretics fall into two main groups:

(a) Those inhibiting proximal tubular sodium reabsorption, causing potassium depletion.

(b) Those antagonising aldosterone either directly, or indirectly by affecting the $Na^+ : K^+$ transport mechanism, causing potassium retention.

# APPENDIX

## POTASSIUM CONTAINING PREPARATIONS

1 g of potassium chloride contains 13 mmol of potassium.

### FOR INTRAVENOUS USE

For use in serious depletion, or where oral potassium cannot be taken or retained. In most cases oral potassium is preferable. Intravenous potassium should be given with care, especially in the presence of poor renal function and the following rules should be observed:

1. Intravenous potassium should not be given in the presence of oliguria unless the potassium deficit is unequivocal and severe.

2. Potassium in the intravenous fluid should not exceed 40 mmol/l.

3. Intravenous potassium should not usually be given at a rate of more than 20 mmol/hour.

### Potassium Chloride Injection B.P.

20 mmol of potassium and chloride in 10 ml.

*WARNING.* This should *never* be given undiluted. It should be added to a full bottle of other intravenous fluid. (10 ml added to a bottle containing 500 ml of fluid gives a concentration of 40 mmol/l.)

### Potassium Chloride and Dextrose Injection B.P.C.

5 per cent dextrose with 40 mmol/l. of potassium and chloride. This is hyperosmolal.

### FOR ORAL USE

1. Potassium Chloride Tablets B.P.—6·5 mmol K and Cl per tablet.
2. Potassium Effervescent Tablets B.P.—6·5 mmol K per tablet.
3. "Slow K" (Ciba)—8 mmol K and Cl per tablet.

## TREATMENT OF HYPERKALAEMIA

### Emergency Treatment

*Calcium chloride (or gluconate).* A 10 per cent solution is given intravenously with ECG monitoring. This treatment antagonises the effect of hyperkalaemia on heart muscle, but does not alter potassium levels.

*WARNING.* Calcium should never be added to bicarbonate solutions, because calcium carbonate is insoluble.

*Glucose 50 g with 20 units of soluble insulin* by intravenous injection lowers plasma potassium rapidly by increasing entry into cells. If the situation is less urgent 10 units of soluble insulin may be added to a litre of 10 per cent dextrose.

If acidosis is present, bicarbonate may be used as an alternative to glucose and insulin injection. 44 mmol (one ampoule) may be injected over 5 minutes.

### Long-Term Treatment

*Sodium polystyrene sulphonate* ("Resonium A"; "Kayexalate") 20–60 g a day by mouth in 20 g doses, *or* 10–40 g in a little water as a retention enema every 4–12 hours. This removes potassium from the body.

# ADDENDUM TO CHAPTERS II AND III

## INVESTIGATION OF ELECTROLYTE DISTURBANCES

In the last two chapters we have outlined conditions in which sodium or potassium concentrations *may* be abnormal. We have pointed out that an abnormal result, particularly of sodium, may not be clinically significant, while "normal" ones do not guarantee "normal" balance. Before making a request some assessment should be made as to whether the result of an estimation will aid diagnosis or treatment.

Sodium and potassium estimations provide the numerical bulk of the workload of most chemical pathology departments; because of this, they are often estimated simultaneously. However, potassium is more often useful than sodium.

The student should bear the following point in mind:

1. *Plasma sodium* **should** *be estimated regularly*

    (*a*) *in the unconscious patient and infants losing fluid* because of the danger of hypernatraemia,

    (*b*) *in the dehydrated patient,* or those with *abnormal losses,* to help diagnosis and to indicate the type of replacement fluid.

2. *Plasma potassium* (*and bicarbonate*) **should** *be estimated regularly* on any patient in whom there is a cause for abnormal levels, because these must be treated:

    (*a*) *in patients with abnormal losses* from the gastro-intestinal tract or kidneys (especially due to *diuretic, or steroid or ACTH therapy*),

    (*b*) *in patients on potassium therapy,*

    (*c*) *in patients in renal failure,*

    (*d*) *in patients in diabetic coma or precoma.*

Groups 1 and 2 are numerically small compared to estimations actually requested.

3. **In the conscious, normally hydrated patient, with no abnormal losses plasma sodium estimation rarely helps.** Mild hyponatraemia is common (p. 57), but treatment in such subjects is usually contra-indicated. Unless renal failure is present, potassium estimation is also unhelpful in such subjects.

# Chapter IV

# HYDROGEN ION HOMEOSTASIS:
# BLOOD GAS LEVELS

THE pH of the extracellular fluid is normally maintained within about 0·05 units of 7·4. Energy production from metabolism is linked to a series of stepwise reactions in which dehydrogenation is of great importance, the hydrogen being transferred to coenzymes such as NAD. Some of the resulting reduced coenzyme (for example $NADH_2$) supplies hydrogen for synthetic reactions. If oxygen is available most of the rest is dehydrogenated again, the hydrogen ultimately combining with the oxygen to form water during oxidative phosphorylation, the oxidized coenzyme being released for re-use. In the absence of oxygen an adequate supply of coenzyme can only be maintained if the hydrogen is passed on to some intermediate product of metabolism; for instance, pyruvate is reduced by $NADH_2$ to form lactate, which cannot be metabolised until oxygen is available. Some free hydrogen ion results from, for instance, adenosine triphosphate (ATP) hydrolysis; in the presence of oxygen most of this is reconverted to water.

Thus, in most tissues, generation of hydrogen ions is linked with the energy production necessary for life, but, if the oxygen supply is adequate, much of the excess is converted to water and pH changes are minimal. However, it has been shown that subjects on a high protein diet pass an acid urine: this seems to be due to the constituent sulphur amino acids. The dehydrogenation step linked to the oxidation of the sulphur to sulphate is not fully understood, but certainly releases hydrogen ions. Excess hydrogen ions (as well as lactate) may also be released during a sudden burst of muscular exercise, and the supply of oxygen may not be adequate to deal immediately with such a load. To prevent a significant change in body pH these unoxidised hydrogen ions must be inactivated until oxidation or elimination from the body. Under normal circumstances homeostatic mechanisms are so effective that *blood* pH varies very little.

Since hydrogen, and not hydroxyl, ions are produced by metabolism the tendency to acidosis is greater than to alkalosis. Homeostatic mechanisms are more effective in dealing with hydrogen than with hydroxyl ions.

## DEFINITIONS

An *acid* is a substance which can dissociate to produce hydrogen ions (protons: $H^+$): a *base* is one which can accept hydrogen ions. Table X includes examples of acids and bases of importance in the body.

### TABLE X

| 1. Acid | | 2. Conjugate Base |
|---|---|---|
| Carbonic acid $H_2CO_3$ $\leftrightarrows H^+$ | $+ HCO_3^-$ | Bicarbonate ion |
| Lactic acid $CH_3CHOHCOOH$ $\leftrightarrows H^+$ | $+ CH_3CHOHCOO^-$ | Lactate ion |
| Ammonium ion $NH_4^+$ $\leftrightarrows H^+$ | $+ NH_3$ | Ammonia |
| Dihydrogen phosphate | | |
| $H_2PO_4^-$ $\leftrightarrows H^+$ | $+ HPO_4^=$ | Monohydrogen phosphate ion |
| Acetoacetic acid | | |
| $CH_3COCH_2COOH$ $\leftrightarrows H^+$ | $+ CH_3COCH_2COO^-$ | Acetoacetate ion |
| $\beta$-hydroxybutyric acid | | |
| $CH_3CHOHCH_2COOH$ $\leftrightarrows H^+$ | $+ CH_3CHOHCH_2COO^-$ | $\beta$-hydroxybutyrate ion. |

An *alkali* is a substance which dissociates to produce hydroxyl ions ($OH^-$). Alkalis are of relatively little importance in the present discussion.

A *strong acid* is highly dissociated in aqueous solution: in other words it produces many hydrogen ions. Hydrochloric acid is a strong acid, and in solution is almost entirely in the form of $H^+Cl^-$. However, the examples given in the above list are, chemically speaking, *weak acids*, little dissociated in water and yielding relatively few hydrogen ions. In the body even very small changes of pH are important and result in disturbances of physiology.

*Buffering* is the term used for the process by which a strong acid (or base) is replaced by a weak one, with a consequent reduction in the number of free hydrogen ions ($H^+$); the "shock" of the hydrogen ions is taken up by the buffer with a change of pH smaller than that which would occur in the absence of the buffer.

For example:

$$H^+Cl^- + \quad NaHCO_3 \leftrightarrows H_2CO_3 \quad + NaCl$$
Strong acid   Buffer       Weak acid       Neutral salt

*pH* is a measure of hydrogen ion activity. It was originally defined as $\log_{10}$ of the reciprocal of the hydrogen ion concentration ($[H^+]$); although it is now known that this is not strictly true, the definition suffices for present purposes. The $\log_{10}$ of a number is the power to which 10 must be raised to produce that number. Thus $\log 100 = \log 10^2 = 2$ and $\log 10^7 = 7$.

Let us suppose $[H^+]$ is $10^{-7}$ ($0 \cdot 000\ 000\ 1$) mol/l

Then $\log [H^+] = -7$

But $pH = \log \dfrac{1}{[H^+]} = -\log [H^+] = 7$

For the non-mathematically minded only a few points need be remembered.

Since at pH 6 $[H^+] = 10^{-6}$ ($0 \cdot 000\ 001$) mol/l
and at pH 7 $[H^+] = 10^{-7}$ ($0 \cdot 000\ 000\ 1$) mol/l

therefore a change of *one pH unit* represents a *tenfold change in $[H^+]$*. This is a much larger change than is immediately obvious from the change in pH values. Although changes of this magnitude do not occur in the body during life, in pathological conditions changes of $0 \cdot 3$ of a pH unit can take place. $0 \cdot 3$ is the log of 2. Therefore a *decrease of pH by* $0 \cdot 3$ (e.g. $7 \cdot 4$ to $7 \cdot 1$) represents a *doubling of $[H^+]$*. Here again the use of pH makes a very significant change in $[H^+]$ appear deceptively small. (Compare the situation if the plasma sodium concentration had changed from 140 to 280 mmol/l). Urinary pH is much more variable than that in the blood: $[H^+]$ can increase 1000-fold (a fall of 3 pH units).

**The Henderson–Hasselbalch equation.**—We have already seen that a buffer absorbs the "shock" of the addition of $H^+$ to a system by replacing a strong acid by a weak one. It will be seen that when the bases in column 2 of Table X buffer $H^+$, the corresponding acid in column 1 is formed. This weak acid and its conjugate base form a *buffer pair*. In aqueous solution the pH is determined by the ratio of this acid to its conjugate base.

Let us take the bicarbonate pair as an example. Carbonic acid ($H_2CO_3$) dissociates into $H^+$ and $HCO_3^-$ until equilibrium is reached (in this case very much in favour of $H_2CO_3$), and the ratio of the two forms will now remain constant (K). We can therefore write

$$K\ [H_2CO_3] = [H^+] \times [HCO_3^-]$$

(that is, at equilibrium, the concentration of $H_2CO_3$ is K times that of the product of $[H^+]$ and $[HCO_3^-]$).

Transposing, $[H^+] = K\ \dfrac{[H_2CO_3]}{[HCO_3^-]}$

But we are interested in $\log \dfrac{1}{[H^+]}$, or pH.

Taking logs (when multiplication becomes addition) and reciprocals throughout

$$\mathrm{Log}\ \frac{1}{[H^+]} = \log \frac{1}{K} + \log \frac{[HCO_3^-]}{[H_2CO_3]}$$

$$\text{Log} \frac{1}{K} \text{ is called pK}$$

$$\text{Therefore } \boxed{pH = pK + \log \frac{[HCO_3^-]}{[H_2CO_3]}}$$

This equation (an example of the Henderson–Hasselbalch equation) is valid for any buffer pair. It is important to notice that the pH depends on the *ratio* of the concentrations of base (in this case $[HCO_3^-]$) to acid (in this case $[H_2CO_3]$).

In practice it is not possible to measure the very low carbonic acid concentration directly. It is in equilibrium with dissolved $CO_2$, the concentration of which ($[CO_2]$) can be calculated (p. 91). If $[CO_2]$ is inserted into the equation in place of $[H_2CO_3]$, the overall dissociation constant is now the sum of those of the two reactions

$$K_1 [H_2CO_3] = [H^+] \times [HCO_3^-]$$
$$\text{and } K_2 [CO_2] \times [H_2O] = [H_2CO_3]$$

This combined constant is usually written as $K'$. The Henderson–Hasselbalch equation for the bicarbonate system then becomes

$$pH = pK' + \log \frac{[HCO_3^-]}{[CO_2]}$$

This is the form that we shall use in the rest of the chapter.

### HYDROGEN ION HOMEOSTASIS

The following points should be noted:

1. *Hydrogen ions can be incorporated in water*, maintaining normal pH.

(*a*) This is the normal mechanism of complete oxidation of $H^+$ during oxidative phosphorylation. Any $H^+$ not inactivated has to be dealt with by homeostatic mechanisms.

(*b*) $H^+$ is converted to water during the conversion of $H_2CO_3$ to carbon dioxide ($CO_2$) and water. As this is a reversible reaction $H^+$ will only continue to be so inactivated if $CO_2$ is removed. This results in bicarbonate depletion.

2. *Hydrogen ions can be lost from the body only through the kidney and the intestine.*—This mechanism, unlike that of 1(*b*) is coupled, in the kidney, with regeneration and reabsorption of bicarbonate ion ($HCO_3^-$) and is therefore the ideal method of eliminating any excess $H^+$.

3. *Buffering of hydrogen ions is a temporary measure.*—The $H^+$ is still in the body, and the presence of the weak acid of the buffer pair causes a small change in pH (see the Henderson–Hasselbalch equation). If

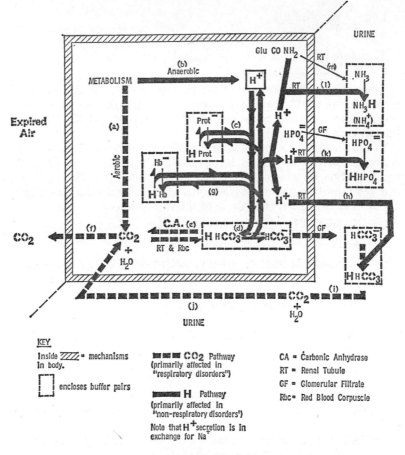

FIG. 3—Cycle of hydrogen ion homeostasis.

$H^+$ is not completely neutralised, or eliminated from the body, and if production continues, buffering power will eventually be used up and the pH will change abruptly.

The normal mechanisms of $H^+$ homeostasis are summarised in a simplified form in Fig. 3. By considering each mechanism in turn we shall see how they are linked. The key to Fig. 3 explains the symbols and initials.

(a) In the presence of equivalent amounts of oxygen most of the $H^+$ liberated by metabolism is converted to water and the carbon to $CO_2$ (aerobic metabolism).

(b) Any excess $H^+$ is liberated into the cell (conveniently referred to as "anaerobic" metabolism).

These two steps can be summarised diagrammatically:

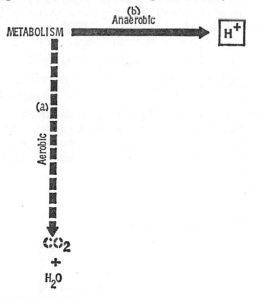

### Anaerobic ($H^+$) Pathway
#### *Tissue Cells*

(c) Some of the unoxidised $H^+$ is buffered by cell proteins, which can act as bases at body pH. The Henderson–Hasselbalch equation for this buffer pair may be written:

$$pH = pK + \log \frac{[\text{Prot}^-]}{[\text{HProt}]}$$

This step may be represented diagrammatically as:

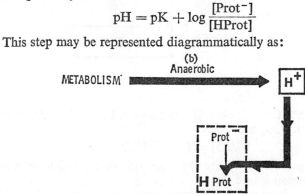

As $\text{Prot}^-$ is converted to HProt there will be a slight fall of cell pH.

### Extracellular Fluid

(*d*) $H^+$ diffuses into the extracellular fluid (ECF) where the *bicarbonate buffer pair* is of by far the greatest importance, and accounts for over 60 per cent of the blood buffering capacity (for acids other than carbonic acid). This $H^+$ is buffered by $HCO_3^-$, and carbonic acid ($H_2CO_3$) and $CO_2$ are produced, with a slight fall in extracellular pH.

$$H^+ + HCO_3^- \rightarrow H_2CO_3 \leftrightharpoons CO_2 + H_2O$$

$$pH = pK' + \log \frac{[HCO_3^-]}{[CO_2]} \begin{array}{l} \leftarrow \text{used up in buffering} \\ \leftarrow \text{produced during buffering} \end{array}$$

Plasma proteins can, like cell proteins, act as buffers but are of little quantitative importance in this respect when compared with bicarbonate.

As $H^+$ is removed from the ECF a flow out of the cell is maintained, reconverting HProt to Prot$^-$, and therefore returning cell pH to normal.

Adding this step to our diagram, we have:

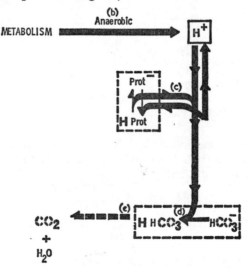

### The Lungs

(*e*) The $H_2CO_3$ formed when $H^+$ is buffered by $HCO_3^-$ dissociates spontaneously to $CO_2$ and water.

$$H_2CO_3 \rightleftharpoons CO_2 + H_2O$$

The $H^+$ is now incorporated in the water. However, if $CO_2$ were not removed the action would stop.

(*f*) The lungs have a large reserve capacity to eliminate $CO_2$. The $CO_2$ concentration is kept low and the above reaction continues to the right: thus the $H_2CO_3$ is removed as it is formed and the extracellular pH changes little. The flow of $H^+$ from cells is maintained, so that cell buffering capacity is restored. However, it should be noted that extracellular buffering is depleted by this mechanism.

We can now complete the diagram for the anaerobic pathway.

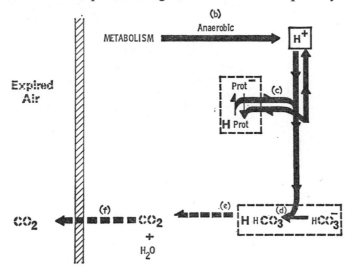

## The Aerobic ($CO_2$) Pathway

### The Erythrocytes

(g) $CO_2$ and $H_2O$, apart from being products of buffering by $HCO_3^-$, are also constantly generated aerobically in the TCA cycle; again, much of the metabolic $H^+$ is converted to water and the $CO_2$ is eliminated in expired air (see diagram over page).

If, as in pulmonary disease, $CO_2$ is not eliminated the reaction discussed above will be reversed, effectively generating $H^+$ from $H_2O$.

$$CO_2 + H_2O \rightarrow H_2CO_3 \rightarrow H^+ + HCO_3^-$$

This reaction is slow, and only a minute proportion of the $CO_2$ dissolved in plasma forms $H_2CO_3$; $CO_2$ itself is toxic. Ideally there should be a mechanism by which the removal of $CO_2$ is speeded up, but by which the acid generated in the process can be buffered.

The erythrocyte contains high concentrations of the enzyme carbonic anhydrase*, a catalyst of the reaction between $CO_2$ and water, and haemoglobin, an important buffer; it thus provides just such a mechanism. $CO_2$ diffuses into red cells, where the formation of $H_2CO_3$ is rapid. Much of the $H^+$ is buffered by haemoglobin.

* Carbonate dehydratase.

$$pH = pK + \log \frac{[Hb^-]}{[HHb]}$$

As this happens the concentration of $HCO_3^-$ increases and it diffuses into the extracellular fluid, electrochemical neutrality being maintained

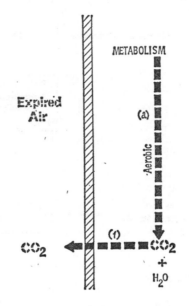

by diffusion of chloride into the cell (the "chloride shift"): the rise of extracellular $[HCO_3^-]$ tends to compensate for any rise in $CO_2$ concentration thereby correcting the *ratio* $[HCO_3^-]/[CO_2]$ and therefore the pH.

As the lungs eliminate $CO_2$ all these reactions (represented diagrammatically on the facing page) are reversed.

Although haemoglobin accounts for only about a third of the buffering power of the blood, it works co-operatively with the bicarbonate system to maintain a normal ratio of $HCO_3^-$ to $CO_2$.

### Elimination of Hydrogen Ions and Bicarbonate Repletion: The Kidney in Hydrogen Ion Homeostasis

*The final restoration of $H^+$ balance can be effected only by the kidneys.* In the pathways discussed above $H^+$ produced anaerobically has been converted to water, but at the expense of bicarbonate buffering power. That generated as a result of $CO_2$ retention is buffered by haemoglobin, which has a limited capacity. Unless the kidneys are functioning, buffering

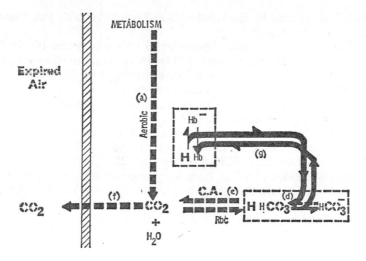

power will be exceeded and the pH will drop significantly. The kidney has two important functions in hydrogen ion homeostasis:

1. To regenerate and reabsorb bicarbonate.

2. To eliminate any $H^+$ not converted to water, thus "recharging" buffers other than bicarbonate.

Carbonic anhydrase, present in high concentration in the tubular cells, is important for both these processes. Both depend on $[CO_2]$ in these cells: the $CO_2$ comes both from the blood and from the urine: the latter is important in bicarbonate reabsorption.

### Bicarbonate Regeneration

The whole length of the renal tubule can actively secrete $H^+$ into the urine. Although the mechanism is not known for certain, the net effect is exchange of one $H^+$ for one sodium ion ($Na^+$). Moreover, in the distal nephron the potassium ion ($K^+$) appears to compete with $H^+$ for secretion. The $H^+$ probably comes from $H_2CO_3$ formed in the cell from metabolic $CO_2$ and water (e). As $H^+$ is secreted, $HCO_3^-$ is "left behind" and diffuses back into the ECF: the $HCO_3^-$ of carbonic acid is thus regenerated, just as it is in the erythrocyte when $H^+$ is buffered by haemoglobin. *This process is accelerated by a high circulating $P_{CO_2}$.*

### Bicarbonate Reabsorption—(h), (i), (j)

Bicarbonate reabsorption from the urine is closely linked with, and dependent on, $H^+$ secretion by the renal tubule. $HCO_3^-$ is filtered at the glomerulus in the same concentration as that in plasma.

1. $H^+$ *is secreted* by the tubule (*h*) and combines with this filtered $HCO_3^-$ to form $H_2CO_3$.

2. Urinary carbonic acid dissociates to $CO_2$ and water (*i*). In the proximal tubule this reaction is catalysed by carbonic anhydrase on the luminal surface of the tubular cells: in the distal nephron where the pH may be lower, it probably occurs spontaneously.

3. The partial pressure of $CO_2$ ($Pco_2$) in the urine rises.

4. $CO_2$ diffuses back into the tubular cell (*j*).

5. In the tubular cell much of this $CO_2$, catalysed by carbonic anhydrase, combines with water to form carbonic acid.

6. $H^+$ *is secreted* and $HCO_3^-$ *is "left behind"* as in 1.

This cycle reclaims $HCO_3^-$ from the glomerular filtrate, thus tending to restore extracellular buffering. It does *not* eliminate $H^+$, because the $H^+$ secreted in 1 is replaced by one formed in reaction 5.

We can summarise this cycle as follows:

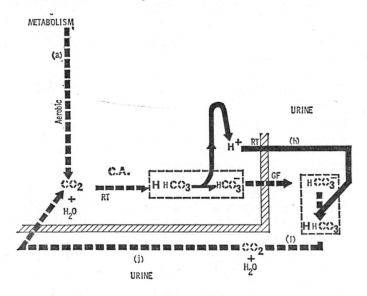

### Elimination and Urinary Buffering of Hydrogen Ions

Under physiological conditions almost all the filtered bicarbonate is reabsorbed from the urine. As this occurs and $[HCO_3^-]$ falls, further $H^+$ secretion causes net loss from the body into the urine; as always, this secretion is linked with $HCO_3^-$ regeneration in the tubular cells. If urine were unbuffered urinary pH would fall very low. In fact, it is rarely below 4·5. There are two important buffer pairs in the urine, other than bicarbonate.

(k) *The phosphate buffer pair.*—At pH 7·4 most of the phosphate in the glomerular filtrate is in the form of monohydrogen phosphate ($HPO_4^=$), and this can accept $H^+$ to become dihydrogen phosphate ($H_2PO_4^-$).

$$pII = pK + \log \frac{[HPO_4^=]}{[H_2PO_4^-]}$$

As in even mild acidosis bone salts are ionised more than at normal pH (p. 235), requirement for increased urinary secretion of $H^+$ is linked with increased buffering capacity in the glomerular filtrate, due to an increase of phosphate liberated from bone.

(l) *Buffering by ammonia.*—The enzyme *glutaminase* is present in renal tubular cells (m), and catalyses the hydrolysis of the terminal amino group of glutamine to form glutamate and the ammonium ion.

$$H_2O + GluCONH_2 \rightarrow GluCOO^- + NH_4^+$$

Ammonia and ammonium ion form a buffer pair

$$pH = pK + \log \frac{[NH_3]}{[NH_4^+]}$$

but at pH 7·4 the equilibrium is overwhelmingly in favour of $NH_4^+$. However, $NH_3$ can diffuse out of the cell into the urine much more rapidly than $NH_4^+$, and, if the urine is acid, will be prevented from returning by avid combination with $H^+$. Thus, dissociation of $NH_4^+$ in the cell is maintained by removal of $NH_3$.

$$GluCONH_2 \rightarrow NH_4^+ \rightarrow NH_3 \rightarrow Urine$$
$$+$$
$$H^+$$

This chain of events liberates $H^+$ in the cell. It seems of no advantage to buffer an $H^+$ in the urine if, simultaneously, another is produced. However, it has been generally assumed that there *is* an advantage, and a tentative hypothesis compatible with this assumption is suggested.

Ammonium, produced in all cells by deamination of amino acids, combines with glutamate to form the non-toxic glutamine. Most of this is reconverted to glutamate in the liver, the ammonium nitrogen forming urea. The amount of acid produced is the same whether this nitrogen is converted to urea, or whether it diffuses into the urine as ammonia. However, in the first case $H^+$ enters the circulation and depletes bicarbonate; in the second an $H^+$ can be lost in the urine in the ammonium ion to balance the $H^+$ produced. Urea synthesis has been shown to fall, and urinary ammonium to increase, in acidosis, favouring renal utilisation of glutamine, and the production of $NH_4^+$ hydrogen ion at a site where it can be secreted. Moreover glutamate, the other product of glutinamase activity, can be further deaminated, and the resulting α-oxoglutarate converted to glucose. Since this process requires $H^+$, an amount approximately equivalent to that generated may be incorporated into glucose, while the ammonia buffers $H^+$ secreted into the urine.

Diagrammatically, the urinary mechanisms may be summarised:

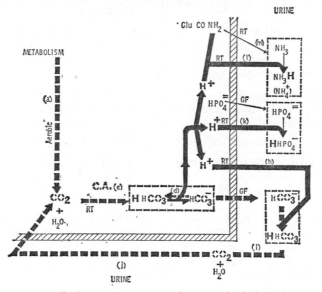

If we now put all these systems together we find that they form the integrated system, shown in Fig. 3.

### Why Bicarbonate?

As we have seen, the bicarbonate buffer pair is the most important one in the extracellular fluid. Ideally the pK of the pair should be near to that of the required pH: the optimum extracellular pH is 7·4, and yet the pK' of the bicarbonate system is 6·1.

Bicarbonate does not behave exactly like any other buffer system since $H_2CO_3$ dissociates readily to $CO_2$ and water: in the body this is catalysed by carbonic anhydrase. $CO_2$ can then be lost in expired air keeping $[H_2CO_3]$ low and the pH above the pK'. Respiratory rate (and $CO_2$ loss) is controlled by the pH (or $Pco_2$) of blood flowing through the hypothalamic respiratory centre, so that elimination can be increased as required. This mechanism is so sensitive that, if lung function is normal, $Pco_2$ rarely rises.

$CO_2$ is a product of aerobic metabolism and can combine with water to produce carbonic acid: thus the material for formation of $HCO_3^-$ is abundantly available when required.

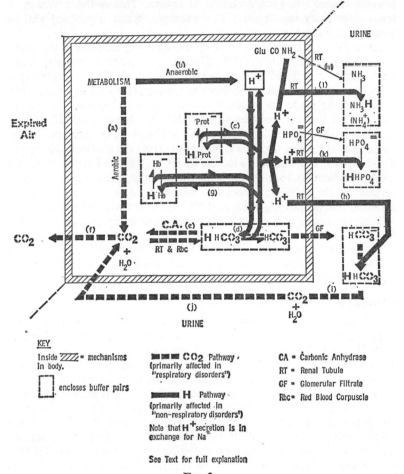

KEY

Inside ▨▨ = mechanisms
in body.

⬜ encloses buffer pairs

▬▬▬ $CO_2$ Pathway
(primarily affected in
"respiratory disorders")

▬▬▬ H  Pathway
(primarily affected in
"non-respiratory disorders")
Note that $H^+$ secretion is in
exchange for $Na^+$

See Text for full explanation

CA = Carbonic Anhydrase
RT = Renal Tubule
GF = Glomerular Filtrate
Rbc= Red Blood Corpuscle

FIG. 3.

## Role of Sodium, Potassium and Glomerular Filtration Rate in Hydrogen Ion Homeostasis

**Sodium and the glomerular filtration rate.**—Secretion of $H^+$ takes place in exchange for $Na^+$. $H^+$ secretion cannot occur if the amount of $Na^+$ in the tubular fluid is too low for quantitatively adequate exchange. Sodium is an abundant cation, and inadequate amounts of it in the tubular lumen are usually the result of a low glomerular filtration rate (GFR): however concentrated the sodium is in the glomerular filtrate (and it cannot differ from plasma sodium levels), a low flow may provide inadequate total amounts of cation. This is the situation in renal circulatory insufficiency (for example, water depletion) and in glomerular disease.

Since $H^+$ is a normal product of metabolism, acidosis is the usual accompaniment of a low GFR. However, in pyloric stenosis, water depletion is associated with abnormal loss of $H^+$ by an extrarenal route, and the low GFR is therefore accompanied by alkalosis. The $H^+$ deficit can be made up by the reaction between $CO_2$ and water (e in Fig. 3), but this can only continue if $HCO_3^-$ is removed. $HCO_3^-$ is present in the glomerular filtrate and will not be reabsorbed in the absence of $H^+$ secretion, thus tending to correct the alkalosis. However, if the GFR is low the total amount of $HCO_3^-$ filtered is not quantitatively sufficient to keep reaction (e) to the right, even if none is reabsorbed by the tubules: moreover, late in pyloric stenosis the urine becomes inappropriately acid, and $HCO_3^-$ is reabsorbed (see below).

Thus *correction of either acidosis or alkalosis is impaired if the GFR is low.*

**Potassium.**—Potassium competes with $H^+$ for secretion in exchange for $Na^+$. If it is deficient *inside the distal tubular cell,* more $H^+$ will be secreted; if it is in excess at the same site less $H^+$ will be secreted.

### Chloride in Hydrogen Ion Homeostasis

*Chloride depletion.*—It has been suggested that if chloride depletion occurs without equivalent loss of sodium (for instance, when it is lost with $H^+$ or $K^+$), the chloride deficiency itself can adversely affect hydrogen ion homeostasis. Sodium is normally actively reabsorbed in the proximal tubule, and because electrochemical neutrality must be maintained, chloride, the predominant anion, follows passively. Deficiency of chloride limits this proximal sodium reabsorption, because the other anion present in significant amounts, $HCO_3^-$, cannot easily pass the tubular cell wall. More sodium will therefore be available for *exchange* with $H^+$ in the distal nephron and the urine becomes acid. More $HCO_3^-$ will be regenerated as more $H^+$ is secreted. This mechanism is thought to aggravate the alkalosis in two conditions:

1. Pyloric stenosis (p. 100).

2. Some cases of potassium depletion (note that this is an extracellular alkalosis only) (p. 65).

Infusion of chloride helps to correct such alkalosis.

In most circumstances plasma chloride estimation is of little value in assessing the state of hydrogen ion balance.

"*Hyperchloraemic acidosis*".—Bicarbonate and phosphate are *buffer anions*—they can accept $H^+$ at physiological pH, and play an important part in hydrogen ion homeostasis. Chloride, by contrast, cannot accept $H^+$ at physiological pH: it is therefore not directly concerned in $H^+$

homeostasis and its level does not usually alter in hydrogen ion disturbances. However, plasma chloride and $HCO_3^-$ levels do sometimes vary inversely.

Hyperchloraemic acidosis occurs in three conditions:

1. Transplantation of the ureters into the colon (p. 99).
2. Inhibition of carbonic anhydrase with loss of bicarbonate and reduced $H^+$ secretion (p. 99).
3. Isolated failure of $H^+$ secretion by the kidney without anion retention (renal tubular acidosis, p. 95).

As in all cases of metabolic acidosis these conditions are associated with low plasma $[HCO_3^-]$. However, unlike most such conditions, the concentration of anions other than chloride does not rise (for instance, in renal glomerular failure there is an increase in urate and phosphate, and in diabetic ketoacidosis acetoacetate and $\beta$-hydroxybutyrate are produced in excess). As electrochemical neutrality must exist, it is obvious that, unless $[Na^+]$ falls, the concentration of some anion must rise.

The colon has been shown to reabsorb chloride from the lumen in exchange for bicarbonate. The mechanism of the hyperchloraemia in the other two situations is less clear. The low plasma bicarbonate concentration results from deficient $H^+$ secretion and therefore deficient $HCO_3^-$ reabsorption and regeneration. $Na^+ : H^+$ exchange is impaired for the same reason and sodium reabsorption is almost entirely by isosmotic proximal tubular mechanisms: this can only occur if chloride accompanies the sodium.

### Disturbances of Hydrogen Ion Homeostasis

Disturbances of hydrogen ion homeostasis involve the bicarbonate buffer pair. In 'respiratory' disturbances abnormalities of $[CO_2]$ are primary, while in so-called 'metabolic' disturbances $[HCO_3^-]$ is affected early and changes in $[CO_2]$ are secondary.

### Measurements Used to Assess Hydrogen Ion Balance

Measurement of blood pH indicates only whether there is overt acidosis or alkalosis. If the pH is abnormal this may be due to a primary abnormality of the $CO_2$ pathway or to one of the $H^+$ pathway (Fig. 3). A normal pH, however, does not exclude a disturbance of these pathways; compensatory mechanisms may be maintaining it. Assessment of these factors is made by measuring components of the bicarbonate buffer system. The concentration of $CO_2$ is calculated from its measured partial pressure ($P_{CO_2}$) by multiplying by the solubility constant of the gas ($0{\cdot}03$ if $P_{CO_2}$ is in mm Hg, or $0{\cdot}225$ if it is in kPa).

**pH and $P_{CO_2}$.**—There is a significant arteriovenous difference in pH and $P_{CO_2}$, and measurement of both must be made on *arterial blood*. Whole blood (heparinised) must be used because estimation of $P_{CO_2}$ may depend on the presence of erythrocytes. Sample collection is important, and details of this are given in the Appendix (p. 109).

**Measurements of blood $[HCO_3^-]$.**—One of two methods may be used to estimate circulating bicarbonate levels. A third value is used to elucidate the cause of the abnormal bicarbonate level.

1. *Plasma total $CO_2$ ($T_{CO_2}$); "plasma bicarbonate concentration"*.— This is probably the most commonly estimated index of hydrogen ion homeostasis and is easily performed on plasma. If pH and $P_{CO_2}$ are not required it has the advantage that venous blood can be used. It is an estimate of the sum of plasma bicarbonate, carbonic acid and dissolved $CO_2$. At pH 7·4 the ratio of $[HCO_3^-]$ to the other two components is about 20 to 1 and at pH 7·1 is still 10 to 1. Thus, if the $T_{CO_2}$ were 21 mmol/l, $HCO_3^-$ would contribute 20 mmol/l at pH 7·4 and just over 19 mmol/l at pH 7·1. Only 1 mmol/l and just under 2 mmol/l respectively would come from $H_2CO_3 + CO_2$. Thus $T_{CO_2}$ is effectively a measure of plasma bicarbonate concentration.

2. *"Actual" bicarbonate*.—This estimate is made on whole arterial blood. It is calculated from the Henderson–Hasselbalch equation, using the measured values of pH and $P_{CO_2}$.

$$pH = 6·1 + \log \frac{[HCO_3^-]}{0·03 \times P_{CO_2} \text{ (mm Hg)}}$$

It represents whole blood $[HCO_3^-]$ and for reasons discussed above, usually agrees well with plasma $T_{CO_2}$. It is the estimate of choice if the other two parameters are being measured.

3. *"Standard" bicarbonate*.—The difference between this value and the "actual" bicarbonate is used to assess the relative contributions to the latter made by abnormalities of the erythrocyte and the renal mechanisms. It is the bicarbonate level which would result from equilibrating whole, arterial blood *in vitro* with a normal $P_{CO_2}$ of 5·33 kPa (40 mm Hg). It is discussed more fully on p. 97.

### Acidosis

The causes of acidosis can be summarised as follows:

1. Hydrogen ion excess with normal homeostatic mechanisms
   (*a*) Ketoacidosis.
      (i) Diabetic.
      (ii) Starvation and tissue damage.
   (*b*) Absolute hypoxia.
   (*c*) Relative hypoxia.

    (i) Muscular exercise (really physiological).

    (ii) Starvation with increased catabolism.

  (*d*) Excessive intake of hydrogen ion.

2. Failure of homeostatic mechanisms

  (*a*) Failure of the kidney to secrete hydrogen ion.

    (i)   Generalised renal failure (p. 18).

    (ii)  Renal tubular failure and renal tubular acidosis (p. 95).

    (iii) Low glomerular filtration rate (p. 12).

    (iv) Acetazolamide therapy.

  (*b*) Retention of carbon dioxide in pulmonary disease.

3. Relative hydrogen ion excess in bicarbonate depletion

  (*a*) Loss of intestinal secretions containing bicarbonate.

  (*b*) Transplantation of the ureters into the colon.

### 1. Hydrogen Ion Excess with Normal Homeostatic Mechanisms

An excess of metabolic $H^+$ accumulates in the conditions listed above.

(*a*) In ketoacidosis acetyl CoA condenses to form acetoacetate which is reduced to form $\beta$-hydroxybutyrate.

(*b*) and (*c*) More $H^+$ may be produced by metabolism than can be converted to water, or otherwise inactivated. True tissue hypoxia, usually due to poor blood flow, may cause it. It may be a temporary phenomenon when the supply of oxygen is normal, but the rate of catabolism is increased by muscular exercise. Excess of $H^+$ may be also produced in starvation, when catabolism exceeds anabolism, and this adds to the acidosis associated with ketone production.

In "shock", when poor tissue perfusion causes local hypoxia, and during muscular exercise, lactate is produced as well as hydrogen ions. This combination is known as "lactic acidosis".

(*d*) Excess ingestion of $H^+$ is rare, and usually iatrogenic in origin. Ammonium chloride can produce acidosis because $NH_4^+$ is an acid.

The normal "anaerobic" or $H^+$ pathway is followed (p. 81). However, the rate of addition of $H^+$ to the system is such that *depletion of $HCO_3^-$ by buffering is more rapid than its regeneration and reabsorption by the kidney*. Thus *the constant finding is a low blood* [$HCO_3^-$].

$$\boxed{\text{Added } H^+} + HCO_3^- \xrightarrow{a} H_2CO_3 \xrightarrow{e} H_2O + CO_2 \xrightarrow{f} \text{lungs}$$

Blood pH depends mainly on the *ratio* of [$HCO_3^-$] to [$CO_2$]. During buffering the former falls as the latter is produced ((*d*) and (*e*)). $CO_2$ removal ((*f*)) is accelerated in acidosis, due to stimulation of the respiratory centre, and $P_{CO_2}$ remains normal or falls. Whether pH remains normal depends on whether $CO_2$ can be removed at a rate so much higher than its production that the fall in $[HCO_3{}^-]$ is balanced.

*In metabolic acidosis with functioning homeostatic mechanisms, the findings in arterial blood are:*

$[HCO_3{}^-]$ *always low*
pH normal or low, depending on severity
$Pco_2$ normal or slightly low

As soon as the rate of $H^+$ production falls to a level at which $HCO_3{}^-$ reabsorption and regeneration by the kidney can compensate for its utilisation in buffering, all parameters become normal.

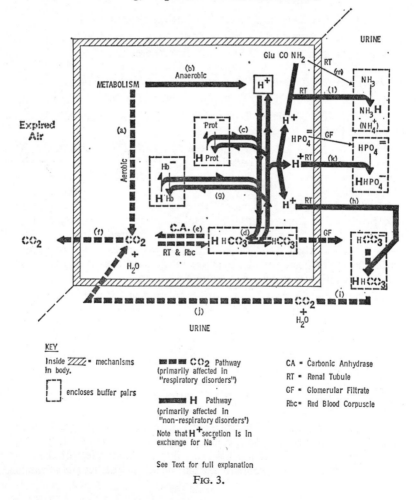

KEY

Inside ////= mechanisms in body.

[_] encloses buffer pairs

■■■ $CO_2$ Pathway (primarily affected in "respiratory disorders")

▬▬ H Pathway (primarily affected in "non-respiratory disorders')

Note that $H^+$ secretion is in exchange for $Na^+$

CA = Carbonic Anhydrase
RT = Renal Tubule
GF = Glomerular Filtrate
Rbc = Red Blood Corpuscle

See Text for full explanation

FIG. 3.

## 2. Failure of Homeostatic Mechanisms

(a) **Failure of the kidney to eliminate hydrogen ions.**—Although the lungs can temporarily restore pH to normal by $CO_2$ elimination, final restoration of $H^+$ and $HCO_3^-$ balance can only take place in the kidney.

### (i) Generalised renal failure.

Commonly in renal failure both glomeruli and tubules are damaged and the disturbance of hydrogen ion homeostasis is a consequence of both these factors. More rarely, one of these lesions occurs in isolation.

### (ii) Renal tubular failure.

If there is a reduction in the number of functioning renal tubular cells there is impairment of several factors concerned in renal tubular secretion of $H^+$.

　　1. $H^+$ secretion itself, in exchange for $Na^+$, is impaired (h), (k) and (l).
　　2. The formation of $NH_4^+$ from glutamine (m) is impaired.
　　3. $HCO_3^-$ reabsorption (i), (j) and (e) and regeneration (which depend on $H^+$ secretion) are impaired.

Bicarbonate concentration falls:

　　1. Because it is lost as $CO_2$ as it buffers $H^+$.
　　2. Because of impaired reabsorption in the kidney.
　　3. Because of the impaired carbonic anhydrase mechanism in the renal tubular cells.

The findings are the same as those already described.

*Renal tubular acidosis.*—This group of conditions may be inborn errors of metabolism, but are more commonly acquired tubular lesions. There is a failure to acidify the urine normally, even after ingestion of an acid load such as ammonium chloride (see Appendix, p. 109). They are probably either due to a deficit of the $H^+$ secreting mechanism itself, or an abnormal permeability of the distal tubular wall to the secreted $H^+$, allowing its diffusion back into the blood: unlike the situation in generalised tubular failure, the tubular cells retain the ability to form ammonia. Because there is no primary glomerular lesion the plasma urea and creatinine levels are often normal. However, prolonged acidosis increases ionisation of calcium and its release from bone (p. 235): this calcium is often precipitated in the renal tubules and the subject may present with uraemia due to nephrocalcinosis and fibrosis. The increased breakdown of bone salts also partially explains the phosphaturia often accompanying renal tubular acidosis. This is one of the conditions in which hyperchloraemia may occur. Since normal $H^+$ secretion cannot take place, $K^+$ is lost in the urine: the resulting association of acidosis and hypokalaemia is rare (see also p. 100).

(iii) *Glomerular failure.*

As explained on p. 89, a low GFR, whether due to water depletion or to glomerular disease, impairs the ability of renal tubular cells to eliminate the $H^+$ produced in normal metabolism. Added to this the $H^+$ load is often increased if the patient is not eating normally, when catabolism will exceed anabolism.

(b) **Retention of carbon dioxide in pulmonary disease.**—We have already seen that severe tissue hypoxia can sometimes be a cause of lactic acidosis. However, this type of acidosis is usually included in the "non-respiratory" ("metabolic") group of disturbances, and the "respiratory" group is usually defined as that in which the primary defect is in $CO_2$ (and carbonic acid) metabolism. To avoid confusion these groups should perhaps be classified as disturbances due to carbonic acid ("carbonic acidosis") and "non-carbonic acidosis". However, as the terminology "respiratory" and "non-respiratory" (or "metabolic") is generally accepted it will be used in this book.

In true respiratory acidosis the primary defect is carbon dioxide retention. As we shall see, this type of acidosis is accompanied by significantly different findings in the blood from those in non-respiratory disturbances.

Let us now follow the consequences of $CO_2$ retention.

$$\text{HHb} \underset{\text{Hb}}{\overset{g}{\longleftarrow}} H^+ + HCO_3^- \overset{d}{\longleftarrow} H_2CO_3 \overset{e}{\longleftarrow} H_2O + CO_2 \overset{f}{-\!\!\!/\!\!\!\longrightarrow} \text{Lungs}$$

Because removal of $CO_2$ by the lungs ($f$) is impaired the $P_{CO_2}$ *of the blood rises.* This is the constant finding in respiratory acidosis (just as a low $[HCO_3^-]$ is a constant finding in non-respiratory ("metabolic") acidosis).

In *acute respiratory failure* the rise in $P_{CO_2}$ increases the rate of reaction ($e$), and therefore the production of $H_2CO_3$.

The $CO_2$ diffuses into erythrocytes, where carbonic anhydrase catalyses its removal by reaction ($e$): much of the $H_2CO_3$ is buffered by haemoglobin ($g$) and $HCO_3^-$ diffuses into the ECF, tending to correct the pH by bringing the ratio of $[HCO_3^-]$ to $[CO_2]$ towards normal. There will be a slight rise in blood $[HCO_3^-]$. However, this correction is usually inadequate to stop a fall in pH.

In *chronic respiratory failure* the bicarbonate regeneration by the kidney, due to increased $P_{CO_2}$ in tubular cells, may raise blood $[HCO_3^-]$ sufficiently to correct the pH (reactions $h$, $i$, $j$, in Fig. 3).

$$pH = pK' + \log \frac{[HCO_3{}^-]}{[CO_2]}$$

Correction of pH by renal mechanisms takes several days to reach its maximum. In many cases of chronic respiratory disease compensation may be complete: if $CO_2$ retention is severe the raised $[HCO_3{}^-]$ may not be sufficient to prevent the fall in pH.

"*Standard bicarbonate*".—The level of the circulating blood bicarbonate (or "actual bicarbonate") is maintained by three mechanisms, the first of which affects "actual", but not "standard", bicarbonate values.

1. Formation of $H_2CO_3$ in erythrocytes from $CO_2$ and water, which is catalysed by carbonic anhydrase, the hydrogen ion being buffered by haemoglobin.

2. A similar mechanism in renal tubular cells, the hydrogen ion being secreted into the urine (regeneration of bicarbonate).

3. Reabsorption of bicarbonate from the renal tubular fluid as $H^+$ is secreted into it. Because, under physiological conditions, urine contains little $HCO_3{}^-$, this mechanism is normally working at near maximum capacity and cannot alter significantly in pathological conditions.

1. The erythrocyte mechanism is *rapid*, and *depends directly on the circulating* $P_{CO_2}$. Because of the vast reserve capacity of the normal respiratory pathway to control $P_{CO_2}$, significant abnormalities of the latter only occur when the lungs or the response of the respiratory centre are abnormal: only in such circumstances will the erythrocyte significantly affect circulating $[HCO_3{}^-]$. It is of great importance as a short-term response to respiratory disturbances, but is limited in the long term by the buffering capacity of haemoglobin.

This mechanism only requires erythrocytes. *It can therefore occur* in vitro *if whole blood is equilibrated with gases of differing* $P_{CO_2}$. *In acute respiratory disturbances*, much of the change in blood $[HCO_3{}^-]$ is due to the erythrocyte mechanism, and can be reversed by equilibrating the blood with a "normal" $P_{CO_2}$ (5·33 kPa; 40 mm Hg). If $[HCO_3{}^-]$ is now measured again the result is the "standard" bicarbonate.

2. The renal mechanism increases bicarbonate regeneration by the kidney in all acidotic states if renal function is normal. It too depends on blood $P_{CO_2}$ and is therefore affected in respiratory disturbances. If acidosis persists the kidney can continue to affect blood bicarbonate levels indefinitely. *In the long term* it is the most important mechanism increasing blood buffering capacity, but the cumulative effect takes some time to establish.

This mechanism, which depends entirely on the kidney, does *not* occur *in vitro*. In chronic respiratory disturbances much of the change in circulating $[HCO_3^-]$ is due to this mechanism. Thus, although the "standard" $[HCO_3^-]$ will be nearer normal than the "actual" $[HCO_3^-]$, it will still be abnormal. (In metabolic disturbances, when $Pco_2$ is near normal the "standard" and "actual" bicarbonate are not significantly different.) Measurement of "standard" $[HCO_3^-]$ is most useful in *acute* respiratory disorders, when a metabolic component is suspected (see below).

*The findings in the arterial blood in respiratory acidosis are:*

*$P_{CO_2}$ always raised*

In *acute* respiratory failure

pH low
"Actual" $[HCO_3^-]$ high normal or slightly raised
"Standard" $[HCO_3^-]$ normal, and lower than "actual" $[HCO_3^-]$

In *chronic* respiratory failure

pH normal or low depending on severity
"Actual" $[HCO_3^-]$ raised
"Standard" $[HCO_3^-]$ raised, but lower than "actual" $[HCO_3^-]$

## Mixed Disturbances

If there is metabolic acidosis, and $CO_2$ retention is also present, some of the compensatory increase in blood $[HCO_3^-]$ is used to buffer acid other than $H_2CO_3$. The rise of actual $[HCO_3^-]$ is impaired, more $CO_2$ is produced, and the pH falls to lower levels than during $CO_2$ retention alone. This situation most commonly occurs in the "respiratory distress syndrome" of the newborn, when there is added "lactic acidosis" due to hypoxia; it may also be due to coexistence of respiratory failure with, for example, ketoacidosis or renal failure.

In such cases the "actual" bicarbonate concentration may be high, normal or low, depending on the relative contributions from the respiratory and metabolic components, and interpretation may be difficult. It is in the *acute* situation of this type that estimation of standard $[HCO_3^-]$ is most useful. If the $Pco_2$ is known to be high the red cell mechanism is brought into play and the "standard" bicarbonate *must* be nearer normal than the "actual" bicarbonate: if there is no metabolic component it will be within the normal range. However, if there is superadded metabolic acidosis some of the "standard" $[HCO_3^-]$ will have been used in buffering, and this value will be low. If the respiratory disease is chronic, "standard" bicarbonate is less easy to interpret, although a low level with a raised $Pco_2$ would indicate severe superimposed metabolic acidosis.

### 3. Relative Hydrogen Ion Excess in Bicarbonate Depletion

If $HCO_3^-$ is lost, buffering capacity for $H^+$ is reduced. Reaction ($d$) goes to the right, increasing production of $H^+$. Compensatory mechanisms are secretion of $H^+$ in the urine and $HCO_3^-$ regeneration and reabsorption. If this is not possible, or is ineffective, the respiratory centre is stimulated, $CO_2$ is lost, and reaction ($e$) is shifted to the left; as usual, this mechanism does not replete $HCO_3^-$.

(a) **Loss of intestinal secretions.**—Many intestinal secretions are alkaline and have concentrations of $HCO_3^-$ above that of plasma; for instance, the bicarbonate concentration of duodenal juice is about twice that of plasma. Excessive loss of these through fistulae, or in severe diarrhoea, may reduce plasma $HCO_3^-$ concentrations and cause acidosis. If the kidney is able to excrete the relative excess of $H^+$ it may be able to restore plasma $[HCO_3^-]$ to normal.

Because electrolyte and water are also lost in the intestinal secretions such disturbances are accompanied by the changes described in Chapters II and III.

(b) **Transplantation of the ureters into the colon.**—The cells of the colon, like those of the renal tubules, are capable of active transport of ions. Normally reabsorption of water and electrolytes in this part of the intestinal tract is almost complete: however, if fluid which contains chloride enters the lumen, the cells tend to reabsorb this chloride in exchange for $HCO_3^-$. Bicarbonate depletion may therefore occur if urine is delivered into the colon, as after transplantation of the ureters at this site. This operation may be performed, with total cystectomy, for carcinoma of the bladder. Unless large doses of oral bicarbonate are given, the result is a very low plasma $[HCO_3^-]$ and very high plasma chloride concentration—an example of "hyperchloraemic acidosis" (p. 90). Acidosis is aggravated by conversion of urea to ammonia (by the action of some intestinal bacteria). The ammonia, after absorption, accepts $H^+$ to form $NH_4^+$ and this is reconverted to urea in the liver. This may also, therefore, be a contributory cause for mild uraemia after this operation.

(c) **Acetazolamide therapy.**—Acetazolamide is a drug, formerly used as a diuretic, and still used in the treatment of glaucoma, which inhibits the enzyme carbonic anhydrase. Since reaction ($e$) is inhibited in the renal tubular cell, bicarbonate reabsorption is impaired, and large amounts of bicarbonate are lost in the urine (the concomitant loss of sodium accounts for the diuretic action). This condition is another example of acidosis which may be accompanied by hyperchloraemia. Because normal $H^+$ secretion cannot occur, $Na^+$ exchange for $K^+$ in the urine is increased and hypokalaemia may occur.

**Secondary effects of acidosis on plasma potassium levels.**—As discussed on p. 64, $H^+$ and $K^+$ compete for exchange with $Na^+$, and acidosis usually causes hyperkalaemia. The exception to this rule occurs when the acidosis is due to failure of the mechanism for renal $H^+$ secretion. The rare combination of acidosis and hypokalaemia occurs in:

Predominant renal tubular failure (p. 14).

Renal tubular acidosis (p. 95).

Acetazolamide therapy (p. 99).

It can also occur due to severe diarrhoea, when both $K^+$ and $HCO_3^-$ are lost in the stools.

### Alkalosis

Because hydrogen, and not hydroxyl, ions are produced by metabolism, alkalosis is a relatively rare condition. The causes of alkalosis are:

1. Alkalosis with normally functioning homeostatic mechanisms.

(*a*) Ingestion of large amounts of soluble base (for instance $HCO_3^-$ as sodium bicarbonate in the treatment of "indigestion", or during intravenous bicarbonate infusion).

(*b*) Loss of hydrogen ion in an unbuffered form (pyloric stenosis).

(*c*) Potassium depletion (this alkalosis is extracellular only).

2. Alkalosis due to abnormal homeostatic mechanisms (respiratory alkalosis).

Alkalosis may present clinically as tetany in spite of normal total plasma calcium levels: this is due to a reduced ionisation of calcium salts in a relatively alkaline medium (p. 235).

**Pyloric stenosis.**—The vomiting of pyloric stenosis is probably the only condition in which an excess of unbuffered $H^+$ is lost. Gastric juice consists largely of hydrochloric acid—a strong acid—and if vomiting is due to an obstruction between the stomach and the duodenum this acid is lost, accompanied by relatively little sodium and potassium. By contrast, vomiting with free communication between the stomach and duodenum results in additional loss of duodenal juice containing relatively high $HCO_3^-$ concentrations and this loss tends to "correct" the potential alkalosis; such vomiting is accompanied by relatively less hydrogen ion, and relatively more electrolyte disturbance than that of pyloric stenosis. Water, of course, is lost in either type, and the electrolyte disturbances of pyloric stenosis are secondary partly to the reduced extracellular fluid volume, and partly to the $H^+$ disturbance.

Loss of $H^+$ in pyloric stenosis causes reaction (*d*) (Fig. 3) to go to the right: $HCO_3^-$ is liberated in a manner analogous to its regeneration in the kidney during $H^+$ secretion. Reaction (*e*) also goes to the right as $[H_2CO_3]$ is reduced, and more $H_2CO_3$ is formed, only to lose $H^+$ and become $HCO_3^-$. Plasma $HCO_3^-$ levels rise, and this rise is accompanied

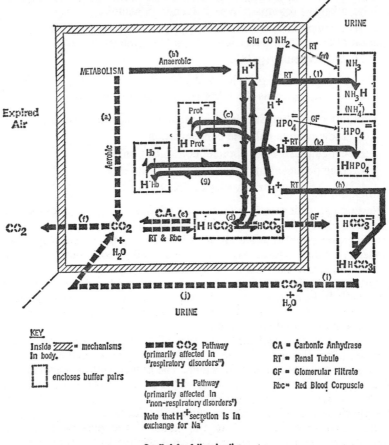

KEY

Inside ////= mechanisms
in body.

[ ] encloses buffer pairs

▬▬▬ $CO_2$ Pathway
(primarily affected in
"respiratory disorders")

▬▬▬ H Pathway
(primarily affected in
"non-respiratory disorders")

Note that $H^+$ secretion is in
exchange for $Na^+$

See Text for full explanation

CA = Carbonic Anhydrase
RT = Renal Tubule
GF = Glomerular Filtrate
Rbc= Red Blood Corpuscle

FIG. 3.

by hypochloraemia due to loss of chloride in the vomitus: this chloride
depletion may aggravate the alkalosis by preventing proximal sodium
reabsorption (p. 90). The low GFR of the accompanying water depletion
reduces urinary bicarbonate loss (p. 90), which would otherwise tend to
correct the alkalosis.

Gastric juice contains relatively little potassium. Subjects with pyloric
stenosis become hypokalaemic because of urinary loss. This loss is
secondary to:

1. $H^+$ deficiency with increased passage of $K^+$ into all cells and

secretion of $K^+$ in the renal tubule, an effect common to all forms of cellular alkalosis (compare hyperkalaemia in acidosis,.

2. Water depletion, and stimulation of aldosterone production, with sodium retention and potassium loss in the urine (p. 37).

3. Hypochloraemia (p. 90).

The $K^+$ depletion may be masked when the patient is dehydrated but should be anticipated as the GFR is increased during treatment.

Thus, after vomiting due to pyloric stenosis the findings are as follows:

1. A low plasma chloride concentration.
2. A high plasma bicarbonate concentration.
3. Haemoconcentration (due to loss of water).
4. A low normal or low plasma $[K^+]$.

Severe hypochloraemic alkalosis is relatively rare because pyloric stenosis is usually treated before this occurs. Nevertheless, when present the diagnosis could (but should not) be made on biochemical grounds alone. The chloride level may be as much as 80 mmol/l lower than that of sodium, rather than the usual 40 mmol/l.

Late in the disease the urine may be relatively acid in reaction, despite severe alkalosis, because of the chloride depletion (p. 90). Two factors hinder renal correction of the alkalosis of pyloric stenosis: fluid depletion with a low GFR and chloride depletion.

*Treatment of the biochemical disorder of pyloric stenosis.*—Treatment consists (if renal function is normal) of replacing water and chloride by administering large amounts of saline. The kidney will then correct the alkalosis. The fluid (which, because of the pyloric obstruction must, of course, be given intravenously) should be at least isosmolar saline. If plasma chloride levels are very low hyperosmolar saline should be given. Potassium should be added to the saline if the plasma potassium concentration is low normal or low (see Appendix to Chapter III).

**Extracellular alkalosis of potassium depletion.**—As discussed on p. 64, potassium depletion causes extracellular alkalosis with intracellular acidosis. If chloride depletion accompanies that of $K^+$ the alkalosis may be aggravated.

**Alkalosis due to abnormal homeostatic mechanisms.**—Overbreathing, whether due to hysteria or because of brain stem lesions, or due to excessive artificial ventilation, increases $CO_2$ loss from the body. Reaction (*e*) (Fig. 3) goes to the left in both renal tubular cells and erythrocytes, and $H^+$ is converted to water with a fall in its concentration and rise in pH. Reaction (*d*) also goes to the left and the concentration of plasma $HCO_3^-$ falls. Because the $Pco_2$ of the renal tubular cell is also kept at low levels, $HCO_3^-$ reabsorption is impaired.

*The arterial blood findings in respiratory alkalosis are:*

$P_{CO_2}$ always reduced.

Actual [bicarbonate] low normal or low.

Standard [bicarbonate] normal in acute state.

pH normal or raised.

*Salicylates*, by stimulating the respiratory centre directly, initially cause respiratory alkalosis. However, in overdosage, ketosis due to vomiting and starvation may superimpose a non-respiratory acidosis on this picture, and this may develop rapidly especially in small children; there is also some evidence that uncoupling of oxidative phosphorylation by salicylates may contribute to the acidosis. Both respiratory alkalosis and non-respiratory acidosis result in low blood $HCO_3^-$ levels, but the pH may be high if respiratory alkalosis is predominant, normal if the two "cancel each other out", or low if metabolic acidosis is predominant. Only by measurement of blood pH can the true state of hydrogen ion balance be assessed.

*The effect of alkalosis on potassium metabolism* has been discussed on p. 64. The combination of alkalosis and hypokalaemia can be a problem in salicylate poisoning.

### INVESTIGATION OF HYDROGEN ION BALANCE

The measurement of the level of plasma bicarbonate *alone* tells us nothing about the state of hydrogen ion balance. For instance, a low concentration may be associated with compensated or uncompensated non-respiratory ("metabolic") acidosis, or respiratory alkalosis. The pH is determined by the *ratio* of $[HCO_3^-]$ to $[CO_2]$ according to the Henderson–Hasselbalch equation.

Nevertheless, in many uncomplicated non-respiratory disturbances (in which the change in bicarbonate is the primary one) a careful clinical history and examination, together with a knowledge of plasma $[HCO_3^-]$, yields adequate information for clinical purposes. For instance, the low bicarbonate levels of renal failure or diabetic keto-acidosis are almost certainly associated with non-respiratory acidosis: that of renal failure is corrected temporarily by dialysis, but infusion of bicarbonate is usually contra-indicated because it involves adminis-tration of fluid and sodium with the risk of overloading the circulation: that of diabetic ketosis responds to insulin and rehydration, and a knowledge of its exact degree is usually unimportant.

Similarly, a subject with chronic bronchitis and a high plasma concentration of bicarbonate undoubtedly has a respiratory acidosis which may or may not be fully compensated. If there is no possibility of improving air entry into the lungs by use of physiotherapy, expectorants and antibiotics, treatment on a respirator is contra-indicated, because a reduction of $P_{CO_2}$ will lead to a bicarbonate loss (by reversal of the changes described on p. 96): unless the patient continues to be respired artificially for life, removal from the respirator will result in return of the $P_{CO_2}$ to its initial high levels, but with a delay of the compensatory increase in bicarbonate for some days. Nothing is to be gained, from the therapeutic point of view, by a knowledge of pH and $P_{CO_2}$ in such a case.

The situation is different if there has been an acute exacerbation of the chronic bronchitis, or if the pulmonary disease is of acute onset and potentially reversible (especially as anoxia may superimpose a "lactic acidosis"). In such cases vigorous therapy or artificial respiration may tide the patient over until the pulmonary condition improves, and more

### TABLE XI

SUMMARY OF FINDINGS IN ARTERIAL BLOOD IN DISTURBANCES OF HYDROGEN ION HOMEOSTASIS

| | | pH | $P_{CO_2}$ | Actual $HCO_3^-$ | Std $HCO_3^-$ | K levels (plasma) |
|---|---|---|---|---|---|---|
| **Acidosis** | | | | | | |
| Non-respiratory | Initial state | ↓ | N | ↓ (primary) | ↓ (primary) | Usually ↑ ( ↓ in renal tubular acidosis and acetazolamide therapy) |
| | Compensated state | N | ↓ (circled) | ↓ (primary) | ↓ (primary) | |
| Respiratory | Acute change | ↓ | ↑ (primary) | N or ↑ | N | ↑ |
| | Compensation | N | ↑ (primary) | ↑↑ (circled) | ↑ | |
| **Alkalosis** | | | | | | |
| Non-respiratory | Acute state | ↑ | N | ↑ (primary) | ↑ (primary) | |
| | Chronic state | ↑ | N or slightly ↑ | ↑↑ (primary) | ↑↑ (primary) | ↓ |
| Respiratory | Acute change | ↑ | ↓ (primary) | N or ↓ | N | |
| | Compensation | N | ↓ (primary) | ↓↓ (circled) | ↓ | ↓ |

Arrows underlined = Primary change.
   „    circled    = Compensatory change.

Note:

1. Generalised potassium depletion can cause extracellular alkalosis.
   Generalised alkalosis can cause hypokalaemia.
   Only the clinical history can differentiate the primary cause of the combination of alkalosis and hypokalaemia.

2. Overbreathing *causes* a [low $HCO_3^-$] in respiratory alkalosis.
   Non-respiratory acidosis with a low [$HCO_3^-$] *causes* overbreathing.
   Only measurement of blood pH and/or $P_{CO_2}$ can differentiate these two.

precise information than the plasma bicarbonate concentration is required for adequate control of treatment. Such information is also desirable in any other clinical situation where mixed respiratory and non-respiratory conditions may be present (for instance, when renal failure is complicated by pneumonia).

## BLOOD GAS LEVELS

In respiratory disturbances associated with acidosis a knowledge of the partial pressure of oxygen (the $Po_2$) is as important as that of pH, $Pco_2$ and $[HCO_3^-]$.

Normal gaseous exchange across the pulmonary alveoli involves loss of $CO_2$ and gain of $O_2$. However, in pathological conditions a fall in $Po_2$ and rise in $Pco_2$ do not always coexist. The reasons for this are as follows:

1. *$CO_2$ is much more soluble than $O_2$ in water*, in which its rate of diffusion is 20 times as high as that of $O_2$. In *pulmonary oedema*, for example, arterial $Po_2$ falls because its diffusion across the alveolar wall is hindered by oedema fluid. Respiration is stimulated by the hypoxia and by pulmonary distension, and $CO_2$ is "washed out". However, the rate of transport of $O_2$ through the fluid cannot be increased enough to restore normal $Po_2$. The result is a *low or normal arterial $Pco_2$* and a *low $Po_2$*. Only in very severe cases is the $Pco_2$ raised.

2. *The haemoglobin of arterial blood is normally 95 per cent saturated with oxygen*, and very little oxygen is carried in simple solution in the plasma: the dissolved $O_2$ is in equilibrium with the oxyhaemoglobin. Overbreathing air with a normal atmospheric $Po_2$ cannot significantly increase the amount of oxygen carried in the blood leaving normal alveoli: it can, however, reduce the $Pco_2$. (Breathing pure oxygen by increasing inspired $Po_2$ can increase arterial $Po_2$, but not the haemoglobin saturation.)

Let us consider the situation in many pulmonary conditions such as *pneumonia, collapse of the lung*, and *pulmonary fibrosis or infiltration*. In such conditions not all alveoli are affected by the disease to the same extent, or in the same way.

(*a*) Some alveoli will be unaffected. The composition of blood leaving these is initially that of normal arterial blood. Increased rate or depth of respiration can lower the $Pco_2$ to very low levels, but not alter either the $Po_2$ or the haemoglobin saturation in this blood.

(*b*) Some alveoli, while having a normal blood supply, may, perhaps because of obstruction of small airways, have little or no air entry. The composition of blood leaving these is near to that of venous blood (it is a right-to-left shunt). Unless increased ventilation can overcome the obstruction it will have no effect on this low $Po_2$ and high $Pco_2$.

(c) Some alveoli may have normal air entry, but little or no blood supply. These are effectively "dead space": increased ventilation will be "wasted", because however much air enters and leaves these alveoli, there is no gas exchange with blood.

Blood from (a) and (b) mixes in the pulmonary vein before entering the left atrium: the result is mixed venous and arterial blood. The high $Pco_2$ and low $Po_2$ stimulate respiration, and if enough unaffected alveoli (a) are present, the reduction of $Pco_2$ in blood leaving these to very low levels may "correct" the high $Pco_2$ from poorly aerated alveoli. For reasons discussed above, neither the $Po_2$ nor the haemoglobin saturation will be significantly altered. The final result is therefore *low or normal arterial* $Pco_2$ with *a low* $Po_2$.

If the proportion of type (b) and (c) alveoli is very high, $Pco_2$ cannot be adequately corrected by overventilation, and the result is a *high arterial* $Pco_2$ *and low* $Po_2$.

In conditions in which almost all the alveoli have a normal blood supply but poor air entry the result will be a *high arterial* $Pco_2$ and *low* $Po_2$. This may be due to *mechanical or neurological defects in respiratory movement*, or to *obstruction of large or small airways*. However, early in an acute asthmatic attack, stimulation of respiration by alveolar stretching may keep $Pco_2$ normal, or even low.

The two groups of findings in pulmonary disease are summarised in the list below. The conditions marked with * can fall into either group.

1. Low arterial $Po_2$ with low or normal $Pco_2$ can occur in such conditions as:

> Pulmonary oedema (diffusion defect)
> Pneumonia*
> Collapse of the lung*
> Pulmonary fibrosis or infiltration*

2. Low arterial $Po_2$ with high $Pco_2$ can occur in such conditions as:

> Chest injury, gross obesity, ankylosing spondylitis (impairment of movement of the respiratory cage)
> Poliomyelitis, lesions of the central nervous system affecting the respiratory centre (neurological impairment of respiratory drive)
> Laryngeal spasm, severe asthma, chronic bronchitis and emphysema (obstruction to airways)
> Pneumonia*
> Collapse of the lung*
> Pulmonary fibrosis or infiltration*

## SUMMARY

### HYDROGEN ION HOMEOSTASIS

1. Any excess of hydrogen ions produced by metabolism can be:
   (a) Buffered as a temporary measure, with a small change in pH,
   (i) By cellular proteins.
   (ii) By bicarbonate in the extracellular fluid (this is quantitatively the most important in non-respiratory ("metabolic") disturbances).
   (iii) By haemoglobin in the erythrocytes (this is a very important buffer in respiratory acidosis).

   (b) Converted to water by the action of carbonic anhydrase in the erythrocytes; the $CO_2$ produced is eliminated by the lungs. While this mechanism at least partially corrects pH it does not replete bicarbonate buffering power.

   (c) Secreted by the kidneys, with reabsorption of bicarbonate from the glomerular filtrate, and regeneration of bicarbonate in the renal tubular cells.

2. Any hydrogen ion secreted into the urine which is not concerned with bicarbonate reabsorption is excreted in a buffered form, mainly as $H_2PO_4^-$ and the ammonium ion. Bicarbonate is regenerated by this process.

3. Renal correction of either acidosis or alkalosis is dependent on a normal glomerular filtration rate.

4. Acidosis may be due to excessive production of hydrogen ion, to failure of the lungs or kidneys, or to excessive loss of bicarbonate.

5. Alkalosis may be due to excessive ingestion of base, excessive loss of hydrogen ion (pyloric stenosis), or to overbreathing.

### BLOOD GAS LEVELS

#### Low Po$_2$ and Normal or Low Pco$_2$

1. Carbon dioxide is very much more soluble than oxygen in water. Its level in the blood is therefore less affected than that of oxygen in pulmonary oedema, in which it may even be low due to respiratory stimulation.

2. Arterial blood is 95 per cent saturated with oxygen. Overbreathing therefore cannot increase oxygen carriage in blood from normal alveoli, but can reduce the Pco$_2$. In ventilation-perfusion defects, where some alveoli have a normal blood supply, but are ventilated poorly, mixture of the "shunted" blood with blood from normal alveoli results in a low Po$_2$ and normal or low Pco$_2$ in the peripheral arterial blood.

**Low Po$_2$ and High Pco$_2$**

3. The Po$_2$ is low and Pco$_2$ is high in total alveolar hypoventilation when neither gas can be adequately exchanged.

## FURTHER READING

CHRISTENSEN, H. N. (1964). *Body Fluids and the Acid-Base Balance*. Philadelphia: W. B. Saunders.

# APPENDIX

## AMMONIUM CHLORIDE LOADING TEST OF URINARY ACIDIFICATION

The ammonium ion ($NH_4^+$) is acidic because it can dissociate to ammonia and $H^+$. If ammonium chloride is ingested the kidneys should normally secrete the excess of hydrogen ion.

### Procedure

No food is taken after midnight.

8 a.m. Ammonium chloride 0·1 g/kg body weight is administered orally.

Hourly specimens of urine are collected between 10 a.m. and 4 p.m., and the pH of each specimen measured immediately with a pH meter (pH paper is not very accurate). If the pH of any specimen falls to 5·2 or below the test can be stopped.

### Interpretation

In normal subjects the urinary pH falls to 5·2 or below between 2 and 8 hours after the dose. In *renal tubular acidosis* this degree of acidification fails to occur. In generalised renal failure the response of the functioning nephrons may give normal results.

## COLLECTION OF SPECIMENS FOR BLOOD GAS ESTIMATION

1. *Arterial* specimens are preferable to *capillary* ones.

2. The syringe should be moistened with *heparin* and the specimen well mixed. Excess heparin may dilute the specimen and cause haemolysis.

3. Gas exchange with the atmosphere should be minimised by *leaving the specimen in the syringe* and expelling any air bubbles which may be present. The needle may be left on the syringe, and bent over to seal the orifice, or the nozzle stoppered.

4. The rate of erythrocyte metabolism should be minimised by keeping the specimen *on ice* or *in the refrigerator* until the estimation is performed.

5. The estimation should be performed *within an hour*.

### Capillary Specimens

In newborn infants arterial puncture may be technically difficult, and it is common to request estimations on capillary specimens (usually heel pricks). The following points should be noted.

1. The area from which the specimen is taken should be warm and pink, because the composition of the capillary blood should be as near arterial as possible. In the presence of peripheral cyanosis results may be dangerously misleading.

2. There should be free flow of blood. Squeezing may cause dilution by ECF.

3. The capillary tube(s) must be heparinised.

4. The tube(s) must be completely filled with blood. Air bubbles will invalidate the results.

5. The ends of the tube(s) should be sealed with Plasticine.

6. Mixing with the heparin should be complete. Usually a small splinter of metal is inserted into the capillary tube, and mixing is performed by passing a magnet up and down the tube.

7. The tube(s) should be put on ice, and the estimation performed at once.

*Whenever possible an arterial specimen should be obtained.*

### Foetal Specimens

In foetal distress, one indication for termination of labour may be foetal acidosis. If the cervix is dilated, and the foetal head presenting, specimens may be taken by puncturing the scalp and allowing blood to flow into long capillary tubes passed up the vagina. Again, air bubbles invalidate the results.

It is preferable to take an arterial specimen from the mother at the same time, and to compare the foetal and maternal results.

# Chapter V

# PITUITARY AND GONADAL HORMONES

## GENERAL PRINCIPLES

THE principles of the laboratory diagnosis of endocrine disorders are outlined in the following chapters. A few general points may be mentioned here.

Endocrine glands may secrete *excessive* or *deficient* amounts of hormone. This may be caused by a *primary* abnormality of the gland itself, or may be *secondary* to an abnormality in a controlling mechanism. In the latter case the endocrine gland itself is normal.

The secretion of most hormones is influenced, directly or indirectly, by the final product or metabolic consequence of its secretion. This is usually a *negative feed-back* whereby a rising final product (hormonal or non-hormonal) inhibits secretion. This concept is used in many dynamic tests of endocrine function.

Whether hormone secretion is excessive or deficient, primary must be distinguished from secondary causes. In addition, borderline abnormal must be differentiated from normal basal hormone levels. If the diagnosis is not obvious on clinical grounds, nor on the results of other laboratory tests, *dynamic tests* are required. These fall into two main categories—*suppression tests* and *stimulation tests*.

(a) *Suppression tests* are used mainly for the diagnosis of abnormal *excessive hormone secretion*. The substance (or an analogue) normally suppressing hormone secretion by negative feed-back is administered and the response measured. *Failure to suppress* implies secretion that is not under feed-back control.

(b) *Stimulation tests* are used mainly for the diagnosis of *deficient hormone secretion*. The trophic hormone that normally stimulates secretion is administered and the response measured. A normal response excludes a primary abnormality of the gland whereas *failure to respond* confirms it.

Responses other than the above all-or-none ones are considered in the relevant sections.

## PITUITARY

The pituitary gland, like the adrenal, consists of two physiologically distinct parts, the anterior pituitary (adenohypophysis) and the posterior pituitary (neurohypophysis).

### Anterior Pituitary

The anterior pituitary gland has a special role in the normal functioning of the endocrine system as many of its hormones control other endocrine glands. An intact pituitary gland is essential, not only for normal thyroid and adrenal function, but also for normal growth and sexual development and function. The hormones secreted by the anterior pituitary are:

*Acidophil* { Growth Hormone (GH or HGH).
Prolactin.

*Basophil* { Adrenocorticotrophic Hormone (ACTH). } *same cell*
Melanocyte Stimulating Hormone (MSH).
Thyroid Stimulating Hormone (TSH).
Follicle Stimulating Hormone (FSH)
Luteinising Hormone (LH).

FSH and LH are also referred to as pituitary gonadotrophins, to distinguish them from chorionic gonadotrophin (HCG), synthesised by the placenta.

Each hormone is secreted by a single cell type as identified by special staining techniques, ultramicroscopy and immunological studies. Routine histological preparations show only three cell types: acidophil, basophil and chromophobe. On this classification, GH and prolactin are secreted by acidophil cells while ACTH, TSH, MSH and the gonadotrophin secreting cells usually appear as basophils. The exact status of the chromophobe cell is uncertain: secretory granules can be demonstrated by ultramicroscopy and they may represent a phase in the development of normal secretory cells. Classification based on light microscopy is not wholly satisfactory, and tumours of chromophobe cells may be associated with excessive secretion of GH or ACTH.

### Posterior Pituitary

The posterior pituitary gland secretes two hormones:

(*a*) **Antidiuretic hormone (ADH)**, which is concerned with maintaining the osmolality of the ECF. Deficiency leads to diabetes insipidus (see p. 46).

(*b*) **Oxytocin**, which acts on the lactating breast to promote milk secretion, and which has a role in producing uterine contractions during parturition.

Of the hormones of anterior pituitary origin, GH, prolactin and the pituitary gonadotrophins are discussed in this chapter. TSH and ACTH are discussed in the chapters on the thyroid (p. 162) and the adrenal cortex (p. 138) respectively.

In man the same cells apparently secrete ACTH and MSH; plasma levels of these two hormones parallel each other in many conditions. As the clinical importance of MSH is limited, it will not be considered separately.

## CONTROL OF PITUITARY HORMONE SECRETION

The general principles of control are outlined in Fig. 4.

### Anterior Pituitary

The hypothalamus and the anterior pituitary gland form a functional unit. Branches of the superior hypophyseal artery divide into a network of capillary loops that enter the median eminence and are intimately applied to the hypothalamic nerve fibres. This network reforms into vessels that pass down the pituitary stalk into a second network of capillaries in the anterior pituitary. Hypothalamic regulatory factors are transported to the anterior pituitary by this *hypophyseal portal system* and control the secretion of anterior pituitary hormones. In most instances their action is stimulatory (releasing factor or releasing hormone—RH), but in the case of prolactin it is predominantly inhibitory.

Several factors control the secretion of releasing, and therefore of anterior pituitary, hormones (Fig. 4).

(*a*) **Feed-back system.**—The hypothalamus (or in the case of the thyroid hormones, the anterior pituitary itself) is sensitive to the circulating level of the target gland hormone: raised levels suppress release of RH (or, in the case of the thyroid, of thyroid stimulating hormone) and low levels stimulate it (negative feed-back).

(*b*) **Inherent rhythms.**—Many hormones are secreted episodically but several, such as ACTH (and consequently cortisol) are also secreted with a 24-hour (circadian) rhythm sufficiently regular to be of value in diagnosis.

(*c*) **Stimuli arising in higher centres** may override the feed-back centres and any rhythm of hormone secretion.

1. Physical stress such as trauma may stimulate the release of GH and ACTH. The stress of insulin-induced hypoglycaemia is used as a test of anterior pituitary function.

*L-Dopa↑*

*Dopamine antagonists ↓*

*Dopamine ↑*

*opposite*

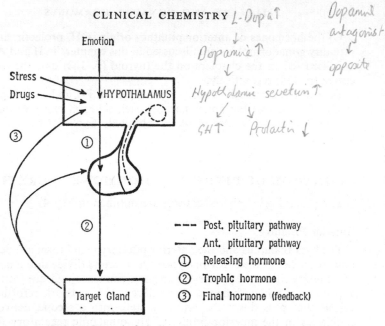

*Hypothalamic secretion ↑*

*GH↑     Prolactin ↓*

Fig. 4.—Control of pituitary hormone secretion.

- - - Post. pituitary pathway
—— Ant. pituitary pathway
① Releasing hormone
② Trophic hormone
③ Final hormone (feedback)

2. Emotional stress may also stimulate GH and ACTH release. Psychiatric illnesses, particularly depression, may mimic primary endocrine disease (p. 145).

(*d*) **Drugs.**—The close interrelationship of neural and endocrine systems is well shown by the effects of certain drugs on the hypothalamic-pituitary unit. Drugs, such as chlorpromazine, which interfere with the action of dopamine, reduce hypothalamic hormone secretion. This results in reduced GH secretion (reduced releasing hormone) and increased prolactin secretion (reduced inhibiting factor). Conversely, L-dopa, which is converted to dopamine, increases hypothalamic hormone secretion which results in increased GH and reduced prolactin secretion.

These effects have been used as pituitary function tests and in the therapy of hypothalamic-pituitary disorders.

**Posterior Pituitary**

The posterior pituitary communicates directly with the hypothalamus via nerve fibres in the pituitary stalk. The hormones of the posterior pituitary (ADH and oxytocin) are synthesised in the hypothalamus and travel down nerve trunks to accumulate in the posterior pituitary, from which they are released under hypothalamic control. Transection of the pituitary stalk or destruction of the pituitary, leading to anterior pituitary insufficiency, does not, therefore, necessarily produce diabetes insipidus. The ADH-producing cells in the hypothalamus are not damaged and may still secrete the hormone into the blood stream in response to the appropriate stimuli.

## GROWTH HORMONE

**Physiology**

There are two facets of growth hormone activity.

**Growth.**—The central role of GH in regulating growth is well illustrated by the dwarfism or gigantism that occur with deficiency or excess of GH in childhood. GH promotes protein synthesis and, in conjunction with insulin, stimulates amino acid uptake by cells. Its effect on skeletal growth however is probably mediated by a small peptide, *somatomedin*. Somatomedin is synthesised in the liver and secretion is dependent on the presence of GH. This mechanism allows a sustained action on skeletal growth despite the marked fluctuations that occur in the plasma levels of GH. Thyroxine is required for normal growth. The growth spurt at the onset of puberty is probably due to androgens.

**Intermediary metabolism.**—GH antagonises the insulin-mediated uptake of glucose. In excess, GH may produce carbohydrate intolerance, but this is usually counteracted by increased insulin secretion. GH also stimulates lipolysis and, as the resultant increase in FFA also antagonises insulin release and its action, the effects are difficult to separate. GH secretion is stimulated by hypoglycaemia and suppressed by hyperglycaemia. The importance of GH in intermediary metabolism and in the maintenance of blood glucose levels is uncertain. Subjects with isolated deficiency of growth hormone may have episodes of hypoglycaemia.

**Control of Growth Hormone Secretion**

Release of GH from the anterior pituitary is controlled by growth hormone-releasing hormone from the hypothalamus. An inhibiting factor has also been described. Secretion is *episodic* and occurs in bursts, usually associated with the onset of deep sleep or exercise, or is a response to a falling blood glucose level after meals. It may be *stimulated* by stress, starvation, hypoglycaemia and by the administration of certain amino acids, or drugs such as L-dopa. Secretion is *suppressed* by hyperglycaemia and is affected by the levels of other hormones. An impaired response to stimuli may occur in obese patients, in hypothyroidism and hypogonadism, in some cases of Cushing's syndrome and in patients receiving large doses of steroids. Oestrogens potentiate GH response.

## Measurement of Growth Hormone

Plasma GH levels are measured by radioimmunoassay, and are expressed in $\mu g/l$ or mU/l. Normal fasting levels at rest are usually below 10 $\mu g/l$. In ambulant subjects the range is much greater. Estimation of GH in urine is not generally used.

### GH EXCESS: ACROMEGALY AND GIGANTISM

The cause of GH excess is a secreting adenoma (usually of eosinophil cells) of the pituitary. The clinical manifestations depend on whether the condition develops before or after fusion of the bony epiphyses.

In childhood GH excess leads to *gigantism*. In these patients epiphyseal fusion may be delayed by accompanying hypogonadism and extreme heights of over 8 feet may be reached. Mild acromegalic features may develop after bony fusion but these giants frequently die in early adult life from infections or progressive tumour growth.

In adults *acromegaly* develops. Bone and soft tissues increase in bulk and lead to increasing size of the hands and other parts due to soft tissue thickening. There may be excessive hair growth and skin secretion. Changes in facial appearance are often marked, due to increasing size of the jaw and sinuses. The gradual coarsening of the features may, however, pass unnoticed for many years. An early complaint is that rings and shoes become too small. The viscera also enlarge and cardiomegaly is common. Although the thyroid enlarges, these patients are usually euthyroid. Menstrual disturbances are common.

A different group of symptoms may occur due to the encroachment of the pituitary tumour on surrounding structures such as the optic chiasma. With progressive destruction of the gland, hypopituitarism develops with consequent deficiency of all anterior pituitary hormones except GH.

### Non-specific Laboratory Findings

Several biochemical abnormalities may occur, but none is constant or specific enough for diagnosis.

*Impaired glucose tolerance* may be demonstrated in about 25 per cent of cases. Only about half of these develop symptomatic diabetes. In most cases, however, the insulin response during the GTT is greater than normal due to the antagonism of insulin by GH. It is probable that only those predisposed to diabetes will develop it under these conditions. In others, the pancreas can secrete enough insulin to overcome the antagonism.

*Plasma phosphate concentration* is *raised* above normal in some severe cases, but may be normal in the later stages, despite activity of the disease.

### Diagnosis

The diagnosis of acromegaly is suggested by the clinical features and radiological findings. It is confirmed by the demonstration of a *raised GH level that is not suppressed by glucose.*

Plasma GH levels may also be used to monitor the effects of treatment.

1. *Basal levels.*—GH levels, in a resting subject after an overnight fast, are usually greater than normal and may reach several hundred μg/l. Because of the wide "normal" range, samples from ambulant patients may fail to distinguish acromegalics with only moderately raised levels from normal subjects.

2. *Glucose tolerance test (GTT) with GH levels.*—During the course of a GTT on a normal subject, plasma GH falls to less than 5 μg/l within 1–2 hours (p. 186). In acromegaly, when secretion is autonomous, this does not occur to the same extent; levels may remain constant or may even increase.

## PROLACTIN

Prolactin is detectable by radioimmunoassay in the blood of even most males and non-pregnant females. Levels rise progressively during pregnancy and remain elevated during lactation. The actions of prolactin in males are still under investigation. It is probable that, acting with placental lactogen, oestrogen and progesterone, it promotes breast development in pregnancy and lactation. Hypersecretion of prolactin other than in pregnancy often causes galactorrhoea. Some cases of galactorrhoea, however, have normal prolactin levels. Gynaecomastia is not usually associated with raised prolactin levels.

Unlike other hormones of the anterior pituitary, which are stimulated by hypothalamic factors, prolactin secretion is normally kept in abeyance by a hypothalamic prolactin-release inhibiting factor (PIF).

*Elevated prolactin levels,* often with galactorrhoea and amenorrhoea, occur:

1. In hypothalamic disorders, whether of functional origin or organic (such as due to tumours), which cause reduced PIF secretion.

2. After surgical section of the pituitary stalk.

3. In some cases of pituitary tumour, either due to secretion of prolactin by the tumour or to mechanical interference with the normal hypothalamic inhibition.

4. In patients taking drugs, particularly phenothiazines, reserpine and methyldopa which reduce PIF secretion. Galactorrhoea is uncommon in this group.

Chlorpromazine administration has been used as a stimulation test of prolactin secretion. Prolactin levels also rise on administration of TRH.

*Low levels* are found in cases of pituitary destruction, for example, due to postpartum infarction and in patients receiving L-dopa.

When prolactin estimations become more generally available the most valuable use will probably be as a sensitive test of hypothalamic disease, where, in contrast to the other anterior pituitary hormones, raised basal levels would be expected.

# HYPOPITUITARISM

Anterior pituitary deficiency usually involves all or several of its hormones. Less commonly, isolated hormone deficiency occurs, particularly of growth hormone (p. 121) or gonadotrophins (p. 126). Total destruction of the anterior pituitary is incompatible with survival, but is uncommon as a result of disease. The anterior pituitary has considerable functional reserve and clinical features are usually absent until destruction of about 70 per cent of the gland has occurred. Provocative tests may demonstrate lesser degrees of impairment. Hypopituitarism may result from disease of the hypothalamus, with loss of releasing hormones, or of the anterior pituitary itself. The end result, anterior pituitary hormone deficiency, is similar and no distinction is made in this section.

## Causes of Hypopituitarism

1. Destruction whether by a primary tumour or by a secondary deposit (especially from breast carcinoma) in the pituitary, hypothalamus or surrounding structures.

2. Infarction, most commonly postpartum (Sheehan's syndrome), or rarely other vascular catastrophes.

3. Granulomatous lesions due to sarcoidosis, tuberculosis or, rarely, fungal infections.

4. Histiocytosis X in children.

5. Pituitary surgery or irradiation.

## Consequences of Hormone Deficiency

In general, with progressively more severe pituitary damage, gonadotrophin and growth hormone deficiency occurs first, followed by that of ACTH and TSH.

The following are features of long-standing cases of hypopituitarism. Mild cases may present with any or all, in varying degree.

1. Gonadotrophin deficiency leads to amenorrhoea, and atrophy of the genitalia. Libido is lost and impotence may develop. Characteristically axillary and pubic hair decrease progressively. In children the onset of puberty is delayed.

2. GH deficiency in children causes dwarfism. In adults and children it may contribute to hypoglycaemia.

3. TSH deficiency produces secondary hypothyroidism which may be clinically indistinguishable from primary myxoedema (p. 171). Lethargy and cold sensitivity are common. In children hypothyroidism is a further cause of retarded growth.

4. ACTH deficiency leads to secondary adrenocortical hypofunction (p. 150). The sodium depletion of Addison's disease is not present, as aldosterone secretion is not significantly under pituitary control. Plasma sodium concentrations may be low, however, because of excessive water retention: cortisol deficiency impairs excretion of a water load. Hypotension may be present. Hypoglycaemia may manifest clinically and there is increased sensitivity to insulin.

5. In postpartum pituitary infarction prolactin deficiency leads to failure of lactation. General apathy and fatigue are common and there may be mild anaemia. Decreased skin pigmentation, due to deficient MSH, leads to striking pallor (a differentiating feature from the hyperpigmentation of Addison's disease).

Fatal coma may develop in hypopituitarism especially after stress, such as that caused by infections. It resembles adrenocortical insufficiency except that the gross sodium depletion does not occur. Other potential hazards are hypoglycaemia, water intoxication and hypothermia.

**Laboratory Findings**

Non-specific laboratory findings may include a low fasting blood glucose level, consequent on GH and cortisol (ACTH) deficiency, and dilutional hyponatraemia, consequent on cortisol (ACTH) deficiency.

Low, or low "normal", levels of urinary gonadotrophins, plasma thyroxine, plasma cortisol and urinary steroids (17-oxosteroids and total 17-oxogenic steroids) are suggestive, but not diagnostic. Unequivocally normal values are against the diagnosis.

**Diagnosis of Hypopituitarism**

The diagnosis of hypopituitarism is best made *directly* by measurement of the anterior pituitary hormones and assessment of pituitary reserve. If these estimations are not available it may be inferred *indirectly* by demonstrating secondary hypofunction of target glands.

TABLE XII

TESTS OF HYPOTHALAMUS AND ANTERIOR PITUITARY RESERVE

| | Test | Secretion assessed | Monitored by |
|---|---|---|---|
| Hypothalamus and anterior pituitary | Insulin hypoglycaemia (p. 130) | GH ACTH (stress centre) | Plasma GH levels Plasma cortisol levels |
| | Metyrapone test (p. 158) | ACTH (feed-back centre) | Plasma 11-deoxycortisol levels Urinary total 17-OGS excretion |
| | Clomiphene test (p. 132) | LH and FSH | Plasma LH and FSH levels Urinary LH excretion |
| | LH/FSH-RH test (p. 132) | LH and FSH | Plasma LH and FSH levels |
| Anterior pituitary | TRH test (p. 177) | TSH | Plasma TSH levels |

17-OGS = 17-oxogenic steroids

**Direct assessment of pituitary reserve.**—The measurement of basal levels of the anterior pituitary hormones is not always sufficient to exclude deficiency in suspected cases. Stimulation tests are often necessary to confirm the diagnosis, particularly in less severely affected patients.

The tests available for the assessment of anterior pituitary reserve are outlined in Table XII and are discussed in greater detail elsewhere in this and the next two chapters. In some instances it is possible to differentiate primary hypothalamic from primary pituitary causes of hypopituitarism.

The most useful, and certainly the least time-consuming test from the patient's point of view, is the simultaneous administration of insulin, LH/FSH-RH and TRH. The response is monitored by the plasma levels of GH, cortisol, LH, FSH and TSH. The use of cortisol as a measure of ACTH secretion assumes a normally functioning adrenal cortex. A reduced cortisol response must be further investigated by adrenal cortex stimulation tests.

**Indirect diagnosis** of hypopituitarism is made by demonstrating target gland hypofunction that responds to administration of the relevant trophic hormone.

(a) *Secondary adrenocortical hypofunction* (p. 150) is diagnosed by the finding of abnormally low plasma or urinary steroid levels that respond to ACTH stimulation.

(b) *Secondary thyroid hypofunction* is diagnosed by low plasma thyroxine levels or by a low radioactive iodine neck uptake that responds to TSH stimulation.

In both these instances it must be remembered that *prolonged* hypopituitarism may lead to secondary atrophy of the target gland with consequent diminished response to stimulation.

Note that these tests establish only the presence or absence of hypopituitarism. The *cause* must be sought by clinical and radiological means.

### GROWTH HORMONE DEFICIENCY

A small percentage of dwarfed children have growth hormone deficiency. In about half of these there is also deficiency of other hormones. In some patients an organic disorder of the anterior pituitary or hypothalamus may be found, and a rare familial, autosomal recessive form is described, but in many cases the cause is unknown. Emotional deprivation may be associated with growth hormone deficiency indistinguishable by laboratory tests from that due to other causes. Growth failure is progressive and may be noted in the first two years of life. GH is the only means of treatment, and as supplies of this are limited it is essential that GH deficiency be proved.

### Diagnosis

**Basal plasma GH levels** in normal children may be low and are rarely diagnostic. Blood taken at a time when physiologically high levels are expected (p. 115) may avoid the more unpleasant stimulation tests. Such times are 60 to 90 minutes after the onset of sleep and about 20 minutes after vigorous exercise.

**Stimulation of GH secretion.**—If GH deficiency is not excluded by the above measurements, one or more stimulation tests are necessary. The simplest is the measurement of plasma GH after oral administration of Bovril (p. 131) (stimulation by the amino acids). If this is negative, intravenous infusion of arginine, or an insulin hypoglycaemia test, is necessary. Levels of GH in excess of 20 $\mu g/l$ in response to any of these tests excludes the diagnosis. Once GH deficiency is established a cause should be sought by appropriate radiological and clinical means.

## GONADOTROPHINS AND GONADAL HORMONES

Hypothalamic, anterior pituitary and gonadal function are closely interrelated. Together they control the normal sexual development and function of both sexes. As with other endocrine systems, abnormalities of function of one gland must be assessed in relation to the other glands of the system.

### GONADOTROPHINS

The two *pituitary gonadotrophins* follicle-stimulating hormone (FSH) and luteinising hormone (LH) are glycoproteins. In the male LH is sometimes referred to as interstitial cell-stimulating hormone (ICSH).

(*a*) FSH stimulates the development of ovarian follicles and, together with LH, stimulates oestrogen secretion by the follicle.

In males FSH stimulates spermatogenesis.

(*b*) LH acts with FSH in stimulating oestrogen secretion, and promotes ovulation in the prepared follicle with formation of a corpus luteum.

In males LH stimulates androgen production by the interstitial cells of the testis.

Pituitary gonadotrophin secretion is controlled by *releasing hormones* from the hypothalamus. A synthetic LH/FSH-releasing hormone is available and may be used to test pituitary function.

Oestrogens, progesterone and testosterone inhibit secretion by negative feed-back. This is the basis of the action of oral contraceptives. Inhibition of gonadotrophin release by high levels of oestrogen or testosterone probably depends on their attachment to hypothalamic binding sites. The drug *clomiphene* may act by displacing these hormones from the sites, and it initiates gonadotrophin release even in the presence of normal circulating levels. It may be used as a pituitary function test or, therapeutically, to induce ovulation.

Gonadotrophin levels are usually low in children and increase at puberty. There is, however, considerable overlap of pre- and post-pubertal values. In the female levels vary during the menstrual cycle (see p. 124) and are high after the menopause.

**Chorionic gonadotrophin** (HCG) is secreted normally by the placenta and also by tumours of trophoblastic origin. It has a structure and action very similar to that of LH.

### Measurement of Gonadotrophins

Bioassay of total urinary gonadotrophins has been replaced in many centres by immunological methods. FSH and LH can be measured in plasma by radioimmunoassay and urinary LH measurement is generally available. Most methods do not distinguish between LH and HCG. Because reference standards and techniques vary, results of gonadotrophin estimations must be compared with normal values issued by the laboratory performing the test.

The high urinary levels of HCG in pregnancy allow the use of simple and rapid immunological techniques for pregnancy diagnosis. Because of their immunological similarity, high levels of LH (such as occur post-menopausally) may give false positive results.

GONADAL HORMONES

The gonadal hormones are steroids synthesised in the ovary and testis. Small amounts are also secreted by the adrenal cortex. They will be considered briefly and current diagnostic uses of their estimation will be discussed.

### Ovary

The main ovarian hormones are oestrogens and progesterone, with small amounts of androgens.

(*a*) The main *oestrogens* are *oestradiol* and *oestrone*. These are metabolised to the relatively inactive *oestriol*. Oestrogens are necessary for the development of female secondary sex characteristics and for normal menstruation. Levels are low or undetectable in children and after the menopause. The changes during the menstrual cycle are considered below.

Oestrogens may be measured in *plasma* by *radioimmunoassay* or in *urine* by *chemical* techniques as *total urinary oestrogens*. It is not necessary, for clinical purposes, to measure the separate oestrogens in urine. The importance of urinary *oestriol* measurement in pregnancy is considered on p. 431.

(*b*) *Progesterone* is synthesised by the corpus luteum in the second half of the ovulatory cycle. It prepares the endometrium to receive the fertilised ovum and is necessary for the maintenance of early pregnancy. Progesterone may be measured in *plasma* or secretion may be assessed by measuring its main *urinary metabolite, pregnanediol*.

(*c*) The main ovarian *androgen* is *androstenedione* which is converted peripherally to *testosterone*. Levels in women are about a tenth of those in men.

## Testis

The Leydig, or interstitial, cells of the testis secrete *testosterone* under the influence of LH. This hormone is responsible for the development, at puberty, of the male physical and mature sexual characteristics. Prepubertal levels are low.

Testosterone is usually estimated in *plasma*, although urine has been used.

### Measurement of Gonadal Hormones

Many *routine* assay methods for these steroids fail to distinguish between closely related hormones. The technique used affects results, and it is important to know what is being measured. Such tests are usually *adequate for diagnostic purposes provided they are interpreted with reference to the normal values of the same laboratory*.

THE NORMAL MENSTRUAL CYCLE

The hormonal changes during a normal ovulatory cycle are shown in Fig. 5.

1. In the first half (*follicular phase*) of the cycle FSH stimulates development of ovarian follicles and, together with LH, stimulates them to secrete oestrogen. Plasma *oestrogen levels are low initially but rise steadily* to a pre-ovulatory peak. FSH levels fall slightly as a result of feed-back suppression.

2. Immediately after the oestrogen peak there is a *sharp burst of LH secretion* and, to a lesser extent, of that of FSH. This is followed by *ovulation*.

3. In the second half (*luteal phase*) of the cycle *progesterone* is secreted and there is a *second rise in oestrogen* levels due to secretion by the corpus luteum.

4. Unless pregnancy supervenes the corpus luteum involutes after about 10 to 12 days. *Progesterone and oestrogen levels then fall* steeply, the endometrium breaks down and menstruation occurs.

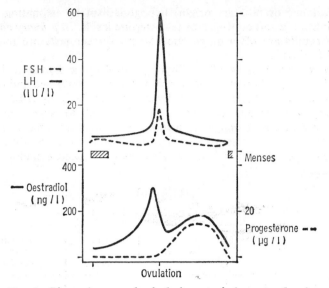

FIG. 5.—Plasma hormone levels during a typical menstrual cycle.

If the ovum is fertilised and implants in the endometrium, the corpus luteum survives and oestrogen and progesterone levels continue to rise, FSH falls and "LH" (HCG measured as LH) rises rapidly.

Interpretation of laboratory findings should be made in relation to the stage of the patient's menstrual cycle.

## DIAGNOSTIC USE OF GONADOTROPHIN AND GONADAL HORMONE ESTIMATIONS

We will briefly consider several clinical situations where estimation of one or more of these hormones is of value.

### Hypogonadism

*Children* with hypogonadism usually present with *delayed puberty*. After puberty the presenting features are usually *infertility* and, in women, *secondary amenorrhoea*. These features are non-specific: gonadal function cannot be assessed in isolation from the general state of the patient, including that of other endocrine glands. Only the laboratory assessment of gonadotrophin and gonadal hormone secretion will be considered here. For a full discussion the reader is referred to the textbooks listed at the end of the chapter.

*Primary* (of gonadal origin) is distinguished from *secondary* (of hypothalamic or pituitary origin) hypogonadism by estimating basal gonadotrophin and oestrogen or testosterone levels. In *primary gonadal failure* results are often diagnostic and no further tests are required (see Table XIII).

In *secondary hypogonadism* ("hypogonadotrophic hypogonadism") the deficient secretion of gonadotrophins must be confirmed and distinguished from low "normal" levels by stimulation tests. Both *LH/FSH-releasing hormone* and *clomiphene* can be used to assess anterior pituitary function: clomiphene also assesses hypothalamic function. Unfortunately the results of these two tests do not usually distinguish between hypothalamic and pituitary causes of reduced gonadotrophin secretion.

### TABLE XIII

LABORATORY FINDINGS IN HYPOGONADISM

| Test | Normal | Usual findings in hypogonadism | |
| --- | --- | --- | --- |
| | | Primary | Secondary |
| *Basal hormone levels* | | | |
| Testosterone or oestrogen | N | Low | Low |
| Gonadotrophins | N | Raised* | Low |
| *Response to stimulation by* | | | |
| HCG (males) | Rapid rise of plasma testo- sterone | No rise* | Normal or slow rise |
| Clomiphene | Rise in plasma gonadotrophins | High basal levels do not usually respond | Impaired response if due to pituitary destruction |
| LH/FSH-releasing hormone | Rise in plasma gonadotrophins | Exaggerated response | Impaired or normal (depending on cause) |

* See text for findings in male infertility

If generalised anterior pituitary insufficiency is thought to be the cause of the hypogonadism, other pituitary function tests should be performed (p. 130).

If *male infertility* is due to *predominant seminiferous tubular failure* (failure of spermatogenesis), Leydig cell function, testosterone secretion and response to HCG stimulation may be normal; LH levels are therefore also normal. FSH levels, however, are raised, possibly due to interruption of a spermatogenesis-hypothalamic feed-back inhibition.

Urinary 17-oxosteroids are largely of adrenal origin (p. 141), and their excretion has been used as an index of androgen secretion. This is an insensitive index of testicular function.

## Detection of Ovulation

It may be necessary to establish whether or not a patient has ovulated, either spontaneously or as a result of therapy to induce ovulation. *Plasma progesterone* should be measured on a blood sample taken during the second half of the menstrual cycle. A value within the "normal range" for that time of the cycle is good presumptive evidence of ovulation. *Luteal function* may be assessed by estimating progesterone in serial samples, and values within the normal range for a week or longer indicate that such function is adequate.

## Monitoring of Gonadotrophin Therapy

Patients with "hypogonadotrophic hypogonadism", not ovulating in response to clomiphene therapy, may be induced to do so by administration of gonadotrophins to mimic the changes of the normal cycle. A course of human menopausal gonadotrophin (predominantly FSH, with some LH) is given to induce development of follicles. Because of the risk of hyperstimulation, this stage must be monitored by daily oestrogen estimations (in plasma or urine) until levels reach those of the pre-ovulatory oestrogen peak. Administration of HCG (to simulate the LH peak) at this point frequently induces ovulation which can be detected by rising progesterone levels (see above).

## Detection and follow-up of Trophoblastic Tumours

Trophoblastic tumours (hydatidiform mole, chorionepithelioma and certain teratomas) secrete HCG. For diagnostic purposes the ordinary pregnancy tests are usually adequate but they are not sensitive enough for follow-up studies. HCG estimation in plasma or urine by radio-immunoassay is at least 20 times more sensitive and, performed at regular intervals, allows early detection and treatment of recurrence.

## Hirsutism and Virilism

*It is essential to distinguish between hirsutism and virilism.*

**Hirsutism** (excessive growth of facial and body hair) is a common complaint and is often associated with menstrual irregularities. It is probably due to slightly excessive androgen secretion from the adrenal cortex, the ovary or both. A number of stimulation and suppression tests have been used in an attempt to establish the source of androgen. In many cases, particularly those due to polycystic ovaries, the plasma testosterone level is slightly raised.

**Virilism** is uncommon but much more serious. The patient presents with *other evidence of excessive androgen secretion* such as enlargement of the clitoris, increased hair growth of male distribution, deepening of the voice and breast atrophy. The main causes are:

1. *Ovarian tumours* (such as arrhenoblastoma and hilus cell tumour) secreting androgens.

2. *Adrenocortical pathology*
   (*a*) Tumours, usually carcinoma
   (*b*) Hyperplasia—Pituitary-dependent Cushing's syndrome (rarely)
        —Congenital adrenal hyperplasia.

3. *Administration of androgens.*

After clinical assessment, *plasma testosterone* and *urinary 17-oxosteroids* should be estimated. *Normal or slightly raised 17-oxosteroid excretion* and an *elevated plasma testosterone* level usually indicate an *ovarian source* of androgens while a *markedly raised 17-oxosteroid* value indicates *adrenocortical* pathology. Distinction of adrenocortical hyperplasia from tumour is considered in Chapter VI (p. 146).


## SUMMARY

1. The anterior pituitary secretes growth hormone (GH), prolactin, adrenocorticotrophic hormone (ACTH), melanocyte stimulating hormone (MSH), thyroid stimulating hormone (TSH), and the two pituitary gonadotrophins, follicle stimulating hormone (FSH) and luteinising hormone (LH). The posterior pituitary secretes antidiuretic hormone (ADH) and oxytocin.

2. The secretion of anterior pituitary hormones is controlled by releasing factors from the hypothalamus. These in turn are controlled by circulating hormone levels (feed-back), or respond to stimuli from higher cerebral centres. In the case of thyroxine, feed-back control is exerted on the anterior pituitary itself.

3. GH controls growth and has a number of effects on intermediary metabolism. Excessive growth hormone secretion causes gigantism or acromegaly. Laboratory evidence of autonomous GH secretion is obtained by failure of suppression of plasma levels of GH during a glucose tolerance test.

4. Excessive secretion of prolactin may occur due to lesions involving the hypothalamus or pituitary stalk. It may present as galactorrhoea.

5. Anterior pituitary hormone deficiency usually involves several hormones. Less commonly, isolated deficiency occurs. The clinical features of hypopituitarism are those of gonadotrophin and sex hormone deficiency and of secondary hypofunction of adrenal cortex and thyroid. Diagnosis is made by demonstrating reduced anterior pituitary reserve by the appropriate stimulation tests.

6. GH deficiency in childhood leads to dwarfism. Diagnosis of such deficiency is made by demonstrating a subnormal GH response to appropriate stimuli.

7. Estimation of gonadotrophins and gonadal hormones (oestrogens, progesterone, testosterone) is useful in a number of clinical situations. These include:

(a) assessment of hypogonadism and male infertility

(b) detection of ovulation

(c) monitoring of gonadotrophin therapy

(d) follow-up of trophoblastic tumours

(e) assessment of virilism.

## FURTHER READING

*General Endocrinology*

HALL, R., ANDERSON, J., SMART, G. A., and BESSER, M. (1974). *Fundamentals of Clinical Endocrinology*, 2nd edit. London: Pitman Medical Publishing Co.

*Medicine* (1972–1973). (A monthly add-on series of practical general medicine), No. 2—Endocrine Diseases. Medical Education (International) Ltd.

*Hypothalamus and Pituitary*

BESSER, M. (1972–1973). The Hypothalamus and Pituitary Gland. *Medicine*, No. 2, 85. Medical Education (International) Ltd.

EDWARDS, C. R. W., and BESSER, G. M. (1974). Diseases of the hypothalamus and pituitary gland. *Clinics in Endocr. & Metab.*, 3, 475.

*Gonadotrophins and Gonadal Hormones*

RYAN, K. (ed.) (1973). Gynecologic endocrinology. *Clin. Obstet. Gynec.* 16, 167.
This contains articles on induction of ovulation and menstrual abnormalities.

LONDON, D. (1972–1973). The Testis. *Medicine*, No. 2, 145. Medical Education (International) Ltd.

BAKER, H. W. G., and HUDSON, B. (1974). Male gonadal dysfunction. *Clinics in Endocr. & Metab.* 3, 507.

FRANCHIMONT, P., VALCKE, J. C., and LAMBOTTE, R. (1974). Female gonadal dysfunction. *Clinics in Endocr. & Metab.*, 3, 533.

# APPENDIX

## Specimen Collection

Many hormones are stable in drawn blood for several hours or even days. Others, such as insulin, renin and ACTH, are unstable and samples should be kept cool and plasma separated from cells without delay. Your laboratory should be consulted for details of specimen collection and handling.

## Procedure for Repeated Sampling

Many "dynamic" tests of endocrine gland function require several blood samples over a short period of time. Repeated venepuncture is unpleasant for the patient and is also undesirable because it introduces an element of stress which may interfere with the results. An indwelling needle helps to avoid these problems. The following is a suitable procedure:

(i) An indwelling needle (a "Braunula", a "Butterfly" or similar needle of at least 19G) is introduced into a suitable forearm vein. It is secured in position with adhesive strapping.
(ii) Isotonic saline is infused *slowly* to keep the needle open.
(iii) At least 30 minutes is allowed between insertion of the needle and sampling to allow any stress-induced elevation of hormones to return to normal.
(iv) All samples may be taken through this needle:

(a) Disconnect the saline infusion (this can be avoided by the use of a disposable 3-way tap between the infusion set and the needle).
(b) Aspirate and *discard* 1–2 ml of saline-blood mixture.
(c) *Using a separate syringe*, aspirate the required volume of blood.
(d) Re-establish the saline flow.

If there is an untoward reaction such as severe hypoglycaemia after insulin administration, intravenous therapy can be given without delay through the indwelling needle.

## TESTS OF PITUITARY AND GONADAL FUNCTION

The tests outlined below are those in use in our laboratory. The references to newer tests, on which these descriptions are based, should be consulted for further details of procedure and interpretation.

## Insulin Stress Test

This test is used to assess the ability of the anterior pituitary to secrete growth hormone, and indirectly ACTH, in response to the stress of hypoglycaemia. As plasma cortisol is measured, the whole hypothalamic-pituitary-adrenal cortex axis is tested.

After an *overnight fast*:

1. Insert indwelling intravenous cannula (see above).
2. Wait at least 30 minutes for hormone levels to return to basal levels.
3. Take *basal blood* samples.
4. Inject soluble insulin intravenously (for dose see below).
5. Take *further blood samples* at 30, 45, 60, 90 and 120 minutes after insulin administration.

**Notes.**—1. Glucose, 11-hydroxycorticosteroids ("cortisol") and GH are estimated on all blood samples. The samples for hormone estimation should be kept at 4°C until delivered to the laboratory.

2. *The dose of soluble insulin* must be sufficient to lower blood glucose levels to less than 2·2 mmol/l (40 mg/dl). The usual dose is 0·15 U/kg body weight. If pituitary or adrenocortical hypofunction is probable, or if a low fasting blood glucose has previously been found, the dose should be reduced to 0·1 or 0·05 U/kg. Conversely if there is resistance to the action of insulin (Cushing's syndrome, acromegaly or obesity) 0·2 or 0·3 U/kg may be required.

3. The test is potentially dangerous and should be done only under *direct medical supervision*. Glucose for intravenous administration should be *immediately available* in case severe hypoglycaemia develops. At the conclusion of the test the patient should be given something to eat.

4. If it is necessary to administer glucose, *continue with the blood sampling*. The stress has certainly been adequate.

**Interpretation.**—*Provided the blood glucose has fallen below 2·2 mmol/l (40 mg/dl)*, preferably with clinical signs of hypoglycaemia, a normal response is a rise in plasma GH of more than 26 mIU/l (about 13 μg/l) to exceed 40 mIU/l (about 20 μg/l) and in plasma 11-OHCS of more than 200 nmol/l (7μg/dl) to exceed 550 nmol/l (20μg/dl). Failure to do so indicates hypothalamic or anterior pituitary hypofunction. Primary adrenocortical hypofunction as a cause of failure of plasma 11-OHCS to rise must be excluded by atetracosactrin stimulation test (p. 157). In Cushing's syndrome neither 11-OHCS nor GH levels rise (p. 146).

### Combined Anterior Pituitary Function Test

(Harsoulis, P., Marshall, J. C., Kuku, S. F., Burke, C. W., London, D. R., and Fraser, T. R. *Brit. med. J.*, 1973, **4**, 326; and Mortimer, C. H., Besser, G. M., McNeilly, A. S., Tunbridge, W. M. G., Gomez-Pan, A., and Hall, R. *Clin. Endocr.*, 1973, **2**, 317).

Thyrotrophin-releasing hormone (TRH) and LH/FSH-RH can be administered with insulin to provide a single test assessing anterior pituitary reserve.

The procedure is identical with the standard insulin stress test, except that 100 μg of LH/FSH-RH and 200 μg of TRH are injected intravenously immediately after the insulin. The responses to the combined stimuli are the same as when they are applied separately.

### Bovril Test of Growth Hormone Secretion

Bovril stimulates growth hormone secretion in children, most women, but not usually in adult males. It is an ideal screening test for possible growth hormone deficiency in children who have low sleeping or post-exercise plasma GH levels. It is safe, and does not require intravenous infusion or special supervision.

1. After an overnight fast, blood is taken for estimation of basal GH levels.

2. The child drinks Bovril (13·3 g/m² body surface or 0·33 g/kg body weight) dissolved in warm water.

3. Further blood samples are taken 30, 60, 90 and 120 minutes after the Bovril.

As little as 0·5 ml of blood may be needed for the estimation. If this is so in your laboratory, the specimen may be obtained from a finger prick, obviating the need for repeated venepuncture.

**Interpretation.**—A plasma GH level of 40 mIU/l (20 μg/l) or more excludes GH deficiency. Poor responses must be confirmed by another test, such as the insulin stress test.

### Clomiphene Test

Clomiphene stimulates the secretion of pituitary gonadotrophins in both males and females past puberty. The side-effects of clomiphene include flickering of vision, minor mental symptoms and, in women, possible hyperstimulation of the ovaries. Because clomiphene may induce ovulation, unwanted pregnancy could be described as a complication, and the patient should be forewarned.

(*a*) **Males** (Anderson, D. C., Marshall, J. C., Young, J. L., and Fraser, T. R. *Clin. Endocr.*, 1972, **1**, 127).

Clomiphene citrate (3 mg/kg/day to a maximum of 200 mg/day) is given orally in divided doses for 10 days.

(i) Normal subjects show a rise of plasma LH to about twice basal levels by day 10.

(ii) Prepubertal patients and those with pituitary insufficiency fail to respond.

(iii) High basal gonadotrophin levels of primary gonadal failure do not usually increase further.

(*b*) **Females.** Because of the risk of possible hyperstimulation of the ovaries at high dosage, the test is usually performed initially with 50 or 100 mg daily for 5 days. A definite rise of LH above basal levels indicates a functioning hypothalamic-anterior pituitary unit.

In many cases the LH/FSH-releasing hormone test provides the required information and is simpler, quicker and possibly safer than clomiphene.

### LH/FSH-releasing Hormone Test

(Mortimer, C. H., Besser, G. M., McNeilly, A. S., Marshall, J. C., Harsoulis, P., Tunbridge, W. M. G., Gomez-Pan, A., and Hall, R. *Brit. med. J.*, 1973, **4**, 73).

The synthetic LH/FSH-releasing hormone stimulates the release of gonadotrophins from the normal anterior pituitary. 100 μg of LH/FSH-RH is given by rapid intravenous injection. *Plasma LH and FSH levels are measured in blood drawn before and at 20 and 60 minutes after* the injection. No side-effects have been described.

In general, patients with *primary gonadal* failure have an *exaggerated* response, and those with *hypopituitarism* an *impaired* response. Most patients with low basal gonadotrophin levels show a variable but definite response.

## HCG Stimulation Test

(Anderson, D. C., Marshall, J. C., Young, J. L., and Fraser, T. R. *Clin. Endocr.*, 1972, **1**, 127).

Human chorionic gonadotrophin resembles LH in its action and stimulates the Leydig cells to secrete testosterone.

The test extends over 5 days. HCG (2000 IU) is given by intramuscular injection on days 0 and 3. *Blood samples* are taken daily for *plasma testosterone* estimation.

(*a*) *Normal subjects* respond with an approximately twofold rise in plasma testosterone levels within the first 3 days, with little further rise after the second injection.

(*b*) Patients with *gonadotrophin deficiency* respond more slowly and levels increase further after the second injection. This is a common type of response of organs with reversible atrophy due to lack of trophic stimulation.

(*c*) Patients with *primary testicular failure* fail to respond.

# Chapter VI

# ADRENAL CORTEX: ACTH

THE adrenal glands are divided into the functionally distinct cortex and medulla: however, although the medulla is part of the sympathetic nervous system, glucocorticoids are probably required for the final stage of adrenaline (epinephrine) synthesis. The cortex is part of the hypothalamic-pituitary-adrenal endocrine system. Histologically the adult adrenal cortex has three layers. The outer thin layer (*zona glomerulosa*) is functionally distinct and secretes aldosterone. The inner two layers, the *zona fasciculata* and the *zona reticularis*, are apparently different forms of the same functional unit and secrete the bulk of the adrenocortical hormones. A wider fourth layer is present in the foetal adrenal gland but disappears after birth.

## THE ADRENAL STEROIDS

### CHEMISTRY

All the adrenal cortical hormones are steroids with the same chemical skeleton, each carbon atom of which is numbered according to inter- national agreement (Fig. 6). The usual chemical classification is based on the number of carbon atoms in the molecule. Most contain 21 of these (for example cortisol) and are therefore referred to as $C_{21}$ steroids. They originate only in the adrenal cortex. The androgens (for example, androstenedione) contain only 19 carbon atoms ($C_{19}$ steroids) and are derived both from the adrenal cortex and gonads. Oestrogens from the ovary are $C_{18}$ steroids. Various groups are attached to the molecule: for example, cortisol has hydroxyl (—OH) groups at positions 11, 17 and 21 while androstenedione has an oxo (=O) group at position 17.

Such apparently small chemical differences produce a variety of biological actions. Synthetic steroids more potent than the natural hormones (e.g. dexamethasone), or with one physiological action accentuated (e.g. anabolic steroids), can be produced by chemical means.

### PHYSIOLOGY

The adrenocortical hormones can be classified according to their main physiological actions: there is, however, functional overlap between groups.

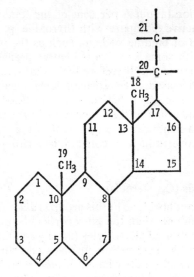

Numbering of the Steroid Nucleus

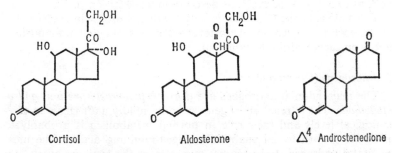

Cortisol       Aldosterone       $\Delta^4$ Androstenedione

FIG. 6.—The more important adrenal steroids.

## 1. Glucocorticoids ($C_{21}$ steroids)

Cortisol and corticosterone are the major glucocorticoids. Their action on carbohydrate metabolism opposes that of insulin and they favour gluconeogenesis. Protein breakdown and lipolysis are stimulated. Glucocorticoid excess impairs glucose tolerance, and alters adipose tissue distribution. Cortisol enhances the ability of the kidney to excrete a water load; it helps to maintain extracellular fluid volume and normal blood pressure.

Cortisol in the blood is mostly protein bound to a specific *cortisol-binding globulin* (CBG, transcortin) and to albumin. Only the unbound free fraction (about 6 per cent of the total at normal levels) is physiologically active (compare with thyroxine, p. 163). Changes in the circulating level of binding protein, such as the increase occurring in pregnancy or with some oral contraceptive preparations, can produce high protein bound and therefore high total levels with a normal, or only slightly increased, free cortisol level (again compare with thyroxine, p. 165). The glucocorticoids are metabolised in the liver, conjugated with glucuronic acid and excreted in the urine.

*Cortisone* is not a major secretion product of the adrenal cortex and is biologically inactive until it is converted *in vivo* to hydrocortisone (cortisol).

## 2. Mineralocorticoids ($C_{21}$ steroids)

The most important hormone in this group is *aldosterone*. It is present in much smaller amounts than the glucocorticoids, but has a powerful effect on the distribution of electrolytes (mostly sodium and potassium) across cell membranes. This action involves all cells but is clinically most evident in the renal tubule where it promotes sodium reabsorption and potassium excretion. Aldosterone is also metabolised in the liver and excreted in the urine, mostly conjugated with glucuronic acid. The actions of aldosterone are considered in more detail on p. 37.

The actions of the $C_{21}$ steroids may overlap. Cortisol in excess has a mineralocorticoid action. Aldosterone has a small and unimportant effect on carbohydrate metabolism.

## 3. Androgens ($C_{19}$ steroids)

The main adrenal androgens are *dehydroepiandrosterone* and *androstenedione*. The adrenal androgens are only mildly androgenic at physiological levels and their role in normal metabolism is probably to promote growth by an anabolic nitrogen-retaining action. The most powerful androgen, testosterone, comes from the testis or ovary, not the adrenal cortex.

There is extensive interconversion of the androgens from the adrenal cortex and the testis. The end products, *androsterone* and *aetiocholanolone*, together with *dehydroepiandrosterone*, are conjugated in the liver and excreted in the urine as glucuronides and sulphates.

### CONTROL OF ADRENAL STEROID SECRETION

The hypothalamus, anterior pituitary and adrenal cortex form a functional unit (the hypothalamic-pituitary-adrenal axis—Fig. 7).

*Cortisol* is secreted in response to *adrenocorticotrophic hormone* (*ACTH*) from the anterior pituitary. ACTH secretion is in turn depen-

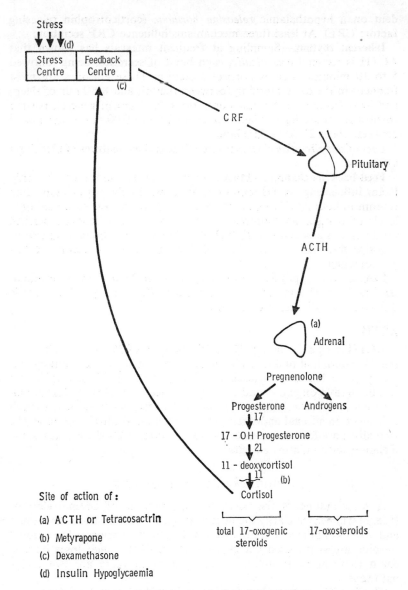

Site of action of :

(a) ACTH or Tetracosactrin

(b) Metyrapone

(c) Dexamethasone

(d) Insulin Hypoglycaemia

FIG. 7.—The control of cortisol secretion.

dent on a hypothalamic *releasing hormone* (corticotrophin releasing factor; CRF). At least three mechanisms influence CRF secretion.

**Inherent rhythm.**—Sampling at frequent intervals has shown that ACTH is secreted *episodically,* each burst of secretion being followed 5 to 10 minutes later by cortisol secretion. These episodes are most frequent in the early morning (between the 6th and 8th hour of sleep) and least frequent in the few hours prior to sleep; in general, plasma cortisol levels are *highest* between about 7.00 to 9.00 hours and *lowest* between about 23.00 to 4.00 hours.

*Loss of this circadian rhythm* is one of the earliest features of Cushing's syndrome.

**Feed-back mechanism.**—Plasma cortisol levels cannot be the only factor influencing ACTH secretion, as shown by the parallel circadian rhythm in both ACTH and cortisol levels. Nevertheless sustained high levels of cortisol do suppress ACTH secretion and deficient cortisol production is associated with high ACTH levels, providing evidence of *negative feed-back* control. Several dynamic tests are based on this phenomenon.

**Stress.**—Stress, physical or mental, can override the first two mechanisms and cause sustained ACTH secretion. The consequent diagnostic difficulties are discussed on p. 145.

## ACTH

ACTH is a single chain polypeptide made up of 39 amino acids. Only the 24 N-terminal amino acids are required for biological activity and this peptide has been synthesised (*tetracosactrin*; "Synacthen") and can be used in diagnosis and therapy in place of ACTH. ACTH *stimulates cortisol* synthesis and secretion by the adrenal cortex. It has much less effect on adrenal androgen production and, at physiological levels, virtually no effect on aldosterone production. ACTH also has some melanocyte-stimulating activity.

### Synthesis of Adrenal Steroids

A brief outline of the pathways of synthesis of adrenal steroids (Fig. 8) will help to clarify the interrelationship of the various hormones and groups of hormones estimated for diagnostic purposes. The adrenal steroids are synthesised via cholesterol (p. 219), which loses its side chain to form pregnenolone. From this point there are two major pathways.

The first ($C_{21}$ pathway) leads, via a series of hydroxylations (additions of OH groups) at positions 17, 21 and 11 to cortisol. Each step is controlled by a specific enzyme and absence of one of these enzymes gives rise to the condition of congenital adrenal hyperplasia (p. 153).

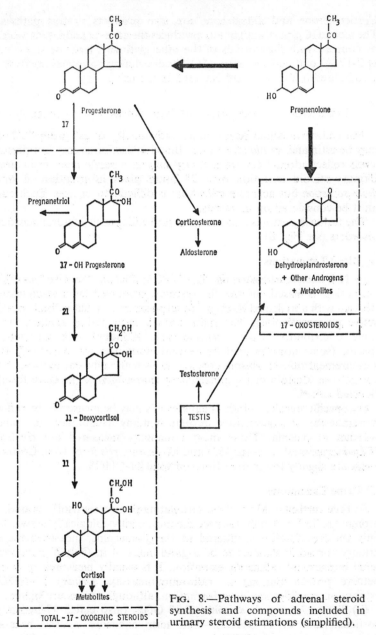

Fig. 8.—Pathways of adrenal steroid synthesis and compounds included in urinary steroid estimations (simplified).

Corticosterone and aldosterone are also products of this pathway. The second ($C_{19}$ pathway) finally produces the adrenal androgens which, by contrast with the steroids of the other pathway, lack the side chain at C-17. As outlined earlier (p. 136), the circulating hormones are further metabolised in the liver and excreted in the urine.

### LABORATORY INVESTIGATION OF ADRENOCORTICAL FUNCTION

Adrenal steroids may be estimated individually, or as "groups". They may be estimated in *blood* or *urine*. Blood collection is easy but plasma levels reflect adrenal cortical activity only at a *single moment* in time. Urinary steroid excretion over 24 hours gives information of *total daily secretion* but accurate collection is difficult to ensure. Each case must be considered on its merits.

The following estimations are used in the diagnosis of adrenocortical disorders (see Fig. 8).

### (a) Blood Estimations

(i) **11-hydroxycorticosteroids** (11-OHCS; Plasma "cortisol").—This estimation is based on the fluorescence produced by unconjugated steroids with a hydroxyl (OH) group at position 11 (that is, biologically active glucocorticoids). The main steroid measured is *cortisol*, with some contribution from *corticosterone*. Although not completely specific (some non-steroidal fluorescent substances are measured) the measurement reflects plasma cortisol levels with sufficient accuracy for clinical use. Certain drugs, particularly *spironolactone*, produce falsely elevated values.

(ii) **Specific steroids.**—Individual steroids may be measured, by radioimmunoassay or competitive protein binding techniques, on small volumes of plasma. Those most commonly measured are *cortisol*, *17-hydroxyprogesterone* (p. 153) and *11-deoxycortisol* (p. 143). Cortisol levels are slightly lower than those of total 11-OHCS.

### (b) Urine Estimations

(i) **Free cortisol.**—Most of the circulating plasma cortisol is bound to protein (p. 136) but only the free fraction is physiologically active. As only the free fraction is filtered at the glomerulus, measurement of urinary cortisol is claimed to be a good index of it. Only if gross may renal impairment reduce its excretion. It is usually measured by competitive protein binding or radioimmunoassay. *Urinary 11-OHCS* estimation provides similar information, although values are higher.

(ii) **Total 17-oxogenic steroids** (T-17 OGS) (also called 17-hydroxycorticosteroids or 17 OHCS).—The term oxogenic is used because during estimation the 17-OHCS are converted to the more stable and easily measured 17-oxo form. This group includes cortisol, its precursors and its metabolites and is used as an index of cortisol output.

(iii) **17-oxosteroids** (previously called 17-ketosteroids) measure most of the products of the androgen pathway, both from the adrenal and from the testes, and their metabolites. In *females* most of the 17-oxosteroids are of adrenal orgin while in *males* 10–20 per cent is derived from the testis from metabolites of testosterone (not itself a 17-oxosteroid). 17-oxosteroid excretion is reduced non-specifically in many illnesses: the main clinical value of this estimation is in the differential diagnosis of virilisation, in some forms of which high levels may be found (p. 128).

(iv) **Steroid "patterns".**—Urinary steroids may be separated by gas or liquid chromatography. Patterns of steroid excretion may be of value in the differential diagnosis of states of hypersecretion, and if necessary can be performed in special centres.

**Plasma ACTH** may be measured by radioimmunoassay. Because the hormone is extremely labile, special collection of samples is necessary and it is not a practical test for routine use. Usually similar information is obtained by measuring plasma cortisol. ACTH estimation is most valuable in the differential diagnosis of Cushing's syndrome (p. 148). Although it distinguishes between primary and secondary adrenocortical hypofunction, simpler tests are available for this purpose.

### Tests of the Hypothalamic-Pituitary-Adrenal Axis (Fig. 7)

Each part of the axis and each control mechanism may be assessed. Details of the tests are given in the appendix.

1. **Assessment of secretory rhythm.**—Plasma cortisol estimations at intervals during the 24 hours will establish the presence of an overall rhythm. Because of the episodic secretion, and the effect of stress, results on single samples may be misleading.

2. **Assessment of feed-back control.**—(a) *Dexamethasone suppression test.*—This test is used for the differential diagnosis of adrenocortical *hyperfunction*, and depends on inhibition of ACTH secretion by the potent synthetic steroid, dexamethasone: this steroid does not contribute significantly to results of the usual steroid estimations. The resultant fall of cortisol secretion may be monitored by measuring plasma cortisol levels or urinary T-17 OGS excretion. Failure of cortisol levels to fall implies either

(i) that the *feed-back centre* is abnormally *insensitive*, or

(ii) that *cortisol secretion*, whether from adrenal or ectopic sites (p. 148), is *autonomous*.

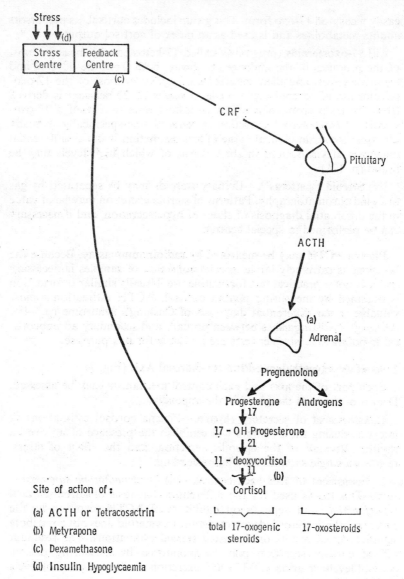

Stress

Stress Centre | Feedback Centre (c)

(d)

CRF

Pituitary

ACTH

(a)
Adrenal

Pregnenolone

Progesterone          Androgens

↓ 17

17 – OH Progesterone

↓ 21

11 – deoxycortisol

↓ 11          (b)

Cortisol

Site of action of :

(a) ACTH or Tetracosactrin

(b) Metyrapone

(c) Dexamethasone

(d) Insulin Hypoglycaemia

total 17-oxogenic steroids          17-oxosteroids

FIG. 7.

(b) *Metyrapone test.*—This test is used primarily in the differential diagnosis of adrenocortical *hypofunction*, and tests the whole pathway (but see p. 148). The drug metyrapone blocks cortisol synthesis by inhibiting the final (11-OH) hydroxylation step. A normal feed-back centre responds to the consequent fall in circulating cortisol by increasing CRF and therefore ACTH secretion. This stimulates cortisol synthesis but, as the final stage is blocked, precursors accumulate and are excreted in the urine to be measured as an increase in the T-17 OGS (Fig. 8). Alternatively, the increase in plasma levels of the immediate precursor of cortisol, 11-*deoxycortisol*, may be measured, thus obviating the need for 24-hour urine collection and considerably shortening the time of the test. Metyrapone tests the whole of the *hypothalamic-pituitary-adrenal axis.* False negative results are considered on p. 159.

3. **Assessment of the "stress" pathway.**—This test is also used primarily in the differential diagnosis of adrenocortical *hypofunction.* A normal response to metyrapone does not necessarily mean that the patient can respond adequately to stress. Ability to do this can only be established by measuring the response to a suitable stress stimulus. The most frequently used stimulus is insulin-induced hypoglycaemia, which can at the same time be used to assess growth hormone secretion.

4. **Assessment of the adrenal cortex.**—A subnormal response to metyrapone or to hypoglycaemia can only be said to be due to hypothalamic or pituitary disease if the adrenal cortex can be shown to be functioning normally. The administration of ACTH or preferably tetracosactrin, if followed in normal subjects by increased adrenocortical steroid secretion: this may be monitored by measuring *plasma cortisol* levels or urinary T-17 OGS output.

(i) *Primary adrenal cortical hypofunction.*—The response is subnormal even after repeated stimulation.

(ii) *Secondary adrenal cortical hypofunction.*—The response is often subnormal initially, but improves after repeated stimulation.

## DISORDERS OF THE ADRENAL CORTEX

The main disorders of adrenocortical function are:

1. **Hyperfunction**

   (*a*) Excessive cortisol secretion—Cushing's syndrome

   (*b*) Excessive aldosterone secretion—Primary aldosteronism (Conn's syndrome) (p. 51).

   (*c*) Excessive androgen secretion—Adrenocortical carcinoma
   Congenital adrenal hyperplasia.

## 2. Hypofunction

(a) Primary—Addison's disease
> Congenital adrenal hyperplasia (deficient cortisol and aldosterone)

(b) Secondary—ACTH deficiency (deficient cortisol)

### CUSHING'S SYNDROME

Cushing's syndrome is produced by an excess of circulating cortisol. The *causes* are:

1. Bilateral adrenal hyperplasia due to:

(a) excessive ACTH secretion by the pituitary (with or without a pituitary tumour);

(b) ectopic ACTH production by a non-endocrine tumour. The most frequent cause of this is bronchogenic carcinoma (see Chapter XXI).

2. Adenoma or carcinoma of the adrenal cortex.

3. Administration of cortisone or its analogues. This should not present a diagnostic problem.

**Clinical features.**—The most usual form of Cushing's syndrome, pituitary-dependent adrenal hyperplasia, usually occurs in women of reproductive age.

The commonest presenting feature is obesity, limited usually to the trunk. Other clinical features are a round plethoric face (moon-face), purple striae on the abdomen, hypertension, muscular weakness, osteoporosis and frequently hirsutism. Menstrual disturbances, especially amenorrhoea, are common.

The most prominent metabolic abnormalities in Cushing's syndrome reflect the *glucocorticoid* action of cortisol. About two-thirds of the patients have *impaired glucose tolerance* as shown by a diabetic glucose tolerance test and of these many have hyperglycaemia and consequent glycosuria. As cortisol has the opposite action to insulin the resultant picture resembles, in certain aspects, that of diabetes mellitus. There is true resistance to the action of insulin, and a diabetic on cortisone therapy will require an increased dose of insulin. Protein breakdown is accelerated and the liberated amino acids are either converted to glucose (gluconeogenesis) or lost in the urine leading to a negative nitrogen balance. This is evident clinically as muscle wasting and osteoporosis. The mechanism underlying the characteristic distribution of adipose tissue is not known.

The *mineralocorticoid* effect of excess cortisol is seen in some, but not all, cases of Cushing's syndrome. *Sodium retention* by the renal tubules leads to an increase in total body sodium and therefore body water

(p. 51). Plasma sodium concentration is usually normal, however, because of this parallel water retention. *Potassium loss* through the renal tubule may lead to severe depletion, reflected usually as hypokalaemic alkalosis.

*Androgen* secretion may be increased and may be a factor in the hirsutism, but virilisation is uncommon.

Certain clinical features may suggest the cause. Marked virilisation suggests carcinoma of the adrenal cortex. Non-endocrine tumours producing ACTH usually present with severe hypokalaemic alkalosis, with none of the physical stigmata of the common form. These patients may also have increased skin pigmentation due to high ACTH levels. Adrenocortical carcinoma is the usual cause of Cushing's syndrome in children.

One of the main difficulties in the diagnosis of Cushing's syndrome is the fact that stress, whether from mental or from physical causes (such as severe illness), can increase secretion and mimic the laboratory findings of the syndrome. Two conditions pose particular diagnostic problems:

(*a*) Simple obesity, often with hirsutism and hypertension, may be very difficult to distinguish from early Cushing's syndrome both clinically and on the results of laboratory tests.

(*b*) Psychiatric disorders, particularly endogenous depression, may have the laboratory features of Cushing's syndrome.

**Diagnosis.**—Investigation of a suspected case of Cushing's syndrome is best performed in hospital and must aim to answer two questions:

(*a*) is there excessive cortisol secretion?

(*b*) if so, what is its cause?

(a) *Is there excessive cortisol secretion?*—In all forms of Cushing's syndrome, urinary excretion of free cortisol (or 11-OHCS) is increased, reflecting the elevated free plasma cortisol level. Furthermore, as the *normal circadian rhythm is lost* (p. 138), free cortisol is excreted into the urine for longer periods within the 24 hours. These two facts make the estimation of *urinary free cortisol* a sensitive index of adrenocortical hyperfunction.

All types of Cushing's syndrome *fail to reduce cortisol* secretion with a small dose of *dexamethasone*. In pituitary-dependent adrenal hyperplasia, the feed-back mechanism is insensitive and ACTH continues to be secreted at this dose level. In the other forms, cortisol secretion is autonomous and output is already maximally suppressed.

The preliminary investigation of Cushing's syndrome is therefore:

Measurement of *urinary free cortisol.*
Assessment of *circadian rhythm.*
Assessment of *dexamethasone suppressability.*

A recommended protocol for in-patients is:

1. Collect urine for two 24-hour periods for estimation of free cortisol (or 11-OHCS) and total 17-oxogenic steroids.

2. Take blood for plasma cortisol or 11-OHCS estimation at about 9.00 hours and 23.00 hours on both days.

3. After completion of the urine collections give 2 mg dexamethasone orally at 23.00 hours, and take blood for plasma cortisol estimation at 9.00 hours the next morning.

*Interpretation.*—In a *normal* subject plasma 11-OHCS levels are between about 160 and 600 nmol/l (6 and 22 µg/dl) at 9.00 hours, are less than about 200 nmol/l (7 µg/dl) at 23.00 hours, and *suppress* to less than about 200 nmol/l (7 µg/dl) on the morning after dexamethasone. Urinary free cortisol levels are normal. These findings virtually exclude the diagnosis of Cushing's syndrome. In *Cushing's syndrome* the 9.00 hours plasma 11-OHCS levels may be normal but there is *loss of* the *normal rhythm*, and *urinary free cortisol* levels are *raised*. Plasma 11-OHCS *fail to suppress* after dexamethasone. In *obese patients* there may be an *abnormal rhythm* and resistance to dexamethasone but *urinary free cortisol* levels are *usually normal*. Patients with *psychiatric disorders* may mimic the changes of Cushing's syndrome in all respects and distinction depends on treatment of the psychiatric disorder.

In difficult cases, measurement of the cortisol secretion rate may be of value but is only available in special centres. In both obesity and psychiatric disorders it may be elevated.

4. If there is still doubt, an insulin stress test (p. 130) may be done. Most patients with *Cushing's* syndrome *fail to show* the normal 11-OHCS response despite adequate hypoglycaemia, whereas most patients with *simple obesity* do *respond*.

(b) *What is the cause?*—Estimation of total 17-oxogenic steroids and 17-oxosteroids on the urines collected may help.

(i) *Total 17-oxogenic steroid* excretion may be normal in pituitary-dependent Cushing's syndrome. Very high values suggest ectopic ACTH production or adrenocortical carcinoma.

(ii) *17-oxosteroids.*—Results of this estimation are usually within normal limits in hyperplasia and adenoma of the adrenal cortex, when cortisol is the main secretion product. Adrenocortical carcinomata, however, are less specific in their action and frequently secrete a variety of steroids, including androgens, in large quantities. It is these patients who show virilisation and the urinary 17-oxosteroid excretion (reflecting androgens) may be very high. Excretion of a large number of steroids is shown by gas chromatography.

Other useful investigations may be:

## TABLE XIV
### TEST RESULTS IN CUSHING'S SYNDROME

| Test | Adrenocortical hyperplasia | | Adrenocortical tumour | |
|---|---|---|---|---|
| | Pituitary dependent | Ectopic ACTH production | Carcinoma | Adenoma |
| *Basal plasma cortisol levels* | | | | |
| 9.00 hours | Raised, may be normal | Both usually very high | Both usually very high | Raised, may be normal |
| 23.00 hours | Raised | | | Raised |
| *Urinary steroid excretion* | | | | |
| Free cortisol (or 11-OHCS) | Raised | Usually very high | Usually very high | Raised |
| Total 17-oxogenic steroids | May be normal | Usually very high | Usually very high | Raised or normal |
| 17-oxosteroids | Usually normal | Usually raised | Usually very high | Normal |
| Plasma ACTH levels | High normal or moderately raised | Usually very high | Very low or undetectable | Very low or undetectable |
| ACTH stimulation | Normal or exaggerated response | Usually responds | Usually no response | Usually no response |
| Metyrapone test | Normal or exaggerated response | May or may not respond | Usually no response | Usually no response |
| Dexamethasone suppression (8 mg/day) | Suppression | Usually no suppression | Usually no suppression | Usually no suppression |

(iii) *Plasma ACTH levels.*—These are moderately raised in most cases of pituitary-dependent Cushing's syndrome, often very high in ectopic ACTH-secreting tumours and very low in cases of adrenocortical tumours (secretion is suppressed by the high cortisol levels).

If ACTH estimation is not available, the following tests should be done; the selection depends on the provisional diagnosis. In all of them interpretation is based on the fact that cortisol secretion by tumours is neither dependent on, nor stimulated by, ACTH whereas secretion from hyperplastic tissue may be.

(iv) *ACTH or tetracosactrin stimulation* elicits a normal or exaggerated response in most cases of adrenocortical hyperplasia, whereas adrenocortical tumours usually, but not always, fail to respond.

(v) *The metyrapone test* (p. 158) produces a rise in urinary total 17-oxogenic steroids in most cases of hyperplasia, but not usually if the cause is an adrenocortical tumour. Levels may in fact fall due to general inhibition of steroid synthesis.

(vi) *A two-dose dexamethasone test* (p. 158) may be done if the diagnosis is still in doubt. Dexamethasone in a dose of 8 mg/day usually reduces T-17 OGS excretion to less than 50 per cent of the basal level in pituitary-dependent adrenal hyperplasia but not in patients with ectopic ACTH secretion or adrenocortical tumour. The feed-back mechanism in the former is set at an abnormally high level, but is responsive to the higher dose of dexamethasone, with a consequent fall in pituitary ACTH secretion. In all other causes of Cushing's syndrome ACTH is already maximally suppressed.

The results of these tests are summarised in Table XIV.

## ADRENOCORTICAL HYPOFUNCTION

### Primary Adrenocortical Hypofunction (Addison's Disease)

Addison's disease is caused by destruction of all zones of the adrenal cortex. Consequently there is a deficiency of glucocorticoids, mineralocorticoids and androgens.

Tuberculosis was once the commonest cause but it has now been superseded in many countries by idiopathic atrophy of the gland, probably due to an autoimmune process. Rare causes of destruction include amyloidosis, mycotic infections and secondary malignancy.

**Clinical features.**—Depending on the degree of adrenal destruction, the patient with Addison's disease may present in one of two ways. He may first be seen in a shocked dehydrated condition (Addisonian crisis); this is a medical emergency requiring immediate treatment. Alternatively the disease may be diagnosed only after a prolonged period of ill health. The clinical features of this form—tiredness, pigmentation of the skin and buccal mucosa, weight loss and hypotension —are common to many severe chronic diseases.

The most serious consequences are due to the *mineralocorticoid deficiency*. *Sodium deficiency* is the cardinal biochemical abnormality of the Addisonian state. Loss of sodium through the kidneys is accompanied by loss of water and while these parallel one another the plasma sodium concentration remains in the normal range (see Fig. 2, p. 40). The dehydration and consequent haemoconcentration are nevertheless evident from the raised haematocrit and total protein concentration. During a crisis, however, there is almost invariably hyponatraemia, as sodium loss exceeds water loss. As the loss is mainly renal the urine may contain considerable amounts of sodium despite the dehydration and hyponatraemia. The Addisonian crisis is therefore the result of massive sodium depletion. Other abnormalities usually include a raised plasma potassium concentration and metabolic acidosis (compare the reverse in mineralocorticoid excess). The fluid depletion leads to a reduced circulating blood volume and renal circulatory insufficiency with a reduced glomerular filtration rate and a moderately raised plasma urea. In the more chronic form of Addison's disease only some of these features may be present.

The absence of *glucocorticoids* is shown by the marked insulin sensitivity of the hypoadrenal state. The glucose tolerance curve is flat and there may be fasting hypoglycaemia, but this only occasionally gives rise to symptoms.

*Androgen* deficiency is not clinically evident because testosterone production by the testis is unimpaired.

The pigmentation that develops in Addison's disease is probably due to high circulating ACTH and MSH levels resulting from the lack of cortisol suppression of the feed-back mechanism.

**Diagnosis.**—If the patient presents in Addisonian crisis, immediate treatment is necessary and there is no time to perform diagnostic tests. Blood can be taken for plasma 11-OHCS estimation but further tests can await recovery from the crisis, as the underlying adrenal insufficiency will not be affected by the therapy.

Urinary steroid estimations are of little use in the diagnosis of Addison's disease for two important reasons.

(*a*) Low values occur non-specifically in many disorders and may be found in normal subjects.

(*b*) If the adrenal destruction is partial, maximum ACTH stimulation of the remnant via the feed-back mechanism may maintain steroid secretion within normal limits. There is, however, no reserve and in a stress situation the residual adrenal cortex is unable to respond to the increased need for cortisol and the patient may develop an adrenal crisis.

*Plasma ACTH* levels will be very high. The same information, however, may be obtained by stimulation tests.

The *essential feature in the diagnosis of Addison's disease* is the demonstration that the adrenal cortex cannot respond to ACTH. In practice, tetracosactrin (for example "Synacthen") is usually used. It has the same biological action as ACTH but as it lacks the antigenic part of the molecule there is much less danger of an allergic reaction.

(a) *Short stimulation tests* (p. 157) are most convenient. A definite rise of plasma 11-OHCS excludes Addison's disease. Lack of response may be due to primary or secondary adrenal atrophy: to distinguish between these a more prolonged stimulus is needed.

(b) *Prolonged stimulation tests.*—These consist of several injections of long-acting tetracosactrin. Failure to respond to this prolonged stimulus confirms the diagnosis of Addison's disease (compare with the TSH stimulation test in primary and secondary hypothyroidism, p. 172).

Dexamethasone, which does not interfere with the usual steroid estimations, may be given during stimulation tests. This is *essential* when adrenal hypofunction is suspected, and the patient has been on other steroids: withdrawal of these may precipitate an adrenal crisis and they contribute to the estimated levels. It is a precaution in such cases, even when not on steroid therapy.

### Secondary Adrenal Hypofunction

As mentioned in the classification of adrenal disorders, adrenocortical hypofunction may be secondary to lack of ACTH.

ACTH release may be impaired by disease of the hypothalamus or of the anterior pituitary, most commonly due to tumour or infarction. Corticosteroid therapy suppresses ACTH release and after such therapy, especially if prolonged, the ACTH-releasing mechanism may be slow to recover. This period of impaired response may last up to a year.

Complete anterior pituitary destruction results in the picture of hypopituitarism (p. 118) but if destruction is only partial there may be sufficient ACTH for basal requirements. As in partial adrenal cortical destruction (p. 149) the deficiency may only become evident under conditions of stress. In these patients, as in those on corticosteroid therapy, stress may precipitate acute adrenal insufficiency. The most usual sources of stress are infections and surgery.

The adrenal crisis due to ACTH deficiency is due to lack of glucocorticoids only. It differs from an Addisonian crisis in that aldosterone secretion (not under the influence of ACTH) is normal; consequently there is not the characteristic sodium depletion and dehydration. The condition is nevertheless potentially fatal and is ushered in by mental disturbance, nausea, abdominal pain and usually hypotension (cortisol deficiency). As in primary glucocorticoid deficiency there may be hypoglycaemia and marked insulin sensitivity. Plasma sodium levels may be low, not because of sodium loss as in Addisonian crisis, but because of dilution due to the delayed water excretion of cortisol deficiency.

If the patient presents in this state blood may be taken for plasma 11-OHCS but treatment should be instituted at once. This single estimation may distinguish acute adrenal insufficiency (low levels) from clinically similar conditions where normal adrenal response to stress produces raised 11-OHCS. It does not distinguish primary from secondary adrenal insufficiency.

**Diagnosis.**—Plasma cortisol levels may be low whether the hypofunction is primary or secondary.

1. *Plasma ACTH* levels clearly distinguish primary (high levels) from secondary (low levels) adrenocortical hypofunction.

2. If this estimation is not available a 5-hour *tetracosactrin test* should be done.

(a) If the response is normal, the hypothalamic-pituitary-adrenal axis may be tested for feed-back (metyrapone test) and stress (insulin stress test) responsiveness.

(b) If there is a definite, but impaired, response two further daily injections of tetracosactrin are given. A further response over this period confirms secondary, reversible, adrenal atrophy.

Prolonged lack of ACTH may cause irreversible cortical atrophy with no response to stimulation. Unlike the situation in Addison's disease, ACTH levels are low or absent.

3. If secondary adrenal hypofunction is diagnosed the "reserve" of other anterior pituitary hormones should be assessed. If the cause of the hypofunction is prolonged steroid therapy, this step is probably not necessary, because only ACTH secretion is likely to have been affected.

### Corticosteroid Therapy

There is the risk of adrenocortical hypofunction when long term corticosteroid therapy is stopped. This may be due either to secondary adrenal atrophy or to impairment of ACTH-releasing mechanisms.

A simple means of testing the ACTH-releasing mechanisms is to estimate the morning plasma 11-OHCS levels two or three days after cessation of steroid therapy. A level within the normal range indicates a functioning pituitary and feed-back centre. It must be emphasised, however, that this does not test the all-important stress pathway (p. 149). Other tests are considered above.

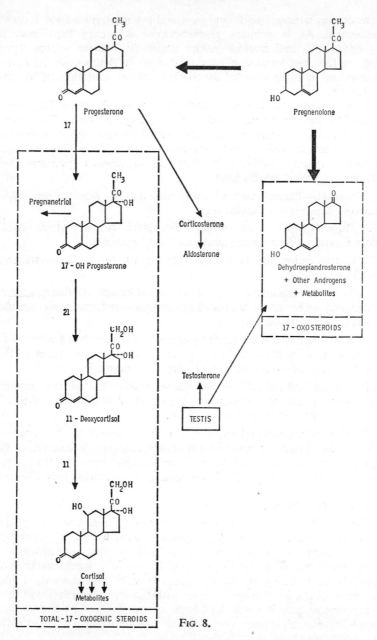

Fig. 8.

## CONGENITAL ADRENAL HYPERPLASIA

This is a rare inherited condition in which there is a deficiency of one of the enzymes involved in the biosynthesis of cortisol (see Fig. 8). The enzyme most commonly affected is that controlling the hydroxylation step at C-21 (21-hydroxylase) which is the penultimate stage of the cortisol pathway. As a result, plasma cortisol levels tend to be low and the feed-back centre is stimulated to secrete greatly increased amounts of ACTH. This, in turn, increases steroid synthesis along both main pathways and may achieve normal cortisol levels if the block is not complete. There will, in any case, be increased production of androgens and of cortisol precursors behind the block. As in Addison's disease a "normal" cortisol level may merely represent a gland working at full capacity but with no reserve to meet stress. In addition, about one-third of the cases seen in infancy have associated impairment of aldosterone synthesis.

With this background of deficient cortisol production leading to excessive ACTH secretion and therefore to androgen overproduction, the clinical manifestations can be explained. The condition may present in several ways.

1. **Female pseudohermaphroditism.**—Because the foetal adrenals start functioning at about the third month of gestation, the abnormal steroid pattern is present *in utero* during the period of development of the external genitalia. In the female child the high levels of androgens can produce pseudohermaphroditism of varying degree.

2. **"Salt-losing" syndrome.**—The cases with impaired aldosterone production can present in the first few days of life with what is, in effect, an Addisonian crisis with the biochemical changes outlined on p. 149.

3. **Progressive virilisation.**—This occurs in males at the age of 2–3 years with enlargement of the penis and rapid osseous development. As bone fusion occurs earlier than normal in these children, they are smaller than their fellows when they become adults.

4. **Milder virilisation.**—This occurs at or after puberty in the female, often associated with amenorrhoea.

5. **Hypertensive form.**—A very much rarer form of congenital adrenal hyperplasia is that due to 11-hydroxylase deficiency. In these patients signs of virilisation are accompanied by hypertension, due most probably to high levels of 11-deoxycorticosterone.

Deficiency of enzymes earlier in the steroid synthetic pathway is excessively rare and usually presents as a severe salt-losing state.

**Diagnosis.**—The immediate precursor to the block in 21-hydroxylase deficiency is *17-hydroxyprogesterone. Plasma levels* of this steroid are raised and may be measured directly, or its breakdown product, *pregnanetriol*, may be measured in the urine.

The increased production of androgens and cortisol precursors results in raised urinary 17-oxosteroids and total 17-oxogenic steroids. All these abnormalities can be corrected by administration of cortisone or dexamethasone, which cut off the excess ACTH secretion.

The major difficulty in diagnosis using the urine tests is the necessity for a 24-hour collection. Not only is this difficult in infants, but in the patient presenting in a salt-losing crisis therapy should not be withheld for 24 hours.

A quicker method of diagnosis is to determine the *11-oxygenation index* using a *random* specimen of urine. Because the final stage of cortisol synthesis is 11-hydroxylation, any block in the pathway (as in congenital adrenal hyperplasia) results in a decreased proportion of total 17-oxogenic steroids with an —OH group at position 11. The ratio of the steroids without an 11-OH to those with 11-OH (the 11-oxygenation index) will be greater than normal.

The urinary chromatographic pattern of corticosteroids and androgens will be abnormal, and may contribute to diagnosis.


## SUMMARY

1. The steroids of the adrenal cortex can be classified into three groups.

(*a*) Glucocorticoids, e.g. cortisol

(*b*) Mineralocorticoids, e.g. aldosterone $\Big\}$ $C_{21}$ steroids

(*c*) Androgens $\qquad$ $C_{19}$ steroids

2. (i) Steroids may be *estimated* in *plasma* by radioimmunoassay or related techniques. Measurement of 11-hydroxycorticosteroids (11-OHCS) is widely used as an assessment of cortisol. *11-deoxycortisol* and *17-hydroxyprogesterone* estimations are sometimes of value.

(ii) *Urinary free cortisol* (or 11-OHCS) estimation is probably the best index of cortisol production. Estimations of excretory products of groups of steroids are:

(*a*) Urinary total 17-oxogenic steroids (T-17 OGS) which includes cortisol, its precursors and metabolites. This test is used as a measure of cortisol production.

(*b*) Urinary 17-oxosteroids which provide a measure of androgens, most of which are derived from the adrenal cortex.

Urinary steroid estimations alone can be very misleading in the diagnosis of adrenocortical disease.

(iii) *Plasma ACTH* estimation is most useful in the differential diagnosis of Cushing's syndrome and in the distinction between primary and secondary adrenocortical hypofunction.

3. Cortisol secretion is increased by ACTH from the pituitary and this in turn is controlled by two hypothalamic centres:

(*a*) A feed-back centre that responds to circulating cortisol levels.
(*b*) A stress centre responding to stresses of various kinds.

There is an overall circadian variation in cortisol secretion. The lowest levels occur around midnight and the highest in the morning.

4. *Cushing's syndrome* is due to excess cortisol. The causes are adrenal hyperplasia due either to excess ACTH from the pituitary or a non-endocrine tumour, or to a tumour of the adrenal cortex. The biochemical features include impaired glucose tolerance and, in some cases, hypokalaemic alkalosis.

There are two stages in diagnosis.

(*a*) Demonstration of excess cortisol production.
(*b*) Determination of the cause.

5. Primary adrenocortical hypofunction (*Addison's disease*) is due to destruction of the adrenal cortex with loss of all its hormones. Biochemically it is characterised by sodium depletion.

Diagnosis is made by demonstrating that the adrenal cortex cannot respond to ACTH by stimulation tests.

6. *Secondary adrenal insufficiency* is caused by diminished ACTH secretion by the pituitary. This may be due to disease of the hypothalamus or pituitary or it may be a result of corticosteroid therapy.

It is usually diagnosed by demonstrating a definite, but impaired, response to tetracosactrin (synthetic ACTH) stimulation. In cases without adrenal atrophy the hypothalamic-pituitary-adrenal axis may be assessed by metyrapone or insulin stress tests.

7. *Congenital adrenal hyperplasia* is due to an inherited enzyme deficiency in the biosynthesis of cortisol. Symptoms are due to deficient cortisol and to excess androgen secretion. Diagnosis is made most rapidly by the 11-oxygenation index, or by estimation of plasma 17-hydroxyprogesterone. Urinary 17-oxosteroids are raised.

## FURTHER READING

MATTINGLY, D. (1968). Disorders of the Adrenal Cortex and Pituitary Gland. In: *Recent Advances in Medicine*, Chap. 5, 15th edit. Eds. D. N. Baron, N. Compston and A. M. Dawson. London: J. & A. Churchill.

HALL, R., ANDERSON, J., SMART, G. A., and BESSER, M. (1974). *Fundamentals of Clinical Endocrinology (2nd edit.).* London: Pitman Medical.

STEINBECK, A. W., and THEILE, H. M. (1974). The adrenal cortex. *Clinics in Endocr. & Metab.*, **3**, 557.

Symposium on the Adrenal Cortex (1972). *American Journal of Medicine*, **53**, pp. 529–685—a number of articles on adrenocortical physiology and pathology, including aldosterone.

BINDER, C., and HALL, P. E. (eds.) (1972). *Cushing's Syndrome: Diagnosis and Treatment*. London: Wm. Heinemann Medical Books.

See also references on p. 129.

# APPENDIX

## PLASMA 11-OHCS

10 ml of blood is collected into a heparin tube and sent to the laboratory as soon as possible. If there is any possibility of delay the specimen should be sent on ice. The method used for estimation depends on fluorimetry of plasma extracts and all these precautions aim to prevent an increase in non-specific plasma fluorescence.

### Interpretation

This is discussed in the relevant sections but in addition it is worth remembering that *raised values not due to an increase in free cortisol* occur with:

1. Improperly collected specimens (non-specific fluorescence).
2. Raised levels of cortisol binding globulin (pregnancy, oestrogen therapy, contraceptive pill).
3. Some drugs, *particularly spironolactone*, which fluoresce and produce falsely high values in both blood and urine.

Prednisolone, fludrocortisone and dexamethasone in usual dosage do not contribute significantly to results.

## TETRACOSACTRIN STIMULATION TESTS

Tetracosactrin is also called Synacthen (Ciba) or Cortrosyn (Organon).

### 1. 30-minute stimulation test

The patient should be resting quietly.

(*a*) 10 ml of blood is taken for basal 11-OHCS.
(*b*) 250 µg of tetracosactrin, dissolved in about 1 ml of sterile water or isotonic saline, is given by intramuscular injection.
(*c*) 30 minutes later 10 ml of blood is taken for 11-OHCS.

Normally plasma 11-OHCS increases by at least 190 nmol/l (7 µg/dl), to a level of at least 540 nmol/l (20 µg/dl).

### 2. 5-hour stimulation test

(*a*) 10 ml of blood is taken for basal 11-OHCS.
(*b*) 1 mg depot *tetracosactrin* is injected intramuscularly.
(*c*) Further blood samples are taken 1 hour and 5 hours after (*b*).

Normally plasma 11-OHCS rise to a level of between 600 and 1300 nmol/l (22 and 46 µg/dl) at 1 hour, and to 1000 and 1800 nmol/l (37 and 66 µg/dl) at 5 hours. Such a response excludes primary (but not secondary) adrenocortical hypofunction. The cause of an impaired response should be investigated by prolonged stimulation.

### 3. 3-day stimulation test

(a) 1 mg depot tetracosactrin is given daily by intramuscular injection.

(b) Plasma 11-OHCS are estimated in blood drawn 5 hours after each injection.

*No further response* after 3 days indicates *primary adrenocortical hypofunction*. An increasing response over this period indicates *secondary adrenal atrophy* (compare thyroid, p. 172).

If there is danger of an adrenal crisis if steroids are withdrawn, dexamethasone, which does not contribute significantly to the 11-OHCS estimation, may be given. Repeated injections of depot tetracosactrin are painful and may, if there is an adrenocortical response, lead to sodium and water retention: *this test is contra-indicated in patients in whom such retention may be dangerous.*

### 2-DOSE DEXAMETHASONE SUPPRESSION TEST

Daily 24-hour urine collections are made for five days.

(a) Day 1—control day;

(b) Days 2 and 3—dexamethasone 0·5 mg, 6-hourly (2 mg/day) is given by mouth;

(c) Days 4 and 5—dexamethasone 2 mg, 6-hourly (8 mg/day) is given by mouth.

Total 17-oxogenic steroids are estimated on the collections of days 1, 3 and 5.

Suppression by either of the two dose levels is shown by a fall to below 14 µmol/24 h (4 mg/24 h) on day 3 (2 mg dose) or to below 50 per cent of the initial level on day 5 (8 mg dose).

For interpretation see p. 148.

### METYRAPONE TEST

#### (i) Using urine estimations

Twenty-four hour urine collections are made for four days.

(a) Days 1 and 2—control urine collection;

(b) Day 3—metyrapone 750 mg is given at 4-hourly intervals for six doses (total 4·5 g);

(c) Day 4—urine collection.

Total 17-oxogenic steroids are measured in all three. A *normal response* is shown by a rise of more than 35 µmol/24 h (10 mg/24 h) above the control level. This may occur on days 3 or 4.

#### (ii) Using plasma 11-deoxycortisol estimation

(a) Blood is taken at 8.00 hours for plasma cortisol and 11-deoxycortisol estimation.

(b) Metyrapone 750 mg is given orally at 4-hourly intervals for six doses (total 4.5g), starting immediately after the basal blood sample.

(c) A further blood sample is taken the following morning at 8.00 hours.

The results depend on the method of estimation used, and the clinician should consult his own laboratory for "normals".

False negative results to metyrapone are usually due to the patient omitting to take one of the doses. This can be assessed by plasma cortisol estimation. If there is incomplete blocking of cortisol synthesis, levels of the hormone do not fall.

Metyrapone administration may be associated with nausea and dizziness.

The *insulin stress test* is described on p. 130.

# Chapter VII

# THYROID FUNCTION: TSH

THE three hormones known to be produced by the thyroid are thyroxine, tri-iodothyronine and calcitonin. Thyroxine and tri-iodothyronine are products of the thyroid follicular cell and influence metabolism throughout the body. Calcitonin is produced by a specialised cell (the C cell) and influences calcium metabolism. As in the case of the adrenal cortex and medulla, this is an anatomical rather than a functional relationship. In lower animals the calcitonin-secreting cells may be completely separated from the thyroid. Calcitonin is considered briefly on p. 239.

## PHYSIOLOGY

### Chemistry

The thyroid hormones are synthesised in the thyroid gland by iodination and coupling of two molecules of the amino acid tyrosine. Thyroxine has four iodine atoms and is referred to as $T_4$; tri-iodothyronine has three and is therefore often known as $T_3$. The chemical structures are shown in Fig. 9.

Iodine is essential for thyroid hormone synthesis and will be considered in more detail.

### Iodine Metabolism

Iodide in the diet is absorbed rapidly from the small bowel. Most natural foods contain adequate amounts of iodide except in regions where the iodide content of the soil is very low. In these areas there used to be a high incidence of goitre, but general use of artificially iodised salt has made this a less common occurrence. Sea foods have a high iodide content and fish and iodised salt are the main dietary sources of the element.

Normally about one-third of the absorbed iodide is taken up by the thyroid and the other two-thirds is excreted via the kidneys.

### Biosynthesis of Thyroid Hormones

The steps in the biosynthesis of the thyroid hormones are outlined below and in Fig. 10. An understanding of the sequence of events is helpful in appreciating both the mechanism of action of the drugs used

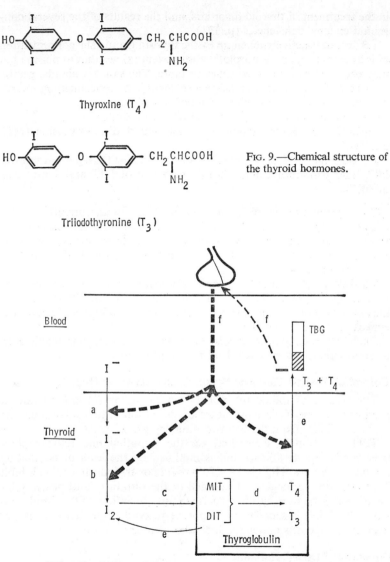

Thyroxine $(T_4)$

Triiodothyronine $(T_3)$

Fig. 9.—Chemical structure of the thyroid hormones.

MIT = Monoiodotyrosine
DIT = Diiodotyrosine
TBG = Thyroxine – Binding Globulin

Fig. 10.—Synthesis and regulation of thyroid hormones.
(See text for explanation of small letters.)

in the treatment of thyroid disorders, and the results of the several congenital enzyme deficiencies (p. 171).

(*a*) Iodide is actively taken up by the thyroid gland. The concentration of iodine in the gland is normally about twenty times that in plasma but may exceed it by a hundred times or more. The salivary glands, gastric mucosa and mammary glands are also capable of concentrating iodine.

Uptake is blocked by thiocyanate or perchlorate.

(*b*) Trapped iodide is rapidly converted to iodine.

(*c*) It is then incorporated into mono- and di-iodotyrosine (MIT and DIT).

(*d*) MIT and DIT are coupled to form thyroxine ($T_4$) (2 molecules of DIT) and tri-iodothyronine ($T_3$) (1 molecule of DIT and 1 molecule of MIT).

The organic binding of iodine (*c*) is inhibited by thiouracil.

The iodotyrosines and $T_3$ and $T_4$ do not exist in a free state in the thyroid gland, but are incorporated in the large protein molecule thyroglobulin. This iodoprotein is the main component of the colloid of the thyroid follicle.

(*e*) Release of $T_3$ and $T_4$ occurs by breakdown of thyroglobulin by proteolytic enzymes. The hormones pass into the blood. Mono- and di-iodotyrosine released at the same time are de-iodinated by another enzyme and the iodine re-utilised.

Each step is controlled by specific enzymes and congenital deficiency of any of these enzymes can lead to hypothyroidism.

### Control of Thyroid Hormone Synthesis and Secretion (Fig. 10)

The uptake of iodide (*a*), conversion to iodine (*b*) and release of thyroid hormones (*e*) are promoted by the action of thyroid-stimulating hormone (TSH) from the anterior pituitary gland.

TSH secretion is dependent on the hypothalamic thyrotrophin-releasing hormone (TRH), but is modified by the level of circulating thyroid hormones. High levels of thyroxine or tri-iodothyronine inhibit TSH secretion by negative feed-back to the pituitary, and possibly inhibit TRH secretion by the hypothalamus. In infants TRH (and consequently TSH) is secreted on exposure to cold. This factor does not seem to be operative after the first year of life.

### Circulating Thyroid Hormone

The hormone secreted in largest amounts is *thyroxine* ($T_4$) and plasma levels are normally about 100 nmol/l ($8\mu g/dl$).

**Thyroxine** in plasma occurs in free and protein-bound forms. Several plasma proteins, including albumin, are capable of binding thyroxine but the main one is an $\alpha$-globulin called thyroxine-binding globulin

(TBG). Almost all the plasma thyroxine is so bound. The free 0·1 per cent or less, is, however, the physiologically active fraction and it is this that controls the pituitary secretion of TSH (compare calcium, p. 235).

Tri-iodothyronine in plasma arises both from secretion by the thyroid and, by conversion, from circulating $T_4$. Normal plasma levels are about 1·5 nmol/l (100 ng/dl). Like $T_4$, more than 99 per cent is carried on the thyroxinebinding proteins but binding is weaker. Despite its lower concentration in plasma, $T_3$ is metabolically more active than $T_4$ and some workers consider $T_4$ to be active only after conversion to $T_3$.

In, for example, iodine deficiency or after treatment of hyperthyroidism, the ratio of $T_3$ to $T_4$ secreted is increased. As a result, the patient may be clinically euthyroid despite low $T_4$ levels.

### Actions of Thyroid Hormones

Thyroid hormones influence and speed up many metabolic processes in the body. They are essential for normal growth, mental development and sexual maturation. They also increase the sensitivity of the cardiovascular and central nervous systems to catecholamines and so influence cardiac output and heart rate. The mechanism of thyroid hormone action is unknown.

## THYROID FUNCTION TESTS

It is essential to confirm a clinical diagnosis of thyroid disease by laboratory tests. Treatment may be prolonged and, in the case of hypothyroidism, life-long. During treatment the original clinical picture disappears and it is not an uncommon problem to be faced with a patient who has taken thyroxine for many years for probable "thyroid trouble". Without an adequately established diagnosis it may be difficult to assess the past history.

Thyroid function tests may conveniently be divided into two groups:

(a) Those measuring circulating thyroid hormones and TSH.

(b) Those measuring radio-iodine uptake by the thyroid gland.

Blood tests are more convenient than radioactive iodine uptake tests. It is necessary to do both only in obscure cases.

### Measurement of Thyroid Hormones and TSH

(a) Plasma thyroxine ($T_4$).—Plasma $T_4$ is preferably measured by competitive protein binding or radioimmunoassay techniques.

An alternative method, still used in some laboratories, is to estimate hormonal iodine, usually as the protein-bound iodine (PBI). These estimations suffer from the major disadvantage that non-hormonal iodine compounds may be included. This produces *falsely raised values* which may either *mimic hyperthyroidism* or, more seriously, *mask*

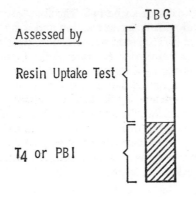

FIG. 11.—Assessment of circulating thyroid hormone.

*hypothyroidism* by elevating low values to within the normal range. The causes of iodine contamination are considered on p. 168.

Plasma $T_4$ levels are high in hyperthyroidism and low in hypothyroidism. However, as more than 99 per cent of measured $T_4$ is protein bound, values depend on the level of thyroxine-binding proteins (see below and Figs. 11 and 13).

(*b*) **Plasma tri-iodothyronine** ($T_3$) may be measured by radioimmunoassay. Estimation is of value in a limited number of situations.

(*c*) **Resin uptake (RU) tests.**—This title applies to any of several methods of assessing *unfilled binding sites on thyroxine-binding proteins* (mainly TBG) and "$T_3$ resin uptake" is *not a measure of circulating $T_3$*. A sample of serum is treated with radioactive $T_3$ or $T_4$ under conditions that ensure saturation of all unbound binding sites (that is, an excess of the radioactive hormone) (a and b, Fig. 12). A resin is then added to the mixture to extract all unbound hormone; the latter is estimated by counting the radioactivity of the resin (c, Fig. 12). It follows that the

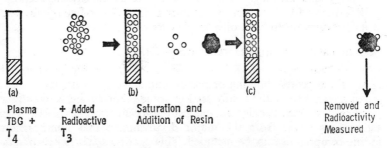

FIG. 12.—Principle of $T_3$ resin uptake test,

greater the number of binding sites available on TBG, the less $T_3$ or $T_4$ will be taken up by the resin and vice versa. Note that the result is expressed as the *resin* uptake, not TBG uptake. Used alone this test is not very sensitive, particularly in cases of hypothyroidism. Its main value is in the calculation of the free thyroxine index (see below).

**Interpretation of $T_4$ and resin uptake** (Fig. 13).—In hyperthyroidism the circulating thyroxine level is increased. This includes both bound and free fractions and the $T_4$ is raised because of the high protein-bound fraction. Because more binding sites on TBG are occupied by the excess thyroxine the resin uptake is also raised (*a*).

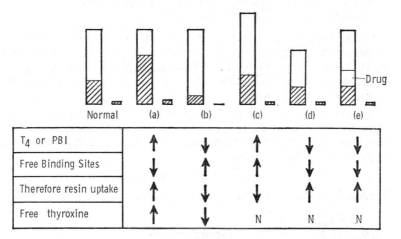

Fig. 13.—Interpretation of tests of circulating hormone (see text).

In hypothyroidism circulating thyroxine is decreased. The $T_4$ level is low and, because fewer binding sites on TBG are occupied, the resin uptake is also low (*b*).

So far we have assumed that measurement of the bound thyroxine is a valid reflection of the important free fraction and therefore a measure of the patient's thyroid status. In three sets of circumstances this is not true and abnormal results may be found with normal free thyroxine levels in a euthyroid patient.

1. *Increased TBG concentration (c)*.—If increased levels of TBG are present, more thyroxine than normal is taken up by it from the plasma. This tendency to reduced free thyroxine results in increased thyroxine synthesis (via the feed-back mechanism) until TBG is again about a third saturated and the free thyroxine is normal. The $T_4$ is therefore raised and, because of the increased number of unsaturated TBG

binding sites, resin uptake is low. This combination of *raised $T_4$ and low resin uptake* occurs with:

    (i) Pregnancy
    (ii) Oestrogen therapy
    (iii) Oral contraceptives.

2. *Decreased TBG concentration* (*d*).—With reduced levels of TBG the reverse process occurs. The $T_4$ is low and, because of reduced binding sites, resin uptake is high. This combination, *low $T_4$ and raised resin uptake,* is seen with:

    (i) Protein loss, such as in the nephrotic syndrome.
    (ii) Congenital TBG deficiency (rare).

3. *TBG binding sites occupied by drugs* (*e*).—Certain drugs, notably salicylates and diphenyl hydantoin ("Epanutin"), occupy binding sites on TBG and displace thyroxine. The resultant low $T_4$ may or may not be accompanied by a raised resin uptake. Low $T_4$ values can only be interpreted with a knowledge of what drugs the patient is taking.

(*d*) **Free thyroxine index (FTI).**—Many formulae have been devised using plasma $T_4$ (or PBI) levels and resin uptake to calculate the free thyroxine. The aim of these formulae is to eliminate changes due only to altered TBG. In principle:

$$T_4 \text{ level} \times \text{resin uptake} = \text{free thyroxine index.}$$

Substituting the results from Fig. 13 we have:

|     | $T_4$ | RU | FTI |     |
|-----|-------|-----|-----|-----|
| (a) | ↑ × | ↑ = | ↑↑ | (hyperthyroid) |
| (b) | ↓ × | ↓ = | ↓↓ | (hypothyroid) |
| (c) | ↑ × | ↓ = | N | (euthyroid) |
| (d) | ↓ × | ↑ = | N | (euthyroid) |

Because of the variable effects of protein binding, $T_4$ estimation should preferably be accompanied by a $T_3$ or $T_4$ resin uptake test and the FTI calculated. This is the *best single indicator* of thyroid secretory function.

(*e*) **"Normalised" thyroxine levels** (for example, effective thyroxine ratio).—A number of tests are becoming available in which the $T_4$ value is corrected for abnormalities of protein binding in a single procedure and is expressed as a ratio of a standard serum. This is analogous to the FTI and may be used as a screening test. Borderline results, especially if low normal, should be confirmed by $T_4$ estimation.

(*f*) **Plasma TSH** may be measured by radioimmunoassay. Estimation is of particular value in the diagnosis of primary hypothyroidism.

Unfortunately most currently available techniques are unable to distinguish between low normal and subnormal levels. Dynamic tests, however, overcome this problem (p. 177).

### Radio-iodine Uptake Tests

Ingested radioactive iodine enters the body iodine pool and is metabolised in the same way as the natural element. For example, after 24 hours about a third of the dose has accumulated in the thyroid gland, about two-thirds has been excreted in the urine and about 0·05 per cent has appeared in the blood as labelled $T_4$ or PBI.

Its appearance at any or all of these sites can be measured to estimate thyroid function. The most commonly used method is direct counting over the thyroid gland. Measurements of urinary loss require a 24-hour collection and assume normal renal function. Measurement of radioactive PBI is technically more difficult than that of neck uptake,

**Neck uptake tests**—The technique is simple. The patient drinks a suitable tracer dose of radioactive iodine and an equivalent dose is set aside as a standard. After the required period of time the radioactivity over the thyroid gland is counted and compared with that of the standard. As physical decay affects both equally, the uptake may be directly expressed as a percentage of the dose.

In hyperthyroidism the uptake is usually above normal and in hypothyroidism it is usually low. Any time period between administration of the dose and counting may be allowed but 4 or 24 hours are commonly chosen. In hyperthyroidism 4 hours is preferable as a significant percentage of the dose may have been discharged from the thyroid as thyroxine within 24 hours. In hypothyroidism, on the other hand, the longer period gives a better separation from the normal.

Two isotopes are in common use. $^{131}I$ has a half-life of 8 days and can be used for prolonged tests such as the 24-hour uptake, and measurement of PB$^{131}I$. Measurable radioactivity from this isotope persists for several weeks. $^{132}I$ has a half-life of 2·3 hours and can only be used for 4-hour uptake tests. It has the advantage that there is a much lower radiation dosage to the patient and that tests can be repeated on the next day if necessary: this is of great advantage when borderline answers are obtained, and virtual absence of radioactivity after 24 hours facilitates the performance of further tests on the next day.

Technetium ($^{99m}Tc$) has also been used. It has a half-life of 6 hours. Radiation to the patient is even lower than with $^{132}I$. It is usually given intravenously, and uptake measured after 20 minutes.

*Extrathyroidal factors affecting neck uptake.*—1. A dose of radioactive iodine enters the body pool of iodine and is diluted by it. The thyroid gland takes up iodine constantly from the body pool and of this a certain proportion will be radioactive. If the body iodine pool is greatly increased, the radioactive iodine is diluted more than normal and a smaller amount will be taken up by the thyroid although the total iodine uptake (radioactive plus non-radioactive) is normal. This will be interpreted as a low uptake. Moreover, large amounts of iodine directly inhibit uptake of iodine by the gland.

Causes of increased body iodine are:

(a) Excess oral iodine intake as, for example, Lugol's iodine and iodides in cough mixture.

(b) Iodine-containing radio-opaque contrast media, as used in, for example, IVPs, cholecystograms and myelograms.

The latter group particularly may invalidate the results of neck up-take for months, especially if injected into a site from which excretion is slow (myelogram). The recommended periods after administration of such substances which should be allowed before performing the test are given in the Appendix (p. 177).

Conversely, raised neck uptakes are found in iodine deficiency, when there is a reduced body iodine "pool".

2. Drugs used in the treatment of thyroid disease alter the uptake so that it no longer accurately reflects thyroxine output. Thiouracil and its derivatives block thyroxine synthesis. They do not directly affect iodine uptake, but the drop in circulating $T_4$ may stimulate TSH secretion and *increase* neck uptake. Administered thyroxine may depress TSH secretion and cause a low neck uptake despite normal or raised levels of circulating thyroxine.

Because of these factors, neck uptake values do not always reflect thyroid function. A full history of drugs and x-rays should be taken, and results interpreted together with the clinical situation. If necessary, stimulation or suppression tests (p. 172 and p. 170) should be performed.

## DISORDERS OF THE THYROID GLAND

Excess or deficiency of circulating thyroid hormone produce characteristic clinical changes. Thyroid disease may exist without hyper- or hypofunction. These changes, and the causes of the abnormal thyroid activity, will be considered only in so far as they influence the results of thyroid function tests.

### HYPERTHYROIDISM (THYROTOXICOSIS)

The syndrome produced by sustained excess of thyroid hormone may be easily recognised or may remain unsuspected for a long time. The key feature is a speeding up of metabolism. Clinical features include tremor, tachycardia (and sometimes arrhythmias), weight loss, tiredness, sweating and diarrhoea. Anxiety and emotional symptoms may be prominent. In some cases a single feature predominates, such as weight loss, diarrhoea, tachycardia or atrial fibrillation.

The *causes* of hyperthyroidism are:

(*a*) Graves' disease (commonest form).

(*b*) Toxic multinodular goitre or single functioning nodule (occasionally an adenoma).

(*c*) Ingestion of thyroid hormones.

(*d*) Rare causes

    (i)   secretion of a thyrotrophin by tumours of trophoblastic origin;

    (ii)  struma ovarii (thyroid tissue in ovarian teratoma);

    (iii) administration of iodine to a subject with iodine-deficiency goitre;

    (iv) TSH-secreting tumour of the pituitary.

**Graves' disease** occurs at any age and is commoner in females. It is characterised by one or more of the following:

    (i)   hyperthyroidism due to diffuse hyperplasia of the thyroid;

    (ii)  exophthalmos;

    (iii) localised (pretibial) myxoedema.

Graves' disease is probably autoimmune in origin. In many cases an immunoglobulin (IgG) capable of stimulating the thyroid is present in the blood. As it acts (in bioassay) later and longer than TSH it is called long-acting thyroid stimulator or LATS. LATS may cross the placenta and produce transient hyperthyroidism in the newborn but its role in adult Graves' disease is controversial. It may be present at high concentrations in the blood of subjects without hyperactivity.

**Toxic nodules,** single or multiple, in a nodular goitre may secrete thyroid hormones *autonomously. TSH is suppressed* by negative feedback and the rest of the thyroid tissue is inactive. This can be detected if a larger dose of radio-iodine is given, and the areas of activity "mapped" by scanning the gland with a counter. A benign secreting *adenoma* produces the same effect and is only distinguished from a "toxic nodule" on histological examination. Toxic nodules are found more commonly in the older age groups, and the patient may present with only one of the features of hyperactivity, most commonly cardiovascular symptoms.

The *feature common to all forms of hyperthyroidism* (with the exception of the *very* rare TSH-secreting pituitary tumour) is that *secretion is not dependent on TSH from the pituitary.*

## Diagnosis of Hyperthyroidism

(a) **Tests on blood.**—The first step is the estimation of *plasma $T_4$*, with a resin uptake test and calculation of the *FTI*. The finding of un-equivocally raised levels in a clinically thyrotoxic patient establishes the diagnosis and no further tests are necessary.

If the results are equivocal, or if there is clinical doubt, the *TRH stimulation test* (see p. 177) is of value. Plasma TSH is often un-detectable in hyperthyroidism and levels *fail to rise* after administration of TRH, the normal response being inhibited by negative feed-back.

(b) **Radio-iodine neck uptake tests.**—Very high uptakes (approaching 100 per cent) may be found in hyperthyroidism, but values may be only moderately raised. The level of neck uptake is not necessarily an index of the severity of the disease, this being essentially a clinical assessment.

If the result is borderline, further tests are used to differentiate "high normal" from pathological results.

In differentiating these conditions it is useful to review the basic cause of the raised uptake. High uptakes may also be due to iodine deficiency, antithyroid drugs or dyshormonogenesis (p. 171). In these situations, the uptake of iodine is a result of TSH stimulation.

The $T_3$ *suppression test* is useful here. The neck uptake is repeated after a short course of tri-iodothyronine in doses about twice that of normal thyroid hormone output (that is, 120 µg/day for six days). This depresses TSH secretion and the uptake will be markedly reduced in all cases other than those due to hyperthyroidism. Lack of suppression is diagnostic of hyperthyroidism.

## $T_3$ Thyrotoxicosis

Plasma $T_3$ levels are raised in virtually all cases of hyperthyroidism, particularly those due to toxic nodules. Rarely $T_4$ levels may be normal and the neck uptake normal or only moderately raised. This rare condi-tion, $T_3$ thyrotoxicosis, should be suspected if a clinically thyrotoxic patient has normal $T_4$ levels. The radio-iodine neck uptake (even if normal) does *not suppress* after $T_3$ administration and there is no TSH response to TRH. The diagnosis is confirmed by measurement of plasma $T_3$ levels. Early clinical hyperthyroidism may pass through a phase with raised $T_3$, but normal $T_4$, levels; $T_4$ levels rise later.

## HYPOTHYROIDISM

Hypothyroidism is due to suboptimal circulating levels of one or both thyroid hormones. The condition develops insidiously and, in its early stages, symptoms are non-specific. Many of the features of frank hypothyroidism are, not unexpectedly, the opposite of those in hyperthyroidism. There is a generalised slowing down of metabolism, with mental dullness, physical slowness and weight gain. The skin is dry, hair falls out and the voice becomes hoarse. The face is puffy and the subcutaneous tissues thickened (myxoedema).

In the most severe form, myxoedema coma with profound hypo-thermia may develop. In children, growth may be impaired and intra-uterine thyroid hormone deficiency leads to cretinism. The fully developed case of myxoedema is easily recognised, but lesser degrees of hypothyroidism with slow onset may be commoner than previously thought.

The causes of hypothyroidism are:

1. **Primary.**—(a) Disease of the gland (autoimmune thyroiditis):
   (i) "Primary" hypothyroidism
   (ii) Hashimoto's disease

(b) The result of treatment:
   (i) post-thyroidectomy
   (ii) post-$^{131}$I therapy for hyperthyroidism

(c) Uncommon causes:
   (i) dyshormonogenesis
   (ii) exogenous goitrogens and drugs

2. **Secondary** to TSH deficiency, due to anterior pituitary or hypo-thalamic disease.

The essential difference between primary and secondary hypothy-roidism is that TSH levels are raised in the former and low in the latter.

Hashimoto's disease and "primary" hypothyroidism are now consid-ered to be different manifestations of the same basic disorder. There is progressive destruction of thyroid tissue and circulating thyroid antibodies are present. The term "dyshormonogenesis" includes the congenital deficiencies of the enzymes involved in thyroxine synthesis. Deficiencies have been described at each of the stages shown in Fig. 10, with differing biochemical features. The end result in each case is reduced thyroxine synthesis and hypothyroidism. Goitre is almost invariable in dys-hormonogenesis, due to continuous TSH stimulation. The commonest form is failure to incorporate iodine into tyrosine (stage (c) of Fig. 10).

If secondary hypothyroidism is long-standing, irreversible atrophic changes occur in the thyroid gland.

### Diagnosis of Hypothyroidism

(a) **Tests on blood.**—Plasma $T_4$ levels and the FTI may be very low but there is often overlap with low normal levels.

*Plasma TSH estimation is the most sensitive test of primary hypo-*

*thyroidism.* In response to even a slight reduction of thyroid hormone levels (which may still be in the "normal" range), TSH secretion is increased and *plasma levels of TSH are raised.* The interpretation of mildly raised levels is controversial but, when combined with a low normal FTI, they probably indicate early hypothyroidism.

*Secondary hypothyroidism* is confirmed by a *TRH stimulation test* as abnormally low TSH levels cannot at present be detected. Intravenous injection of TRH (see p. 177), may produce one of three responses:

(i)  a normal response (which in a case of obvious hypothyroidism probably indicates hypothalamic disease);

(ii) no response (confirming anterior pituitary hypofunction);

(iii) an exaggerated response which is seen in primary hypo-thyroidism.

(*b*) **Radio-iodine neck uptake tests.**—These are of less value in the diagnosis of hypothyroidism than of hyperthyroidism because of considerable overlap with the lower limit of normal, and the widespread use of iodine-containing preparations, but, combined with TSH stimulation tests, can be informative.

If falsely low uptakes due to iodine excess have been excluded (not always possible), results far below normal are diagnostic of hypothyroidism. Borderline results may be due to:

(i)   Primary hypothyroidism

(ii)  Secondary hypothyroidism

(iii) "Low normal" uptake.

To resolve these causes a *TSH stimulation test* is performed. TSH is given and the uptake repeated on the following day; the uptake may, if necessary, be measured at daily intervals after successive injections of TSH. The test in this form is only possible if $^{132}I$ has been given. There are several possible responses to TSH.

1. *A rapid rise of uptake to high normal or high levels* after one dose of TSH indicates a *normally functioning gland,* but does not exclude *early secondary hypothyroidism.*

2. *No response* indicates that the gland cannot increase its uptake in response to TSH, indicating *primary hypothyroidism* or *long-standing secondary hypothyroidism* with atrophy of the thyroid.

3. *A subnormal response with no further increase after a second dose of TSH* indicates that the gland is already almost maximally stimulated by endogenous TSH (*mild primary hypothyroidism*).

4. *A stepwise increase with successive injections* indicates a functional gland with the sluggish response seen in *secondary hypothyroidism*: (compare with adrenal, p. 150). After prolonged $T_4$ therapy, which suppresses the pituitary, a similar effect may be found.

## EUTHYROID GOITRE

Thyroxine synthesis may be impaired by iodine deficiency, by drugs such as para-aminosalicylic acid, or possibly by minor degrees of enzyme deficiency. The consequent slight reduction of circulating thyroxine level results, by the feed-back mechanism, in increased TSH secretion. This stimulates thyroxine synthesis and levels are therefore maintained in the normal range. As a result the thyroid enlarges (goitre) but hypothyroidism is avoided. In areas with low iodine content of the soil, iodine deficiency used to be common (endemic goitre).

Plasma $T_4$ levels are usually normal. Low levels in clinically euthyroid patients are often accompanied by raised plasma $T_3$ levels (see p. 163). TSH levels are often normal.

Inflammation of the thyroid (thyroiditis), whether acute or subacute, may produce marked but temporary aberrations of thyroid function tests. These conditions are relatively uncommon.

### Other Biochemical Findings in Thyroid Disease

1. **Cholesterol level.**—In *hypothyroidism* the synthesis of cholesterol is impaired but its catabolism is even more impaired and plasma cholesterol levels are high. This may help both in diagnosis and in following the effect of treatment but many other factors cause raised cholesterol levels. Although, statistically, low cholesterol levels occur with *hyperthyroidism*, plasma cholesterol estimation is of little value in the individual case because of overlap with the normal range.

Several clinical tests of thyroid function, such as Achilles tendon reflex duration and ECG, depend on the peripheral action of thyroxine (on muscle contraction and on cardiac muscle).

2. **Hypercalcaemia** is very rarely found with severe thyrotoxicosis. There is an increased turnover of bone, probably due to direct action of thyroid hormone (p. 239).

3. **Glucose tolerance tests** may show a "lag" curve in thyrotoxicosis probably due to rapid absorption. In hypothyroidism, by contrast, the glucose tolerance curve may be flat. This test is of little use in diagnosis.

4. **Plasma creatine kinase (CK)** levels are often raised in hypothyroidism but, again, the estimation does not help diagnosis. CK levels may also be raised in thyrotoxic myopathy.

5. **Urinary total 17-oxogenic steroid** and **17-oxosteroid** values are low in untreated hypothyroidism, and this finding does not necessarily indicate a pituitary origin of the disease.

### Thyroid Antibodies

Circulating thyroid antibodies, although not tests of thyroid function, should be sought in the diagnosis of thyroid disease. The two antibodies most useful clinically are:

## TABLE XV
### RESULTS OF THYROID FUNCTION TESTS

| | True T₄ abnormalities | | | | TBG abnormalities | | Iodine "pool" abnormalities | |
| | Hyperthyroidism | | Hypothyroidism | | | | | |
| | True hyperthyroidism | Ingestion of thyroxine | Primary | Secondary | Raised levels | Low levels | Deficiency | Excess |
|---|---|---|---|---|---|---|---|---|
| Plasma T₄ | ↑ | ↑ | ↓ | ↓ | ↑ | ↓ | N | N |
| T₃-resin uptake test | ↑ | ↑ | ↓ | ↓ | ↓ | ↑ | N | N |
| Free thyroxine index | ↑ | ↑ | ↓ | ↓ | N | N | N | N |
| PBI | ↑ | ↑ | ↓ | ↓ | ↑ | ↓ | ↓ | ↑ |
| Plasma TSH levels | ↓ | ↓ | ↑ | ↓ | N | N | usually N | N |
| Radio-iodine neck uptake | ↑ | ↓ | ↓ | ↓ | N | N | ↑ | ↓ |
| (a) Suppressed by T₃ | No | — | — | — | — | — | Yes | — |
| (b) Stimulated by TSH | — | Yes | No | Yes | — | — | — | yes usually |
| Response to TRH | Absent | Probably absent | Exaggerated | Absent | Normal | Normal | Normal | Normal |

(*a*) Complement fixing microsomal antibody.

(*b*) Antibody to thyroglobulin.

Antibodies may be detected in high titre in most cases of Hashimoto's thyroiditis and many cases of "primary" hypothyroidism Their role, if any, in the aetiology of these conditions is uncertain. A high incidence of antibodies in thyrotoxicosis reflects the focal thyroiditis often seen histologically in the gland and correlates with development of post-operative hypothyroidism in these cases.

## SUMMARY

1. There are two circulating thyroid hormones (apart from calcitonin)—thyroxine and tri-iodothyronine. Their synthesis depends on an adequate supply of iodine, and is controlled by circulating thyroid hormone levels via TSH from the anterior pituitary gland.

2. An excess of circulating thyroid hormone produces the syndrome of hyperthyroidism. This may occur in Graves' disease or may be due to a functioning nodule of the thyroid.

3. A decreased circulating thyroid hormone level produces the syndrome of hypothyroidism. This may be primary, due to disease of the thyroid gland, or be secondary to pituitary or hypothalamic disease. TSH levels are raised in the first group and absent in the second.

4. Euthyroid goitre represents compensated thyroid disease. Thyroid function tests may be abnormal.

5. Thyroid function tests may be considered under two headings.

(*a*) Assessment of circulating thyroid hormones.

(*b*) Measurement of iodine uptake by the thyroid.

6. Circulating thyroid hormone levels may be assessed by $T_4$ or PBI estimation and resin uptake test. Changes in TBG may give abnormal results. The free thyroxine index is an expression of thyroid status regardless of TBG changes.

7. Iodine uptake may be measured by radio-iodine neck uptake. Borderline results may be clarified by the $T_3$ suppression test (high) or the TSH stimulation test (low).

8. Many thyroid tests are invalidated by extra-thyroidal factors:

(*a*) Iodine-containing drugs or radio-opaque media invalidate the radioactive neck uptake and the PBI.

(*b*) TBG changes during pregnancy, oestrogen therapy or on oral contraceptives require special interpretation of results of plasma $T_4$ and PBI estimations. They do *not* affect neck uptake.

(c) During and shortly after the treatment of thyroid disease the radio-iodine neck uptake results may not reflect circulating hormone levels.

9. The TRH test is of value in the confirmation of hyperthyroidism, and in the diagnosis of secondary hypothyroidism.

## FURTHER READING

EVERED, D. (1974). Diseases of the thyroid gland. *Clinics in Endocr. & Metab.*, **3**, 425.

ACLAND, J. D. (1971). The interpretation of the serum protein-bound iodine: a review. *J. clin. Path.*, **24**, 187–218.

FISHER, D. A. (1973). Advances in the laboratory diagnosis of thyroid disease —Part I. *J. Pediat.*, **82**, 1–9, Part II, *Ibid*, 187–191.

HAMBURGER, J. I. (1971). Application of the radioiodine uptake to the clinical evaluation of thyroid disease. *Seminars in Nuclear Medicine*, **1**, 287–300.

HOFFENBERG, R. (1973). Triiodothyronine. *Clin. Endocr.*, **2**, 75–87. (A review of physiology and pathological changes.)

Symposium on Graves' disease (1972). *Proc. Mayo Clin.*, **47**, 801–878.

See also references given on p. 129.

# APPENDIX

1. Many factors alter thyroid function tests so that the results no longer reflect the thyroid status of the patient. The commoner (and avoidable) ones are listed below with the time periods that should elapse before measuring the PBI or radioactive iodine uptake. In the case of drugs used in the treatment of thyroid disease the PBI reflects the resultant thyroid status of the patient whereas the neck uptake does not. The times given are approximate—for greater detail on individual drugs or radio-opaque media see:

DAVIS, P. J. (1966). *Amer. J. Med.*, **40**, 918 (for PBI).

MAGALOTTI, M. F., HUMMON, I. F., and HIERSCHBIEL, E. (1959). *Amer. J. Roentgenol.*, **81**, 47 (for radio-iodine uptake).

### TABLE XVI

SUBSTANCES INTERFERING WITH THYROID FUNCTION TESTS

| *Iodine Interference* | PBI | Neck uptake |
|---|---|---|
| Dietary iodide (fish, iodised salt, etc.) | — | 3 days |
| Drug iodide such as KI (suspect in cough mixtures and amoebicides such as "Enterovioform") | 2 weeks | 8 weeks |
| Iodine-containing contrast media | | |
| (a) Rapidly cleared (for example, IVP, angiogram, cholecystogram, salpingogram, bronchogram) | 8 weeks | 8 weeks |
| (b) Slowly cleared (for example, myelogram) | 1 year + (may be indefinite) | 1 year + |
| | | |
| *Interference with Thyroxine Metabolism* | | |
| (a) Blocking uptake (for example, perchlorate, thiocyanate, nitrate) | — | 2 weeks |
| (b) Inhibiting organic binding—thioureas, thiouracils | — | 6 months |
| (c) Thyroid hormones, PAS | — | 8 weeks |

2. **TRH test** (Hall, R., Ormston, B. J., Besser, G. M. Cryer, R. J., and McKendrick, M. *Lancet*, 1972, **1**, 759).

(a) A basal blood sample is taken.

(b) 200 μg of TRH in 2 ml saline is injected rapidly intravenously.

(c) Further blood samples are taken 20 and 60 minutes after the TRH injection.

TSH is measured on all samples.

The maximum response occurs at 20 minutes. Published normal values are:

Males    5·9–18·3 mU/l
Females 6·8–20·0 mU/l.

# Chapter VIII

# CARBOHYDRATE METABOLISM AND ITS INTERRELATIONSHIPS

In most parts of the world carbohydrate is the main source of calorie intake. Under normal circumstances starch is the main dietary carbohydrate, disaccharides contribute significantly and monosaccharides are a minor component of the diet.

## CHEMISTRY

### Monosaccharides

The basic units of carbohydrate are monosaccharides. These compounds are classified by the number of carbon atoms in the molecule. Physiologically the most important groups are the *hexoses* (six carbon atoms) and *pentoses* (five carbon atoms).

The main hexoses of physiological importance are *glucose, fructose* and *galactose*. These are all reducing sugars and therefore reduce Benedict's solution (p. 211). The important pentoses include *ribose* (in RNA) and *deoxyribose* (in DNA).

### Disaccharides

The disaccharides are molecules composed (as the name suggests) of two monosaccharides. The commonly encountered disaccharides are *sucrose* (fructose + glucose), *lactose* (galactose + glucose) and *maltose* (glucose + glucose). Lactose and maltose are reducing sugars, sucrose is not.

### Polysaccharides

Polysaccharides are long chain carbohydrates. *Starch* is found in plants and is a mixture of amylose (straight chains) and amylopectin (branched chains). *Glycogen* is found in animal tissue and has a highly branched structure. Both these polysaccharides are composed of glucose subunits.

## OUTLINE OF INTERMEDIARY METABOLISM

A knowledge of the main pathways of glucose metabolism and their interrelationships with those of fat and protein, is essential to the understanding of disorders of carbohydrate metabolism.

The end products of breakdown of dietary carbohydrates are glucose (from polysaccharides, sucrose and lactose), fructose (from sucrose) and galactose (from lactose). Fructose and galactose are converted to glucose in the liver and will not be considered separately here.

In the following discussion the letters refer to those in Fig. 14.

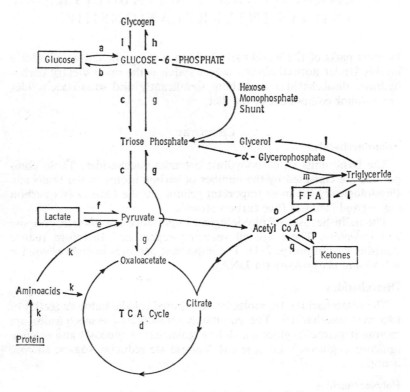

FIG. 14.—Pathways of carbohydrate metabolism.
(See text for explanation of small letters.)

## Glucose Uptake by Cells (*a*) and Formation (*b*)

The entry of glucose into cells other than those of the liver, the central nervous system and erythrocytes, is dependent on *insulin*. Once in the cell, glucose is converted to *glucose-6-phosphate* (*G-6-P*). G-6-P is a central compound of glucose metabolism and can proceed by a number of pathways. Reconversion of G-6-P to glucose requires the enzyme *glucose-6-phosphatase* which is found only in liver and kidney.

### Glycolysis (c) and the Tricarboxylic Acid (TCA) cycle (d)

Some G-6-P is used to provide energy. It is broken down through the glycolytic (Embden-Meyerhof) pathway (c) to pyruvate, which in the presence of oxygen is completely metabolised, via acetyl CoA, in the tricarboxylic acid (TCA, Krebs', citric acid) cycle (d) to carbon dioxide and water. If oxygen is not available glucose is broken down only as far as pyruvate. Unoxidised coenzyme converts this to lactate (e) and there is accompanying acidosis, (see p. 190). When oxygen becomes available, this step is reversed (f). Complete oxidation of glucose produces almost 20 times as much energy as ATP than does glycolysis alone.

Glycolysis is stimulated by *insulin*.

### Gluconeogenesis (g + b)

The glycolytic pathway can proceed in the reverse direction, using some different enzymes, and glucose can be formed from substances in, or entering, the pathway or the TCA cycle. Such new glucose formation (*gluconeogenesis*) (g) can occur only in the liver and kidney, where G-6-P can be converted to glucose (b). The main substrates for gluconeogenesis are the carbon chains of some amino acids (especially alanine) while lactate and glycerol make a small contribution.

Gluconeogenesis is stimulated by *glucocorticoids* and opposed by *insulin*.

### Glycogenesis (h) and Glycogenolysis (i)

G-6-P may be synthesised into glycogen (h), the storage form of glucose (glycogenesis). This reaction takes place mainly in liver and muscle, but can occur in most tissues. On a "normal" diet about 25 per cent of the daily carbohydrate intake is stored as liver glycogen. *Glycogenesis cannot occur in the brain cells or erythrocytes:* as they cannot store glucose they are dependent on an adequate supply from the blood.

Glycogenesis is stimulated by *insulin*.

Glycogen may be broken down again to G-6-P by *glycogenolysis* (i). The G-6-P formed may enter the glycolytic pathway (c) or, in liver and kidney, may be converted to glucose (b).

Glycogenolysis is stimulated by *adrenaline* (which acts on liver and muscle) and *glucagon* (which acts on liver only) and is opposed by *insulin*.

### Hexose Monophosphate Shunt (Pentose Shunt) (j)

This route of glucose metabolism by-passes part of the glycolytic pathway and has two important functions, the *generation of pentoses for nucleic acid synthesis* and the *generation of reduced nicotinamide*

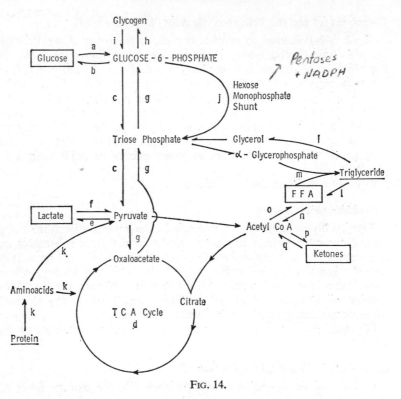

FIG. 14.

*adenine dinucleotide phosphate (NADPH).* NADPH is required, among
other things, for anabolic processes, such as the *synthesis of fatty acids
and steroids* and has a role in maintaining the *integrity of the erythrocyte
membrane.* A key enzyme in the hexose monophosphate pathway is
*glucose-6-phosphate dehydrogenase (G-6-PD),* deficiency of which may
be associated with *haemolytic anaemia.*

### INTERRELATIONSHIPS BETWEEN CARBOHYDRATE AND PROTEIN AND LIPID METABOLISM

As is apparent from Fig. 14, metabolism of glucose, fat and protein
is closely linked.

*Protein breakdown (k)* provides amino acids which may be used for
gluconeogenesis. Such breakdown is promoted by *glucocorticoids* and
inhibited by *insulin. Incorporation of amino acids into protein is promoted by insulin* and *growth hormone.*

*Lipolysis* (triglyceride breakdown) in adipose tissue (*l*) yields free fatty acids (*FFA*) and glycerol. If adequate intracellular glucose is available, FFA are reincorporated into triglycerides via α-glycerophosphate (*m*). Glycerol cannot be utilised by adipose tissue due to a relative lack of the enzyme *glycerokinase* and is carried to the liver where it enters the glycolytic or gluconeogenic pathway. In the absence of adequate amounts of glucose FFA pass into the blood stream and some are metabolised by the tissues via acetyl CoA (*n*). Fatty acids are also synthesised from acetyl CoA (*o*). Acetyl CoA in excess of the capacity of the TCA cycle is converted in the liver to ketones (*p*) which can be used by other tissues for energy (*q*).

Lipolysis is *promoted* by *growth hormone (GH)*, *adrenaline* and *glucocorticoids* and *opposed* by *insulin* and *glucose*.

It is important to stress that glucose is normally the only energy source for brain cells, although they can develop the ability to metabolise ketones after prolonged fasting. As these cells are unable to store glycogen they are dependent on normal blood glucose levels. Hypoglycaemia, particularly of rapid onset, leads to coma and, if untreated, to cerebral damage and death.

## MAINTENANCE OF BLOOD GLUCOSE CONCENTRATIONS

### Hormonal Control

The importance of an adequately high blood glucose concentration for cerebral function and the relative unimportance of hyperglycaemia is reflected in the fact that only insulin reduces blood glucose, while several hormones can raise it. Glucose levels are maintained not only by the formation of glucose from glycogen or by gluconeogenesis, but also by the utilisation of alternative fuel sources such as FFA and ketone bodies which spare glucose.

The subsequent discussion centres on the hormonal control of intermediary metabolism. There is, however, growing evidence that the sympathetic nervous system plays an important role in the control of hormone secretion and in the mobilisation of fat stores. This aspect will not be dealt with in detail and interested readers should consult the references at the end of the chapter.

Three hormones, *insulin, glucagon* and *growth hormone* are intimately concerned with the day-to-day regulation of glucose metabolism and are considered below. The *glucocorticoids* and *adrenaline* are more important in abnormal states such as prolonged fasting and stress. It must be emphasised that the effect of these hormones depends not only on their absolute concentration but also on the concentration of opposing hormones. Thus a falling level of insulin allows the effects of the

TABLE XVII

ACTIONS OF HORMONES ON INTERMEDIARY METABOLISM

| | Insulin | Glucagon | Growth Hormone | Glucocorticoids | Adrenaline |
|---|---|---|---|---|---|
| **Carbohydrate metabolism** | | | | | |
| **(a) in liver** | | | | | |
| — glycolysis | + + | | | | |
| — glycogenesis | + + | | | | |
| — glycogenolysis | | + + | | | + |
| — gluconeogenesis | — | | | + | |
| **(b) in muscle** | | | | | |
| — glucose uptake | + + | | — | — | |
| — glycogenesis | + | | | | |
| — glycogenolysis | — | | | | + |
| **Protein** — synthesis | + | | + | | |
| — breakdown | — | | | + | |
| **Fat** — synthesis | + | | | | |
| — lipolysis | — | | + | + | + |
| Secretion—stimulated by | Hyperglycaemia, Amino acids, Glucagon, G.I.T. hormones, Adrenaline, Fasting | Hypoglycaemia, Amino acids | Hypoglycaemia, Other | Hypoglycaemia, Stress | Stress |
| inhibited by | | Fasting, Insulin | | | |
| Results | Uses and stores available glucose | Provides glucose | Spares glucose | Provides glucose | Provides glucose |
| | | | Provide FFA as alternative fuel | | |
| Blood FFA levels | Fall | | Rise | | |
| Blood glucose levels | Fall | | Rise | | |

+ stimulates      — inhibits

opposing hormones to become apparent without, necessarily, an increase in their secretion.

Table XVII is supplied for reference on the action of the relevant hormones. Insulin and glucagon will be discussed briefly here and the other hormones in their respective chapters.

### Insulin

Insulin is a protein consisting of 51 amino acids arranged as two peptide chains linked by disulphide bonds. The $\beta$-cells of the pancreatic islets of Langerhans synthesise a larger, single chain molecule (pro-insulin), which is split to form the active hormone.

**Insulin secretion.**—1. Probably the most important factor controlling insulin secretion is the *level of blood glucose*. As this rises insulin secretion increases.

2. Insulin levels are higher after an oral dose of glucose than after an equivalent intravenous one: this is probably due to glucose-induced release of an *intestinal hormone* (or hormones) which stimulate insulin secretion. The exact nature of the hormone(s) is unknown.

3. Some amino acids, notably *leucine and arginine*, stimulate insulin secretion. This is of importance in leucine sensitivity (p. 206).

4. *Sulphonyl urea compounds*, such as tolbutamide, are used therapeutically to stimulate insulin secretion.

*Adrenaline inhibits* insulin secretion.

**Actions of insulin.**—1. *Insulin promotes entry of glucose into cells*, especially those of muscle and adipose tissue. It is not required for glucose uptake by cells of the liver or central nervous system, or by erythrocytes.

This action is opposed by
growth hormone
glucocorticoids

2. Insulin *promotes glycogenesis* and *inhibits gluconeogenesis*. Thus, by storing glucose when available, insulin lowers blood glucose concentrations.

3. Insulin inhibits lipolysis and stimulates lipogenesis. It is therefore *antiketogenic* and *reduces blood FFA levels*.

4. Insulin and growth hormone act together to *stimulate protein synthesis*.

**Measurement of plasma insulin.**—Plasma insulin levels are usually estimated by radioimmunoassay. Results of bioassay, which measures "insulin-like activity", usually give adequate agreement with immunoassay. Insulin estimation is most useful in the investigation of hypoglycaemia.

## Glucagon

Glucagon, a single chain polypeptide consisting of 29 amino acids, is synthesised in the α-cells of the islets of Langerhans.

**Secretion of glucagon.**—Glucagon secretion, by contrast with that of insulin, is *suppressed by hyperglycaemia* and *stimulated by hypoglycaemia*. Proteins and amino acids, however, stimulate both glucagon and insulin.

The *actions of glucagon* are:

1. Stimulation of glycogenolysis in liver only.
2. Stimulation of gluconeogenesis.

Both these actions *raise the blood glucose level.* Paradoxically, glucagon stimulates insulin secretion.

Glucagon may be *measured* by *radioimmunoassay*, although the estimation is not routinely available.

## Interrelationship between Insulin and Glucagon

As insulin and glucagon have opposing effects on blood glucose levels, their molar ratio is more important in this respect than the absolute concentration of either: a high insulin/glucagon ratio lowers glucose concentration, while a low one raises it. After a protein meal, when both hormones are increased, insulin stimulates protein synthesis, while glucagon prevents the hypoglycaemia which would result from unopposed insulin action.

## Hormonal Interactions

Normally meals are separated by periods of fasting. Changes in secretion of the hormones already discussed, especially insulin, glucagon and growth hormone (GH), tend to:

1. maintain glucose concentrations within relatively narrow limits;
2. maintain glycogen stores adequate for emergency needs, while converting excess carbohydrate into fat;
3. maintain body protein, again converting excess amino acids to fat, the nitrogen being excreted in the urine.

**Effect of eating or glucose ingestion.**—*During the first 2 hours* glucose is absorbed from the gastro-intestinal tract and blood levels rise. This causes

1. A 10- to 15-fold *rise in insulin levels* (the effect of glucose on insulin secretion is potentiated by amino acids and possibly, by gastro-intestinal hormones).

2. *A fall of glucagon levels and almost complete disappearance of GH* from the plasma.

As a result of this insulin dominated pattern, blood glucose levels, *which reach a peak at about an hour*, then fall as the rate of utilisation exceeds that of absorption, normally to *below 6·7 mmol/l (120 mg/dl) after 2 hours*. Plasma FFA levels also fall.

Failure of blood glucose concentration to fall below 6·7 mmol/l (120 mg/dl) by 2 hours may occur in:

Insulin deficiency (diabetes mellitus)
Growth hormone excess (acromegaly)
Excess of other hormones antagonising insulin:
    Corticosteroids (Cushing's syndrome and stress)
    Adrenaline (phaeochromocytoma and stress)

*Between 2 and 4 hours insulin levels fall*, although remaining well above basal concentrations. *GH and glucagon levels rise*. These changes minimise "overswing" of blood glucose which would occur due to unopposed insulin action. Relative excess of insulin at this time causes *reactive hypoglycaemia* (p. 205).

**Fasting.**—If fasting continues beyond 4 hours insulin levels fall very low and the effects of the opposing hormones (glucagon, GH and glucocorticoids) become evident. This hormonal pattern tends to maintain blood glucose concentrations adequate for brain metabolism, by

1. stimulating glucose production from glycogen by glycogenolysis (glucagon) and from amino acids, liberated by protein breakdown, by gluconeogenesis (glucocorticoids);

2. sparing glucose by releasing FFA by lipolysis (GH and glucocorticoids). FFA can be metabolised by tissues other than brain: moreover, they form acetyl CoA, which, in excess, condenses to form ketones (p. 189): these, too, can provide an energy source for tissues other than brain and liver. High levels of FFA also spare glucose by directly inhibiting glucose uptake by muscle cells.

Thus during fasting *blood glucose levels remain normal until late, but ketosis occurs early*.

*Fasting hypoglycaemia* occurs if there is either insulin excess, or deficiency of the opposing hormones. Note that insulin is the key hormone in the above sequence of events, and that the effects of the opposing hormones are largely dependent on insulin levels.

**Stress**

Under conditions of stress adrenaline is released from the adrenal medulla, and cortisol secretion from the adrenal cortex is stimulated through the stress pathway (p. 138). These hormones both tend to increase blood glucose levels at a time when it is needed by the brain. Under severe stress (for instance "shock") hyperglycaemia and possibly glycosuria may occur. Similarly, hyperglycaemia may occur when there is pathological excess of these two hormones, as in Cushing's syndrome (cortisol) and in phaeochromocytoma (adrenaline).

## DISTURBANCES OF CARBOHYDRATE METABOLISM

The ancient Greeks knew that patients with diabetes mellitus passed sweet urine and the blood sugar concentration was one of the first biochemical estimations to be applied clinically. Because of this, stress has been laid on abnormalities of blood glucose levels in disturbances of carbohydrate metabolism. Low blood glucose levels are indeed dangerous in themselves, because brain cells cannot form glucose from other substances and cannot easily metabolise other foodstuffs. However, *moderate* hyperglycaemia is almost harmless and it is the metabolic consequences of the inability of cells to utilise glucose which are dangerous.

### MEASUREMENT OF BLOOD GLUCOSE

Interpretation of blood glucose levels requires some knowledge of both the method of estimation and of the type of sample used.

(a) *Methods.*—There are two main groups of methods:

(i) Enzymatic methods, using glucose oxidase or other enzymes. The result is considered to represent "true glucose".

(ii) Methods which depend on the reducing property of glucose which (because of the presence of non-glucose reducing substances) may overestimate glucose by 0·3–1·1 mmol/l (5–20 mg/dl) or more. The percentage error is greatest at low levels of glucose.

Either method can be adapted for use by automated techniques. The most commonly used is a reduction method but, when automated, this method gives values closely approaching those of the more specific enzymatic techniques.

(b) *Type of sample.*—Until recently all glucose estimations were performed on whole blood. Some multichannel automated apparatus, however, uses plasma, as this is the type of sample required for other estimations performed simultaneously. Plasma or serum values of glucose are 10–13 per cent higher than those of whole blood.

"True glucose" values on whole blood are given in this chapter.

Either whole blood or plasma must be assayed immediately, or mixed with an inhibitor of glycolysis. *In vitro* metabolism of glucose by the cellular elements of the blood will produce falsely low levels. Tubes containing fluoride or iodoacetate (both of which inhibit glycolysis) with an anticoagulant are used.

## Interpretation of Blood Glucose Levels

(a) **Hyperglycaemia** occurs

1. in the syndrome of diabetes mellitus (p. 191),
2. in patients receiving glucose-containing fluids intravenously,
3. temporarily in severe stress,
4. after cerebrovascular accidents.

(b) **Hypoglycaemia** (blood glucose less than 2·2 mmol/l or 40 mg/dl) also has a number of causes. These are considered on p. 204.

## Glycosuria

Tests for glycosuria are considered in the Appendix. Unless of renal origin, glycosuria indicates hyperglycaemia. Glycosuria usually occurs only when the blood glucose level exceeds about 10 mmol/l (180 mg/dl), and this figure is referred to as the "renal threshold" for glucose. However, the tubules can actually reabsorb a maximum amount of glucose per unit time and this amount corresponds to blood levels of 10 mmol/l only when the glomerular filtration rate (volume per minute) is normal. With a reduced GFR (due to glomerular damage or circulatory insufficiency), and reduced volume of fluid delivered to the tubules, far higher *concentrations* of glucose (but the same amount/minute) may be completely reabsorbed and glycosuria may not occur even with severe hyperglycaemia. The clinical significance of this is discussed on p. 193.

## KETOSIS

Ketosis occurs when there is a deficiency of *intracellular glucose metabolism* such as in:

(a) fasting or starvation. Unless advanced, when hypoglycaemia may occur, blood glucose levels are normal but insulin levels are very low;

(b) diabetes mellitus in which, despite raised blood glucose levels, relative or absolute insulin deficiency prevents cell uptake and metabolism of glucose.

In these circumstances, the rate of triglyceride breakdown in adipose tissue exceeds that of synthesis and FFA are released and used as an alternative energy source. In the liver excess FFA are metabolised to acetyl CoA which is either oxidised in the TCA cycle or, after condensation, forms *acetoacetate*. Acetoacetate is reduced to *β-hydroxybutyrate* and decarboxylated to *acetone*. These three substances are *ketone bodies*, which spare glucose by providing fuel for extrahepatic tissues.

Ketones are normally present in the blood in very small amounts; during *ketosis* they accumulate in the circulation (ketonaemia) and are excreted in the urine (ketonuria). β-Hydroxybutyrate and acetoacetate production is associated with H+ production and, in excess, this contributes to the acidosis in diabetic coma (*ketoacidosis*). Rapid tests are available for the detection of ketonuria (see Appendix) but urinary excretion is variable and the severity of ketosis is assessed, when necessary, by blood ketone levels (p. 215). Although β-hydroxybutyrate, the predominant ketone present in ketosis, is not measured by these tests, they are adequate for clinical assessment.

The treatment of ketosis is that of the cause with, if necessary, correction of the acidosis with bicarbonate.

## "LACTIC ACIDOSIS"

As outlined on p. 181, the end product of glycolysis is pyruvate which, depending on the availability of oxygen, is either further metabolised in the TCA cycle or converted to lactate. As lactic acid is almost completely dissociated at the pH of the body fluids, high blood lactate levels are ineffective in buffering the excess hydrogen ions accumulating under the same circumstances. The result is a metabolic acidosis, usually known as "lactic acidosis".

"Lactic acidosis" is usually due to poor tissue perfusion, with cellular anoxia. It occurs:

1. In circulatory failure and "shock". This is the commonest cause.

2. Rarely in diabetes mellitus without ketosis, or during the treatment of hyperglycaemic ketoacidosis.

3. In a number of severe illnesses, including acute leukaemia, septicaemia, Type I glycogen storage disease (especially after fasting) and ethanol ingestion.

4. It has been described during phenformin therapy.

5. Without identifiable cause ("idiopathic").

The diagnosis of lactic acidosis may be suspected in a patient with metabolic acidosis and one of the above conditions. The patient is usually seriously ill. Other causes for metabolic acidosis such as ketosis, renal failure or poisoning should be excluded. Although a definite diagnosis rests on raised blood lactate levels, this estimation is neither practicable nor necessary in the majority of cases.

The underlying condition should be treated and the acidosis corrected with intravenous bicarbonate: large amounts may be needed to restore the pH to normal. The prognosis of idiopathic lactic acidosis is poor, probably because the underlying abnormality is unrecognised and untreated. Infusion of lactate is theoretically less effective, because bicarbonate is a better buffer. In practice the improved tissue perfusion which follows any type of infusion may correct the acidosis of hypovolaemic "shock" or severe diarrhoea.

## DIABETES MELLITUS

The syndrome of diabetes mellitus is due to absolute or relative insulin deficiency. There are many causes and severity varies. The classification given below, based on the recommendations of the Medical and Scientific Section of the British Diabetic Association, combines aetiology, clinical features and the response to the oral glucose tolerance test.

### (a) Primary, Idiopathic or Essential Diabetes

This is the commonest form, and has a hereditary basis.

Primary diabetes has been classified into several *stages* of severity based on the clinical features and the results of the glucose tolerance test. The terms used are those of the British Diabetic Association with the corresponding American Diabetes Association terminology in parentheses.

i. **Potential diabetic (prediabetic).**—A subject with a normal glucose tolerance test but a strong family history of the disease (e.g. the identical twin of a diabetic, or the offspring of two diabetic parents), or a woman who has given birth to a child of 4·5 kg or over.

ii. **Latent diabetic (suspected diabetic).**—A subject with a normal glucose tolerance test but who has had a diabetic type of glucose tolerance test during pregnancy, after cortisone administration, when obese or at a time of stress such as during a severe infection.

iii. **Asymptomatic diabetic (latent or chemical diabetic).**—A subject with a diabetic glucose tolerance test but without symptoms of diabetes.

iv. **Clinical diabetic (overt diabetic)**

In the British terminology *prediabetic* is reserved for the period in the life of a diabetic before the diagnosis is made and is thus applied retrospectively.

### (b) Secondary Diabetes

i. **Pancreatic**—absolute insulin deficiency due to pancreatic destruction in *chronic pancreatitis, pancreatic carcinoma* or *haemochromatosis* or following *total pancreatectomy.*

ii. **Presence of insulin antagonists**, as in *acromegaly* (excess growth hormone), *Cushing's syndrome* or *steroid therapy* (excess glucocorticoids) and *pregnancy.*

iii. **Inhibition of insulin secretion** due to excessive catecholamine activity (e.g. *phaeochromocytoma, thyrotoxicosis* or *severe stress* such as myocardial infarction) or associated with potassium depletion, and with diuretic therapy.

### The Clinical and Laboratory Features of Primary Diabetes Mellitus

Although the disturbance in carbohydrate metabolism is not the only, and conceivably not the primary, abnormality in diabetes mellitus, its measurement is important in diagnosis and management. A commonly used clinical classification is based on the severity and behaviour of the disease. In young subjects diabetes tends to be severe, management is difficult and the patients are liable to develop ketosis and coma. Plasma insulin levels are very low or insulin may be undetectable (*juvenile onset diabetes or insulin requiring diabetes*). In older persons (*maturity onset diabetes*), frequently associated with obesity, the metabolic disturbances are less severe and may be overshadowed by vascular or other complications. Plasma insulin levels are lower than in corresponding weight matched non-diabetic subjects and there is a delay in insulin secretion in response to glucose, maximum levels being reached later and persisting longer. This classification is only a guide. Mild, obesity-associated diabetes may occur in children and older subjects may present with, or progress to, a metabolically unstable type of disease.

Most of the metabolic changes of diabetes mellitus can be explained as a consequence of insulin deficiency and tend to parallel it in severity. In the *mildest* form, insulin production in the fasting state may be adequate for normal metabolism and *fasting blood glucose levels are normal*. Following a meal, or administration of glucose, as in the glucose tolerance test, less glucose than normal is metabolished by the liver due to delayed insulin response. This results in higher than normal peripheral blood glucose levels which persist at a high level for longer than normal because of reduced utilisation, producing a mild *diabetic type of glucose tolerance curve*.

In more *severe* insulin deficiency, there is *fasting hyperglycaemia and glycosuria*. The osmotic diuresis leads to water and electrolyte depletion with the "classical" symptoms of diabetes, polyuria and polydipsia. There may be protein breakdown, causing muscle wasting. If the insulin deficiency is very severe, the patient, if untreated, progresses steadily to ketoacidosis, coma and death (see later).

Abnormalities in *lipid metabolism* secondary to insulin deficiency also occur. Lipolysis is stimulated and plasma FFA levels rise. In the liver these FFA are converted to acetyl CoA and ketone bodies, or are re-esterified into endogenous triglycerides and incorporated into pre-$\beta$-lipoproteins. Cholesterol synthesis is also increased with an increase in $\beta$-lipoprotein and if insulin deficiency is very severe such as in uncontrolled diabetes, chylomicrons may accumulate in the blood. The hyperlipoproteinaemia of poorly controlled diabetes may be of any type, most commonly IV or V (p. 230).

Vascular disease, due in part to the lipid abnormalities, is common, and may present as cerebrovascular disease or peripheral vascular insufficiency.

The biochemical basis of many of the other clinical features of diabetes mellitus is not clear. Abnormalities of small blood vessels particularly affect the retina and kidney. In the latter, nodular glomerulosclerosis (Kimmelstiel-Wilson kidney) may produce the nephrotic syndrome. Infections are common and may aggravate renal and peripheral vascular disease. Diabetic and potentially diabetic women tend to give birth to large babies.

### Diagnosis of Diabetes Mellitus

The diagnosis of diabetes mellitus depends on the demonstration of carbohydrate intolerance either as frank hyperglycaemia or as an impaired response to a glucose load.

1. **Glycosuria.**—Testing of the urine is the oldest method of diagnosing diabetes and certainly the simplest. It can, however, lead to false conclusions. The renal threshold for glucose is about 10 mmol/l (180 mg/dl) and glycosuria is an indirect indicator of hyperglycaemia. While a positive result requires further investigation a negative test by no means excludes diabetes. There are two main reasons for *false negative* results:

(*a*) The *timing of the urine specimen* tested is important. Early morning specimens of urine are used for many tests because the urine is usually most concentrated at that time of day. However, as it is collected after a period of fasting the test will be positive for sugar only if the blood glucose level has exceeded 10 mmol/l (180 mg/dl) during this period—that is, in severe diabetes. A specimen of urine passed about one hour after a meal is the most valuable one for a screening test.

(*b*) In the presence of renal disease with a *reduced glomerular filtration rate* (p. 189), tubular reabsorption may remove glucose at concentrations well above the usual "renal threshold" of 10 mmol/l. This is seen particularly in elderly people and in cases of undiagnosed diabetes with renal damage. As the "classical" symptoms of diabetes —polyuria and polydipsia—depend on glycosuria, these patients are "asymptomatic".

*False positive* tests may be found in *renal glycosuria* (p. 211).

2. **Random blood glucose level.**—This test is only of value if glucose levels are very high, when they are almost certainly due to diabetes. A normal result does not exclude the diagnosis. In either case a fasting blood glucose estimation should also be performed.

3. **Fasting blood glucose level.**—The blood glucose level after an overnight fast is commonly used to diagnose diabetes. While a glucose level of greater than 6·7 mmol/l (120 mg/dl) is strongly suggestive, and one over 7·2 mmol/l (130 mg/dl) almost diagnostic, many diabetics have fasting glucose levels in the normal range. A high fasting blood

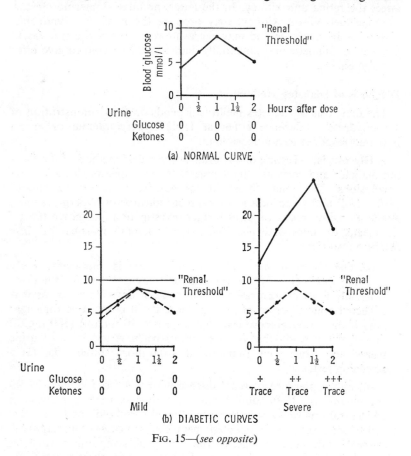

(a) NORMAL CURVE

(b) DIABETIC CURVES

FIG. 15—(*see opposite*)

glucose concentration indicates inadequate insulin output for even basal requirements and no further test is required. A normal fasting level may be present in mild diabetes (Fig. 15) and under these circumstances further tests should be performed.

4. **Two-hour post-glucose blood glucose level.**—This is probably the simplest screening test for diabetes mellitus and is in fact a "modified" glucose tolerance test. By two hours after a meal the blood sugar has

normally fallen below 6·7 mmol/l (120 mg/dl). Although a meal is often recommended as the load, a more standardised stimulus is that given by 50 g of glucose orally, as in the GTT. A normal result almost certainly excludes clinical diabetes.

5. **Glucose Tolerance Test (GTT).**—There are only two indications for performing a GTT in the diagnosis of diabetes:

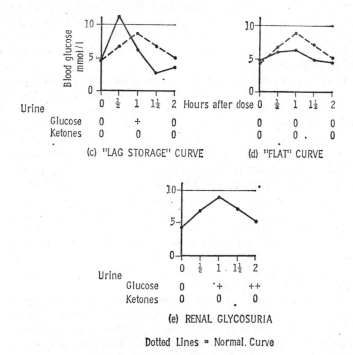

(c) "LAG STORAGE" CURVE    (d) "FLAT" CURVE

(e) RENAL GLYCOSURIA

Dotted Lines = Normal. Curve

Fig. 15.—Glucose tolerance curves.

(a) Detection of mild diabetes when there are normal fasting and random glucose levels. In this respect it is probably no more success-ful than a single estimation two hours after a glucose load.

(b) To ascertain the renal threshold for glucose. Knowledge of this helps the patient to control insulin dosage by testing for urinary sugar.

There is little point in performing a GTT on a subject with markedly raised fasting or random glucose values and similarly, in a patient with fasting hyperglycaemia and glycosuria, no further information on the renal threshold will reward the investigator or the patient for the discomfort of a GTT. Measurement of plasma insulin levels during a routine GTT is of *no* diagnostic value.

It is convenient to consider the glucose tolerance test in more detail here.

**Factors influencing the result of the GTT.**—1. *Dose of glucose.* The use of a dose of 50 g as recommended by the British Diabetic Association is probably adequate. In America a dose of 1·75 g/kg ideal body weight or a standard dose of 100 g is usually given, and is claimed to be a more sensitive test. The larger dose may cause nausea and variable gastric emptying (compare xylose absorption test, p. 286) and does not improve diagnostic accuracy.

2. *Capillary or venous blood.* Although there is very little difference in glucose levels between these two types of sample during fasting, at the high levels found during the GTT, capillary tends to give higher results than venous blood and to return to fasting levels more slowly. These differences should be allowed for in the interpretation of the test.

3. *Age.* There is deterioration of glucose tolerance with increasing age and mild impairment is less significant in older than in younger subjects (compare urea, p. 473). This may, in part, be due to a poor diet.

4. *Previous diet.* No special restrictions are necessary if the patient has been on a normal diet for 3–4 days. If, however, the test is performed after a period of carbohydrate restriction, such as a reducing diet, abnormal glucose tolerance may be shown. The probable reason for this is that metabolism is set in the "fasted" state favouring gluconeogenesis. Mildly abnormal tests should be repeated after appropriate dietary preparation.

5. *Time of day.* Most glucose tolerance tests are performed in the morning and the values quoted are for this time. There is evidence that tests performed in the afternoon show higher blood glucose values and the accepted normals may not be applicable. This phenomenon may be due to a circadian variation in islet cell activity.

**Types of glucose tolerance curve.**—The GTT is performed to assess the response to glucose. This depends not only on insulin and other hormone secretion, but also the rate of absorption of glucose and on the integrity of other compensatory mechanisms such as hepatic glycogenesis.

(a) *The "normal" GTT* (Fig. 15a and Table XVIII).—In Britain the usual criteria employed are those of the Medical and Scientific Section of the British Diabetic Association, for a 50 g glucose load and using whole blood. Table XVIII compares these with the Fajans-Conn criteria commonly used in America which, despite a larger glucose load, are basically similar.

TABLE XVIII

Upper Limits of Normal Glucose Tolerance (mmol/l with mg/dl in brackets)

|  | Fasting | 1 hour | 1½ hours | 2 hours |
|---|---|---|---|---|
| 1. British Diabetic Association: | | | | |
|    Whole blood (capillary) | | 10·0 (180*) | | 6·7 (120) |
|    Whole blood (venous) | | 8·9 (160*) | | 6·1 (110) |
|    Plasma or serum | | 10·3 (185) | | 7·8 (140) |
| 2. Fajans-Conn: | | | | |
|    Whole blood (venous) | 5·6 (100) | 8·9 (160) | 7·8 (140) | 6·7 (120) |
|    Plasma or serum | 6·4 (116) | 10·3 (185) | 9·0 (162) | 7·8 (140) |

*Note:*
1. * This level is the upper limit at any time during the test, but the highest level is found usually at 1 hour.
2. It is *not* necessary that the level at 2 hours should return to that of the fasting patient.

In many cases, if the test is prolonged for 3–4 hours, the level of glucose will be found to fall below the fasting level and return to normal at half to one hour later. This is considered further on p. 206.

Although, usually, only the changes in glucose levels are followed, these are accompanied by alterations in insulin and growth hormone levels as well as in plasma potassium and phosphate concentrations. These latter changes probably reflect the uptake of glucose by cells under the influence of insulin.

(b) *Diabetic glucose tolerance curve* (Fig. 15*b*).—The most significant finding in the diagnosis of diabetes is the *failure of glucose levels to fall below 6·7 mmol/l (120 mg/dl) (capillary) or 6·1 mmol/l (110 mg/dl) (venous) by two hours.* The peak level is frequently above normal and fasting levels may or may not be raised (see p. 194). If the peak concentration exceeds the renal threshold for glucose glycosuria will be present. Note that, because of the delay in urine reaching the bladder, maximum glycosuria is found in the next specimen *after* the highest blood glucose concentration is reached.

(c) *Lag storage curve* (Fig. 15*c*).—The peak blood glucose level may be higher than normal but the 2-hour value is within normal limits, or often low. This implies a delay in early compensatory mechanisms ("lag") without necessarily impairment of insulin response. The hypoglycaemia may be due to outpouring of insulin in response to the initial high glucose levels. Such a curve may be found in:

(i)  Apparently normal individuals (reactive hypoglycaemia).

(ii) After gastrectomy or gastro-jejunostomy when rapid entry of glucose into the intestine leads to rapid absorption and sudden outpouring of insulin with "overswing".

(iii) In very severe liver disease, probably due to a decreased glycogenesis. This is rarely of diagnostic value.

(iv) Rarely in thyrotoxicosis, probably due to rapid absorption of glucose.

If the glucose concentration at the peak of the curve is above the renal threshold level, glycosuria will occur in the next urine specimen.

(d) *Flat glucose tolerance curve* (Fig. 15*d*).—Blood glucose levels fail to rise normally following a glucose load. This type of curve is sometimes considered to indicate malabsorption. It is, however, not uncommonly found in completely healthy individuals and interpretation is difficult. It can also occur due to lack of glucocorticoid in Addison's disease or due to lack of GH in hypopituitarism. It should never be used as the sole test of malabsorption.

(e) *Renal glycosuria* (Fig. 15*e*).—The blood glucose levels follow the normal pattern, but glucose is found in some or all of the urine specimens.

## Other Types of Glucose Tolerance Test

Several variations of the glucose tolerance test have been described. Pretreatment with cortisone has been suggested, to identify latent diabetes, but is of doubtful significance. Intravenous glucose tolerance tests suffer from the major disadvantage that the intestinal phase of glucose homeostasis is bypassed.

### Aetiology of Primary Diabetes Mellitus

It is generally accepted that diabetes mellitus is a genetically determined disease (although the inheritance is complex) and that the metabolic effects are those of insulin deficiency. The nature of the basic abnormality is still a matter of controversy and evidence has been presented for and against several theories. Ideally a theory should explain both metabolic and vascular abnormalities.

1. The basic lesion may be in the $\beta$-cell of the pancreas. Several studies have shown an impaired early insulin response to glucose in the unaffected twins of diabetics (that is, in presumed potential diabetics). This defect may be severe and present as "juvenile onset" diabetes or it may never enter the clinical phase. It is presumed that in most diabetics

it only becomes apparent when some stress is applied to the $\beta$-cell, such as obesity, or the presence of excess cortisol or growth hormone.

2. The abnormality may be primarily extrapancreatic and long-continued antagonism of, or resistance to, insulin action may lead to eventual exhaustion of the $\beta$-cell. Variants of this theory consider the primary fault to be in fat metabolism, excess FFA antagonising insulin. Other circulating insulin antagonists have been suggested. This theory may explain "maturity onset" diabetes in obese subjects. Neither of these theories explains the vascular disease. There is some evidence that in insulin deficiency there is increased glycoprotein synthesis and that this may lead to capillary basement membrane thickening. The theory that vascular changes are primary, interfering with pancreatic secretion of insulin, is controversial.

## Principles of Management of Diabetes Mellitus

The *management* of diabetes mellitus will be only briefly considered.

Juvenile diabetics require *insulin* in amounts that vary according to circumstances. Maturity onset diabetics are frequently controlled by *diet* with *weight reduction*. In this group insulin secretion may be stimulated by the *sulphonylurea* drugs such as tolbutamide. *Biguanides*, such as phenformin, are also used to lower blood glucose; the mechanism of their action is uncertain, but does not depend on the presence of functioning islet cells.

"Resistance" to insulin is found in *pregnancy* (possibly due to high levels of sex hormones) and in *infections* and *trauma* (possibly due to glucocorticoid excess). Should any of these conditions develop in a diabetic patient on insulin therapy, the dosage may have to be increased. A patient is usually termed "insulin resistant" if he needs more than 200 U a day, those needing 60 to 200 U being called "insulin insensitive".

A diabetic may become resistant to administered insulin because of development of antibodies.

## Secondary Diabetes

Many cases of *Cushing's syndrome* or *acromegaly* show impaired glucose tolerance on testing, but in only about 20 per cent is the abnormality severe enough to require treatment. It is probable that these cases are latent diabetics. Generally the disease is mild and ketosis and vascular complications are uncommon. Treatment is directed at the primary disease.

After total *pancreatectomy*, or *extensive pancreatic destruction* by disease, insulin requirements are often less than those of a severe primary diabetic, possibly because glucagon-secreting $\alpha$ cells are lost as well as insulin-secreting $\beta$ cells.

## ACUTE METABOLIC COMPLICATIONS OF DIABETES

The diabetic patient may develop one of several complications requiring emergency therapy. The main ones are:

(a) diabetic ketoacidosis,
(b) hyperosmolal non-ketotic coma,
(c) hypoglycaemia due to overtreatment (p. 204).

### Diabetic Ketoacidosis

Diabetic ketoacidosis is an extension of the metabolic events outlined on p. 192. *Precipitating factors* are usually infections or gastro-intestinal upsets with vomiting. Insulin may mistakenly be withheld by the patient who reasons "no food, therefore no insulin". The consequences are due principally to two factors—*glycosuria* and *ketosis*.

Blood glucose levels are usually in the range of 28–39 mmol/l (500–700 mg/dl) but may be considerably higher. This results in severe *glycosuria* which produces an *osmotic diuresis* leading to loss of water and depletion of body electrolytes. Vomiting is frequently present in ketosis and adds to the fluid and electrolyte depletion. The fluid loss involves both intra- and extracellular fluid with reduction of circulating blood volume, reduced renal blood flow and glomerular filtration rate, and severe cellular dehydration. As tubular reabsorption of glucose depends on volume of glomerular filtrate as well as concentration of glucose therein, reduced GFR allows greater proportional reabsorption of glucose. In some cases this may be complete and glucose is no longer present in the urine despite hyperglycaemia. In such cases the plasma urea level is usually raised and there is evidence of haemoconcentration such as elevated haematocrit and total protein levels. Both these findings are the result of *dehydration*.

As a result of insulin deficiency *lipolysis* is accelerated. More FFA are produced than can be metabolised by peripheral tissues and they are converted in the liver to ketone bodies, or incorporated into endogenous triglycerides. Severe *hyperlipaemia* may be present. Hydrogen ions are produced with ketone bodies (other than acetone) and are buffered by plasma bicarbonate. They are also excreted in the urine and urinary pH drops. The production usually exceeds the ability of the kidney to secrete hydrogen ions and plasma bicarbonate levels fall. The developing metabolic acidosis stimulates the respiratory centre and breathing becomes deeper with a lowering of $Pco_2$. This respiratory compensation may maintain blood pH within the normal range for a while, but at the cost of further lowering of the plasma bicarbonate and very low levels are found. The deep sighing respiration (*Kussmaul respiration*) with the odour of acetone on the breath is a classical feature of diabetic ketoacidosis.

Plasma *potassium levels* are usually *raised* before treatment is started, due probably to the acidosis and to deficient entry of glucose into cells (p. 68) and to the low GFR. It must be realised, however, that urinary potassium loss has been high and that there is a *total body deficiency*. This will be revealed during treatment, as potassium re-enters the cells, with resultant, and possibly severe, hypokalaemia.

The losses in a typical case of diabetic ketoacidosis may amount to 6 or 7 litres of water and 300–500 mmol each of sodium and potassium.

The findings in diabetic ketosis are therefore:

Clinical —overbreathing (acidosis)
dehydration (osmotic diuresis)

In blood—hyperglycaemia
ketonaemia
acidosis with low total $CO_2$ (bicarbonate)
hyperkalaemia (occasionally normokalaemia)
haemoconcentration (p. 41) and mild uraemia

In urine —glycosuria $\Big\}$ while the urinary flow is adequate
ketonuria
low pH (unless there is renal failure)

**Associated findings.**—Changes in plasma phosphate levels parallel those of potassium. Low levels may be found for several days. Amylase concentration may be markedly elevated in both urine and plasma, and should not be interpreted as indicating acute pancreatitis. These may be found coincidently, but are of no importance in diagnosis, prognosis or treatment. They are mentioned to avoid the dangers of misinterpretation.

### Hyperosmolal Non-ketotic Coma

In diabetic ketoacidosis there is always some hyperosmolality due to the hyperglycaemia; the term "hyperosmolal" coma (or "precoma") is, however, usually confined to a condition in which there is marked hyperglycaemia, but no ketonaemia or acidosis. (The reason for these different presentations is not clear.) It is more common in older subjects. Blood glucose levels are very high, often greater than 50 mmol/l (900 mg/dl). The resulting glycosuria produces an osmotic diuresis, with severe dehydration and electrolyte depletion. Hypernatraemia and uraemia are often present due to predominant water loss, and these aggravate extracellular hyperosmolality. Coma is thought to be the result of consequent cerebral cellular dehydration, which may also lead to hyperventilation: the consequent respiratory alkalosis and slight fall in plasma bicarbonate concentration should not be confused with that of metabolic acidosis.

Osmolarity can be calculated with sufficient accuracy for clinical purposes as described on p. 35.

### Investigation of a Diabetic Presenting in Coma

A known diabetic may present in coma, or in a confused state ("precoma"), due to one of the metabolic complications of diabetes (ketoacidosis, hyperosmolal coma or hypoglycaemia), to a cerebrovascular accident, or to an unrelated cause.

The initial assessment is made (and therapy started) on *the clinical findings*. Blood must be taken at once and sent for estimation of *glucose*, *potassium*, bicarbonate or arterial pH, sodium and urea. Urine may be tested for glucose and ketones.

The usual findings are shown in Table XIX.

TABLE XIX

| Diagnosis | Clinical features | Blood glucose | Plasma [$HCO_3^-$] | Urine glucose | Urine ketones |
|---|---|---|---|---|---|
| Ketoacidosis | Dehydration Acidotic breathing | High | Low | +++ | +++ |
| Hyperosmolal coma | Dehydration May be hyper-ventilation | Very high | N or slightly reduced | +++ | Neg. |
| Hypoglycaemia | Non-specific | Low | N | Neg. | Neg. |
| Cerebrovascular accident | Neurological May be hyper-ventilation | May be raised | May be low | May be + | Usually Neg. |
| Other | Variable | Usually N | Variable | Usually Neg. | Usually Neg. |

The following points should be noted:

(*a*) *It is unwise to rely solely on urine testing to diagnose hyperglycaemia.* The urine may have been in the bladder for some time and reflect earlier and very different blood glucose levels. Moreover, glycosuria (and ketonuria) may be absent despite high blood glucose levels if the GFR is much reduced by dehydration or renal disease.

(*b*) Rapid assessment of blood glucose levels by Dextrostix (Ames) is often recommended to confirm the clinical diagnosis before blood glucose levels are available. *It is essential to remember that improperly stored reagent strips may give completely and dangerously fallacious results.* Again, more weight should be given to the clinical findings.

(c) Repeated arterial puncture is undesirable. If arterial pH is measured initially, plasma bicarbonate should also be estimated to provide a base-line for assessing progress.

(d) Mild uraemia and hypernatraemia (often severe) are usually present as a result of dehydration and contribute to the raised osmolality. These results influence therapy after the initial stage.

(e) Blood ketone levels may be assessed semi-quantitatively (see Appendix) but are rarely required for diagnosis. However, in a severely acidotic patient without ketonuria they may give added information. Minimal ketonaemia in such a patient would indicate another cause for the acidosis, such as that associated with lactate production. As the initial therapy is similar, this investigation should not delay treatment.

(f) Cerebrovascular accidents may be associated with hyperglycaemia and glycosuria ("piqûre diabetes"). Stimulation of the respiratory centre may cause overbreathing and consequent respiratory alkalosis, with a low plasma bicarbonate level. This should not be confused with metabolic acidosis. Diagnosis depends on clinical findings.

In summary, institution of therapy in diabetic coma should be based on clinical diagnosis. Blood samples should be sent to the laboratory immediately, but treatment should never be delayed until answers are available. Side-room tests must be interpreted with caution.

### Principles of Treatment of Diabetic Coma

Only the outline is discussed here. For details consult the references given at the end of the chapter. The treatment of hypoglycaemia is considered on p. 209.

**Ketoacidosis.**—1. *Repletion of fluid and electrolytes* should be vigorous. Isotonic saline is used initially and continued if the sodium concentration is normal. If there is hypernatraemia it should be replaced by hypotonic saline. If acidosis is severe bicarbonate may be given, as the action of insulin is impaired by severe acidosis. It is unnecessary to correct the bicarbonate deficit entirely as rapid correction follows adequate fluid and insulin therapy. *Potassium* should be administered as soon as plasma levels start to fall.

2. *Insulin is given immediately*, half intravenously and half subcutaneously. The intravenous dose will act until restored circulation allows absorption from the subcutaneous site. Further dosage will depend on blood glucose levels. Recently continuous, or frequently repeated, infusion of very small doses of insulin has been used with great success.

3. The factor which precipitated the coma (such as infection) should be sought and treated.

4. Frequent monitoring of blood glucose and plasma potassium levels is necessary to assess progress and to detect developing hypoglycaemia or hypokalaemia.

**Hyperosmolal coma.**—Treatment of hyperosmolal coma is similar to that of ketoacidosis, but should be more cautious as sudden reduction of extracellular osmolality may do more harm than good (p. 34). *Relatively small doses of insulin* should be given to reduce blood glucose levels slowly. These patients are often very sensitive to the action of insulin. Fluid should be replaced with hypotonic solutions but, if severe hypernatraemia is present, this, too, should be done cautiously.

## HYPOGLYCAEMIA

Hypoglycaemia is defined as a true blood glucose level of below 2·2 mmol/l (40 mg/dl). The corresponding plasma glucose level is 2·5 mmol/l (45 mg/dl). Symptoms may develop at levels greater than this when there has been a rapid fall from a previously elevated, or even normal, value; by contrast, some people may show no symptoms at levels below 2·2 or even 1·7 mmol/l (30 mg/dl), especially if these have fallen gradually. As discussed earlier, cerebral metabolism is dependent on an adequate supply of glucose from the blood and the symptoms of hypoglycaemia resemble those of cerebral anoxia. Faintness, dizziness or lethargy may progress rapidly into coma and, if untreated, death or permanent cerebral damage may result. If the fall of blood glucose is rapid there may be a phase of sweating, tachycardia and agitation due to adrenaline secretion, but this phase may be absent with a gradual fall. Existing cerebral or cerebrovascular disease may aggravate the condition. Rapid restoration of blood glucose concentration is essential.

### Causes of Hypoglycaemia

There is no completely satisfactory classification of the causes of hypoglycaemia as overlap between different groups occurs. An approach that we have found useful is to consider three main groups:

(a) Insulin or other drug-induced hypoglycaemia;

(b) Fasting hypoglycaemia;

(c) Reactive hypoglycaemia.

**Insulin or other drug-induced hypoglycaemia.**—This is probably the commonest cause. Hypoglycaemia in a diabetic may follow accidental overdosage, be due to changing *insulin* requirements, or to failure to take food after insulin has been given. Self-administration for suicidal purposes is not unknown and homicidal use is a remote possibility. *Sulphonylureas* may also induce hypoglycaemia, especially in the elderly and salicylate poisoning in children may be complicated by hypoglycaemia. Other drugs such as antihistamines have been suspected in some cases and a drug history is an important part of investigation.

**Fasting hypoglycaemia.**—As discussed on p. 183, the maintenance of a normal blood glucose concentration in the fasting subject depends on several factors.

(*a*) Levels are lowered by insulin which stimulates glucose metabolism.

(*b*) Levels are maintained or raised by:

    (i) Glycogenolysis, which depends on *adequate stores of hepatic glycogen*. *Adrenaline* and *glucagon* promote this.

    (ii) Gluconeogenesis, which occurs mainly by conversion of amino acids to glucose in the liver. This process is again dependent on *hepatic integrity* and is promoted by *glucocorticoids*.

    (iii) Uptake of glucose by muscle is hindered by *glucocorticoids* and *growth hormone*.

Alterations in this delicate balance can cause hypoglycaemia.

1. *Hyperinsulinism*—(*a*) islet cell tumour (insulinoma) (p. 207),

                      (*b*) islet cell hyperplasia.

2. *Hepatic disease.*—Despite the central role of the liver in intermediary metabolism, hypoglycaemia occurs only when liver damage is very extensive. This rarely presents diagnostic problems.

3. *Glycogen storage disease.*—This condition is seen primarily in childhood and is considered on p. 210.

4. *Endocrine causes.*—Hypoglycaemia may occur in adrenal or pituitary insufficiency. Rarely it is the sole manifestation and these conditions are considered in Chapters V and VI.

5. *Non-pancreatic tumours.*—Although many types of malignant tumours, carcinoma (especially hepatocellular) as well as sarcoma, have been incriminated in the production of hypoglycaemia, it occurs most commonly in association with mesodermal tumours resembling fibrosarcomata, that occur mainly retroperitoneally. They are slow-growing and may become very large. Hypoglycaemia may be the presenting feature. The mechanism is uncertain and may be secretion of an insulin-like substance or excessive utilisation of glucose (p. 448).

6. *Idiopathic hypoglycaemia of infancy* is considered on p. 209.

**Reactive hypoglycaemia.**—Hypoglycaemia may occur as a reaction to an ingested substance. In such cases there is frequently a recognisable pattern of attacks related to meals or to particular foodstuffs.

1. *Functional hypoglycaemia* (sensitivity to glucose).—Some people develop symptoms of hypoglycaemia 2–4 hours after a meal or after a

glucose load. Unconsciousness usually does not occur. The diagnosis is made by performing a glucose tolerance test and collecting blood at intervals for 5–6 hours. The temporary fall below fasting levels found at $2\frac{1}{2}$–3 hours in many normal people is exaggerated. Interpretation is not easy as patients suspected of having functional hypoglycaemia may not develop symptoms coinciding with this low level and levels as low as 1·7 mmol/l (30 mg/dl) may be found, without symptoms, in glucose tolerance tests on normal patients, particularly young women. A similar "reactive" hypoglycaemia may be seen after *gastrectomy*. These patients show a "lag storage" glucose tolerance curve and excessive insulin secretion (p. 197). A similar exaggerated insulin response is seen in some, but not all, cases of idiopathic functional hypoglycaemia.

Hypoglycaemia 3–5 hours after a carbohydrate meal may indicate early diabetes. It is possibly due to the delayed but exaggerated insulin response seen in the early stages of the disease.

2. *Leucine sensitivity.*—In the first six months of life *casein* may precipitate severe hypoglycaemia. This is due to its content of the amino acid, leucine. Leucine sensitivity is probably due to excessive stimulation of insulin secretion and there is often a familial incidence. The condition appears to be self-limiting and does not usually persist above the age of 6 years. Diagnosis is confirmed by the demonstration of hypoglycaemia within half-an-hour after an oral dose of leucine or casein. Treatment is a low leucine diet.

Normal individuals do not show a significant drop in blood glucose after leucine, but a number of patients with insulinoma do demonstrate leucine sensitivity.

3. *Galactosaemia.*—Reactive hypoglycaemia seen only after milk (lactose) is included in the infant diet, may be due to galactosaemia (p. 371).

4. *Hereditary fructose intolerance.*—This is a *rare* cause of hypoglycaemia. Symptoms start after the introduction of sucrose or fruit juice to the diet. Fructose administration leads to symptoms of hypoglycaemia in half to one hour, accompanied by nausea or vomiting and abdominal pain. There is failure to thrive and progressive liver damage with hepatomegaly, jaundice and ascites. Cirrhosis may develop. Diagnosis is based on the demonstration of fructosuria and hypoglycaemia after oral or intravenous administration of fructose. The basic defect is a deficiency of the enzyme *fructose-1-phosphate aldolase* and accumulation of *fructose-1-phosphate*. The hypoglycaemia is probably due to inhibition of glycogenolysis and gluconeogenesis by fructose-1-phosphate.

5. *Alcohol-induced hypoglycaemia.*—Hypoglycaemia may develop between 2 and 10 hours after ingestion of alcohol, usually in considerable amounts. The incidence of this complication is unknown. It is described

most frequently in chronic alcoholics when starvation or malnutrition is present, but may occur in young persons after first exposure. The probable mechanism is reduced hepatic output of glucose due to suppression of gluconeogenesis during metabolism of the alcohol. This is potentiated in the fasting or starved state when gluconeogenesis is the main source of blood glucose. Clinical differentiation from alcoholic stupor may be impossible and requires blood glucose estimation. Frequent infusions of glucose may be needed in treatment. It is difficult to reproduce the hypoglycaemia after infusion of alcohol unless the patient fasts for 12 to 72 hours.

### Insulinoma

An insulinoma is a tumour of the islet cells of the pancreas. As with other endocrine tumours, hormone secretion is inappropriate and usually excessive. It may occur at any age and is usually single and benign but may rarely be malignant. Multiple tumours may be present and it may be part of the pluriglandular syndrome (p. 281). The tumour involves β-cells although other islet cells are usually present in it.

Symptoms may be vague at first. Hypoglycaemic attacks occur typically at night and before breakfast (fasting hypoglycaemia). Personality or behavioural changes may be the first feature and many of these patients present to psychiatrists.

The *diagnosis* of insulinoma depends on the demonstration of *hypoglycaemia with inappropriately high insulin levels*, usually after a 12 to 14-hour fast. The test may need to be repeated on several occasions. In some patients a fast of up to 72 hours may be necessary.

There are several provocative tests that may be used to assist diagnosis. These are described fully in the Appendix and will only be considered briefly here.

1. **Intravenous glucagon test.**—Following administration of intravenous glucagon there is a rise of plasma insulin levels due to direct stimulation of the islet cells. Blood glucose levels also rise due to stimulation of hepatic glycogenolysis, further stimulating insulin secretion. As they return to normal, insulin secretion ceases. In cases of insulinoma late hypoglycaemia may develop as insulin continues to be secreted. This test is not completely specific, but is positive in over 70 per cent of cases of insulinoma. It is safer than the intravenous tolbutamide test (see below).

2. **L-leucine test.**—Oral administration of l-leucine produces hypoglycaemia and excessive insulin secretion in about 50 per cent of cases of insulinoma. It is the most specific provocative test available for this purpose.

3. **Intravenous tolbutamide test.**—Intravenous tolbutamide produces profound hypoglycaemia and an excessive insulin response in about

80 per cent of cases of insulinoma, but is less specific than the l-leucine test. It is dangerous and it has been suggested that it be restricted to patients in whom an insulinoma is unlikely. *It should never be performed if hypoglycaemia is present.*

The glucose tolerance test has no place in the diagnosis of insulinoma. Any type of response may be found and insulin levels may vary widely.

The *treatment* of insulinoma is by surgical removal if possible.

### Hypoglycaemia in Children

Hypoglycaemia in infancy is not uncommon and is important because permanent brain damage can result, especially in the first few months of life. Only the main causes will be outlined.

(*a*) **Neonatal period.**—Blood glucose levels in the neonatal period as low as 2·2 mmol/l (about 40 mg/dl) or, in the first 72 hours, 1·7 mmol/l (about 30 mg/dl) may be considered "normal". In premature (less than 2·5 kg) infants levels below 1·1 mmol/l (about 20 mg/dl) may occur without clinical evidence of hypoglycaemia. Signs of hypoglycaemia at this age are convulsions, tremors and attacks of apnoea with cyanosis. Neonatal hypoglycaemia occurs particularly:

1. *In babies of diabetic mothers.* If the foetus is exposed to hyperglycaemia during pregnancy, islet cell hyperplasia occurs and the consequent hyperinsulinism may produce hypoglycaemia when the supply of excess glucose from the mother is removed after parturition. Severe hypoglycaemia has been noted in babies whose mothers have been treated with *sulphonylurea drugs* during pregnancy. In some series the incidence of hypoglycaemia in babies of diabetic mothers is 50 per cent or higher. Most of these infants show no clinical signs and the significance of these low levels is controversial. Similar islet cell hyperplasia and neonatal hypoglycaemia may be found with severe erythroblastosis foetalis.

2. *In babies suffering from intra-uterine malnutrition* (for instance, infants of toxaemic mothers, or the smaller of twins). These babies are usually "small for dates" and may show a tendency to hypoglycaemia during the first week, possibly due to poor liver glycogen stores. Prematurity is an aggravating factor as most of the liver glycogen is laid down after 36 weeks. They usually show clinical signs of hypoglycaemia.

(*b*) **Early infancy.**—Soon after birth, or after the introduction of milk or sucrose, hypoglycaemia may be due to:

1. *Glycogen storage disease* (p. 210).
2. *Galactosaemia* (p. 371).
3. *Hereditary fructose intolerance* (p. 206).

(c) **Infancy.**—1. *Idiopathic hypoglycaemia of infancy* consists of a group of cases, probably of multiple aetiology. Symptoms usually develop after fasting or after a febrile illness. The diagnosis is made by excluding other causes. There is a high incidence of brain damage.

2. *Leucine sensitivity* is seen mainly in the first six months of life and is considered on p. 206.

3. *Ketotic hypoglycaemia* is seen in the second year of life and develops after fasting or a febrile illness. These infants are usually "small-for-dates" babies. Ketonuria precedes the hypoglycaemia (as it does in starvation) and the diagnosis can be established by feeding a ketogenic diet (high fat, low calorie) for 48 hours, during which clinical hypoglycaemia occurs. The underlying mechanism is only poorly understood.

(d) **"Adult"** causes, including insulinoma, must be considered at all times.

### Approach to the Diagnosis of Hypoglycaemia

Symptoms can only be attributed to hypoglycaemia if hypoglycaemia is demonstrated. This statement may seem superfluous but it is not unknown for patients to be submitted to multiple tests for the differential diagnosis of hypoglycaemia that does not exist, or worse, to undergo treatment for it.

One of the best diagnostic opportunities is often missed. It should be an absolute rule that *before glucose is administered* to a patient in a suspected hypoglycaemic attack *blood is taken for glucose and insulin.* This step may considerably shorten later investigations. *Hypoglycaemia with high levels of insulin* is seen only with insulinoma or islet cell hyperplasia, administration of insulin or, occasionally, of sulphonylureas, leucine sensitivity and post-gastrectomy hypoglycaemia.

As always, any investigations should be preceded by careful clinical appraisal. Points of special note in the history are times of symptoms in relation to meals or any particular food, drug history and whether or not true unconsciousness has ever occurred. This should help to classify the condition as spontaneous (fasting) or reactive hypoglycaemia, and to suggest any underlying endocrine deficiency.

Unless the cause is clinically evident and confirmed by appropriate tests, insulinoma must be excluded.

Further diagnostic procedures are outlined in the relevant sections and in the Appendix.

*Treatment of hypoglycaemia* is by the urgent intravenous administration of 50 per cent glucose solution *after withdrawal of a blood sample for diagnosis* (see above). Some cases may need to be maintained on a glucose infusion until the cause has been established and treated.

## GLYCOGEN STORAGE DISEASE

Glycogen is synthesised from G-6-P (p. 181) by several specific enzymes and broken down to G-6-P by a different pathway also involving several enzymes. Deficiencies of any of these enzymes may result in the formation of abnormal glycogen, or the accumulation of normal glycogen. At least eight types of glycogen storage disease have been recognised. Only the commonest form, involving the liver, will be discussed here.

### Von Gierke's Disease (Type 1)

This is the classical form of glycogen storage disease and is due to deficiency of the enzyme *glucose-6-phosphatase*. As discussed on p. 180, this enzyme is essential for conversion of glucose-6-phosphate to glucose and patients with a deficiency of the enzyme are liable to profound hypoglycaemia. Small amounts of free glucose may be liberated during breakdown of glycogen to glucose-6-phosphate and contribute to blood glucose. Glycogen accumulates in the liver and kidneys and hepatomegaly is present. The metabolic consequences include:

1. Hypoglycaemia.
2. Raised levels of FFA leading to ketosis and endogenous hyperlipoproteinaemia (secondary to deficient intracellular glucose).
3. Lactic acidosis.
4. Hyperuricaemia.

The diagnosis is confirmed by demonstrating the absence of the enzyme on liver biopsy. If this test is not available, two indirect clinical tests are of value:

(i) Administration of glucagon produces a rise in blood sugar in normal subjects by accelerating glycogenolysis. This will not occur in glucose-6-phosphatase deficiency.

(ii) Infusion of galactose or fructose in normal subjects is followed by a rise in blood glucose after conversion in the liver. When the final enzyme in the pathway, that which converts G-6-P to glucose, is absent, this does not occur.

In both tests there is an excessive rise of blood lactate. Treatment is to take frequent meals to maintain blood glucose levels and prevent cerebral damage.

## REDUCING SUBSTANCES IN THE URINE

Although Benedict's test for urinary reducing substances is still used in many laboratories, it has been replaced in most ward side-rooms by

Clinitest tablets (Ames) or Clinistix (Ames). Clinistix contains the enzyme glucose oxidase, and is specific for glucose; it is used for detection of glycosuria. Clinitest, like Benedict's test, on which it is based, gives a positive result with any reducing substance. It is semiquantitative and is valuable in control of diabetic therapy, whereas Clinistix is unsatisfactory for this purpose. Urine screening in the newborn and infants should be done by Benedict's test or Clinitest because of the diagnostic importance of non-glucose reducing substances, such as galactose, in this age group.

The more important causes of a positive Benedict's test or Clinitest are:

1. Glucose ⎱
2. Glucuronate ⎰ common
3. Lactose —common in pregnancy
4. Fructose ⎫
5. Galactose ⎪
6. Pentoses ⎬ rare
7. Homogentisic acid ⎭

Slight reduction may occur if urate or creatinine is present in high concentration.

Clinistix will confirm the presence or absence of glucose. Non-glucose reducing substances are further identified by chromatography and specific tests.

The significance varies with the substance.

## Glucose

Glycosuria may occur in:

(a) Diabetes mellitus.
(b) Renal glycosuria.

Glucose may be present in the urine when blood levels are normal. This is due to a low renal threshold and may be found in *pregnancy* or as an *inherited* (autosomal dominant) characteristic. The condition is usually benign. It is also found in generalised proximal tubular defects (Fanconi syndrome). Differentiation from diabetes mellitus is made by simultaneous blood and urine glucose estimation or, if necessary, by a glucose tolerance test.

## Glucuronates

A large number of drugs, such as salicylates, and their metabolites are excreted in the urine after conjugation with glucuronic acid in the liver (p. 289). These glucuronates are a relatively common cause of reducing substances in the urine.

**Galactose**

Galactose is found in the urine in galactosaemia (p. 371).

**Fructose**

Fructose may appear in the urine after very large oral doses of the sugar, or excessive fruit ingestion, but in general fructosuria occurs in two rare inborn errors of metabolism, both transmitted by autosomal recessive genes.

(a) *Essential fructosuria* is seen almost exclusively in Jews and is a harmless condition.

(b) *Hereditary fructose intolerance* (p. 206) is a serious disease characterised by hypoglycaemia that may lead to death in infancy.

**Lactose**

Lactosuria may occur in:

(a) Late pregnancy and during lactation.

(b) Congenital lactase deficiency (p. 279).

**Pentoses**

Pentosuria is very rare. It may occur in:

(a) *Alimentary pentosuria* after excessive ingestion of fruits such as cherries and grapes. The pentoses excreted are arabinose and xylose.

(b) *Essential pentosuria* a rare recessive disorder characterised by the excretion of xylulose due to a block in the metabolism of glucuronic acid. It is seen usually in Jews and is harmless.

**Homogentisic Acid**

Homogentisic acid appears in the urine in the rare inborn error alkaptonuria (p. 369). It is usually recognisable by the blackish precipitate formed.

Reducing substances found in the urine during *late pregnancy* may be:

(a) Glucose—due to previously unrecognised diabetes mellitus or to lowered renal threshold.

(b) Lactose.

## SUMMARY

1. Carbohydrates after digestion and absorption are mostly metabolised to glucose.

2. Glucose, after conversion to glucose-6-phosphate, may be stored as glycogen (glycogenesis), or utilised for energy production via the glycolytic pathway (glycolysis) and the tricarboxylic acid (TCA) cycle. An important alternate pathway of glucose metabolism is the hexose monophosphate shunt. The liver plays a central role in glucose metabolism.

3. Insulin is essential for the normal metabolism of glucose. Glucagon, growth hormone, glucocorticoids and adrenaline also play important roles.

4. In the fasting state blood glucose levels are maintained by breakdown of glycogen (glycogenolysis) and formation of glucose, chiefly from amino acids (gluconeogenesis). Alternative energy sources in the fasting state are free fatty acids and ketone bodies.

5. Ketosis develops as a result of deficient intracellular glucose metabolism. If excessive, an acidosis may develop.

6. Lactic acidosis may develop in states of tissue anoxia.

7. Diabetes mellitus is a syndrome developing as a result of relative or absolute deficiency of insulin. It is diagnosed on the basis of hyperglycaemia which may be present in the fasting state (severe diabetes) or may only develop after a glucose load in milder cases. The acute metabolic complications of diabetes mellitus include ketoacidosis and hyperosmolal coma.

8. The commonest cause of hypoglycaemia is excessive insulin administration. Other causes vary with the age of onset and may be classified into fasting or reactive hypoglycaemia. The estimation of plasma insulin is of great value in the diagnosis of insulinoma and a number of provocative tests are described for this purpose.

9. Reducing substances in the urine are not necessarily glucose. The significance of the finding depends on the particular substance.

## FURTHER READING

*Medical Clinics of North America* (1971). **55**, No. 4, 791–1080.
*British Journal of Hospital Medicine* (1972). **7**, No. 2, 151–234.
Both of these issues are devoted to review articles on all aspects of diabetes mellitus.

CAHILL, G. F., Jr. (1971). Physiology of insulin in man. *Diabetes*, **20**, 785.

MARKS, V. (1974). The investigation of hypoglycaemia. *Brit. J. hosp. Med.*, **11**, 731.

### References

ELLENBERG, M., and RIFKIN, H. (Eds.) (1970). *Diabetes Mellitus: Theory and Practice*. New York: McGraw Hill/Blakiston.

This book deals with all aspects of diabetes mellitus with a particularly good section on diagnosis.

STANBURY, J. B., WYNGAARDEN, J. B., and FREDRICKSON, D. S. (Eds.) (1972). *The Metabolic Basis of Inherited Disease*. New York: McGraw Hill/Blakiston.

This contains details of the inherited disorders of carbohydrate metabolism.

MARKS, V., and ROSE, F. C. (1965). *Hypoglycaemia*. Oxford: Blackwell Scientific Publications.

# APPENDIX

## Glycosuria

Glycosuria is best detected by enzyme reagent strips such as Clinistix (Ames). Directions for use are supplied with the strips. Clinistix detects about 5·6 mmol/l (100 mg/dl) of glucose in urine, but is not suitable for the relatively accurate quantitation required for control of diabetic therapy. The concentration of glucose in normal urine varies from 0·1 to 0·8 mmol/l (1–15 mg/dl).

*False negative* results may occur if the urine contains large amounts of ascorbic acid, as may occur from therapeutic doses or after injection of tetracyclines which contain ascorbic acid as preservative.

*False positive* results may occur if the urine container is contaminated with detergent.

## Reducing Substances

Reducing substances in the urine may be detected by:

(*a*) Benedict's qualitative solution. 0·5 ml (about 8 drops) of urine is boiled with 5 ml of reagent for a few minutes. A positive result is the formation of a yellow-red precipitate that discolours the urine from a greenish solution (+) to orange–red with heavy precipitate (++++).

(*b*) Clinitest tablets (Ames)—used as directed. These tablets are more convenient than Benedict's solution for patients controlling their own insulin dosage, but are caustic and require careful handling.

These two tests are less sensitive than Clinistix but are more accurate for quantitation. Benedict's test is slightly more sensitive than Clinitest and therefore has a greater incidence of "false" positive results, e.g. due to drug metabolites.

$$+ \text{ represents } \pm \text{ 10–30 mmol/l (0·2–0·5 g/dl)}$$
$$++ \text{ represents } \pm \text{ 30–60 mmol/l (0·5–1 g/dl)}$$
$$+++ \text{ represents } \pm \text{ 60–120 mmol/l (1–2 g/dl)}$$
$$++++ \text{ represents over 120 mmol/l (2 g/dl)}$$

expressed as glucose

The significance of positive results of these tests is discussed on p. 210.

## Detection of Ketones

(*a*) **Ketonuria.**—Most simple urine tests for ketones detect acetoacetate, because of the greater sensitivity of reagents to this than to acetone, and because acetoacetic acid is present in greater concentration. $\beta$-Hydroxybutyrate is not measured.

**Rothera's test.**—5 ml of urine is saturated with a mixture of ammonium sulphate and sodium nitroprusside (as a powder). 1–2 ml concentrated ammonia is added and mixed. A purple–red colour indicates the presence of acetoacetate (or acetone).

Ketostix and Acetest (Ames) are strips and tablets respectively, impregnated with the reagents of Rothera's test. Instructions are supplied with the reagents. They are more convenient to use.

Rothera's test detects about 0·5 mmol/1 (5mg/dl) acetoacetate while the tablet tests are slightly less sensitive (10–20 mg/dl). After a fast of several hours ketonuria may be detected by these methods. Occasional colour reactions resembling but not identical to that of acetoacetate, may be given by phthalein compounds such as phenolphthalein, bromsulphthalein (BSP) or phenolsulphthalein (PSP).

**Gerhardt's test.**—3 per cent ferric chloride is added drop by drop to 5 ml of urine. A red–purple colour develops if acetoacetate is present in quantities 2·5–5·0 mmol/1 (25–50 mg/dl) or greater.

This test is less sensitive than those described above and, if the latter are negative, need not be done. If however, ketone bodies are found, Gerhardt's test gives additional information about the severity of the ketosis.

A large number of compounds that may be found in urine give a colour with ferric chloride. The commonest are salicylates which give an almost identical colour. Boiling the urine for several minutes will destroy ketone bodies and subsequent testing will give a negative result. Salicylates are unchanged by this treatment.

(*b*) **Ketonaemia.**—The severity of ketonaemia may be rapidly assessed using the reagents for Rothera's test or, more conveniently, using Ketostix (Ames).

1. Serum or plasma is obtained by centrifuging a specimen of blood.

2. Ketostix is dipped into the serum, immediately removed, and excess serum drained against the side of the tube.

3. Exactly 15 seconds later compare any purple colour developing with the colour chart on the reagent bottle. The colour developed is semiquantitative.

$$+ \quad 1\text{–}2 \text{ mmol/1 } (10\text{–}20 \text{ mg/dl}) \text{ acetoacetate}$$
$$+ + \quad 3\text{–}5 \text{ mmol/1 } (30\text{–}50 \text{ mg/dl}) \qquad \text{,,}$$
$$+ + + \quad 8\text{–}12 \text{ mmol/1 } (80\text{–}120 \text{ mg/dl}) \qquad \text{,,}$$

Note that $\beta$-hydroxybutyrate does *not* react.

$+ + +$ results may be further assessed by diluting the serum in water, e.g. 1 drop serum and 3 drops water gives a final dilution of $\frac{1}{4}$. The test result can then be recorded as $+ + + (\frac{1}{4})$.

## Oral Glucose Tolerance Test

The indications for and interpretation of the glucose tolerance test are discussed on p. 195; only the procedure is given here.

1. The patient fasts overnight (at least 10 hours). Water, but no other beverage, is permitted.

2. A fasting blood sample for glucose is withdrawn and a specimen of urine obtained.

3. 50 g glucose (children 1·75 g/kg body weight to a maximum of 50 g) is dissolved in water. The patient drinks this within a period of five minutes.

4. Further samples for glucose are taken 30, 60, 90, 120 minutes after glucose.

5. Urine samples are obtained at 60 and 120 minutes.

During the test the patient should be at rest and may not smoke.

## Provocative Tests for Insulinoma

Several points apply to all the tests:

1. As many blood samples are to be taken, an indwelling needle, kept open by a saline infusion, is required (see p. 130).

2. Samples are taken for both *glucose* and *insulin*. Glucose is taken into a fluoride tube and insulin into a plain tube or one with an anticoagulant. (N.B. Heparin interferes with some insulin estimations—check with your laboratory.) The insulin tube is kept at 4°C until it is delivered to the laboratory.

3. These tests may be dangerous. 50 per cent glucose for intravenous administration *must be immediately available* and given if dangerous hypoglycaemia develops. Should it be necessary to terminate the test, a blood specimen should be taken, if possible, *before glucose administration*.

4. The tests are done with the patient lying down. At the conclusion of the test the patient should be given food.

5. These tests should be preceded by a normal diet for at least three days.

(*a*) **Intravenous glucagon test.**—1. The patient fasts overnight.

2. A *basal blood sample* for glucose and insulin is drawn.

3. 1 mg of glucagon (children 30 $\mu$g/kg body weight to a maximum of 1 mg) is injected intravenously.

4. Further samples for glucose and insulin are taken 5, 10, 15, 20, 30, 60, 120 and 180 minutes after glucagon.

*Response:*

1. *Blood glucose concentration.* The blood glucose rises by about 40 per cent and is followed in some patients with insulinoma, but not in normal subjects, by a fall to below the basal value.

2. *Plasma insulin concentration.* Plasma insulin rises in 5–20 minutes up to 100 mU/l. The response in patients with insulinoma is much greater.

A similar excessive rise may be seen in states of insulin resistance, e.g. obesity. The test is positive in about 70 per cent of cases of insulinoma.

(*b*) **L-leucine test.**—1. The patient fasts overnight.

2. A *basal blood sample* for glucose and insulin is drawn.

3. L-leucine (200 mg/kg body weight) is taken orally mixed with cold water.

4. Further blood samples for glucose and insulin are taken 30, 60, 90 and 120 minutes after l-leucine.

*Response:*

1. *Blood glucose concentration.* Within 30 minutes the blood glucose falls by up to 1·4 mmol/l (about 25 mg/dl) in normal subjects, or may fall to hypoglycaemic levels in cases of insulinoma.

2. *Plasma insulin concentration.* The normal insulin response is an increase of up to 30 mU/l within 30 minutes. An excessive response is seen in about 50 per cent of patients with insulinoma. This response is very specific and is seen only in patients with insulinoma, in subjects who are taking sulphonylureas and in leucine-sensitivity hypoglycaemia in children (p. 206).

(c) **Intravenous tolbutamide test.**— *Warning:* This is the most dangerous of the provocative tests, and should only be used in cases with a suggestive history, but with negative or inconclusive results of the above tests.

1. The patient fasts overnight.

2. A *basal blood sample* for glucose and insulin is drawn. The blood glucose should be estimated *before* proceeding with the test and if it is in the hypo-glycaemic range tolbutamide is *not* given.

3. 1 g tolbutamide (children 20 mg/kg body weight to a maximum of 1 g) is given intravenously.

4. Further blood samples for glucose and insulin are taken 5, 10, 15, 20, 30, 60, 120, 180 and 240 minutes after tolbutamide.

*Response:*

1. *Blood glucose concentration.* The blood glucose falls to about 50 per cent of the fasting value in normal subjects and returns to at least 70 per cent of the fasting value by 180 minutes. The fall in patients with insulinoma is usually greater and, more important, the level at 180 minutes is less than 70 per cent of the fasting value.

2. *Plasma insulin concentration.* In normal subjects the plasma insulin rises by up to 100 mU/l in 5–20 minutes. The rise in patients with insulinoma is very much greater.

This test is positive in about 80 per cent of patients with insulinoma. A similar exaggerated insulin response may be seen with insulin resistance, for example, in obesity and in liver disease but the blood glucose fall is usually within normal limits. Patients with alcohol-induced hypoglycaemia may show a glucose response resembling that of insulinoma but without the excessive insulin response.

# Chapter IX

# PLASMA LIPIDS AND LIPOPROTEINS

THE association of abnormalities of plasma lipids with atheromatosis and ischaemic heart disease has stimulated much research on the subject and has led to the introduction of new terminologies. Older, purely chemical, classifications of lipid disturbances have been replaced by others based on lipoproteins, the form in which lipids circulate in the blood. The first part of this chapter will describe and correlate the terms in current usage.

## TERMINOLOGY AND CLASSIFICATION

### CHEMICAL CLASSIFICATION OF PLASMA LIPIDS

The chemical structures of the four forms of lipid present in plasma are illustrated diagrammatically in Fig. 16.

(a) **Fatty acids** are straight chain compounds, the length of the chain depending on the number of carbon atoms in the molecule. Fatty acids may be *saturated* (containing no double bonds), or *unsaturated* (with one or more double bonds). The main saturated fatty acids in plasma are palmitic (16 carbon atoms) and stearic (18 carbon atoms) acids. Fatty acids may be *esterified* with glycerol to form glycerides, or they may be free, when they are called *free fatty acids* (FFA) or *non-esterified fatty acids* (NEFA). In the blood FFA are carried mainly on albumin. Normal fasting concentrations are very low.

The metabolism of FFA is considered in Chapter VIII and on p. 223.

(b) **Triglycerides** consist of glycerol, each molecule of which is esterified with three fatty acids. 95 per cent of the lipids of adipose tissue are triglycerides.

(c) **Phospholipids** are complex lipids containing phosphate and a nitrogenous base. The major phospholipids in plasma are *lecithin* and *sphingomyelin*. The phosphate and nitrogenous bases are water soluble, a fact that is important in lipid transport.

(d) **Cholesterol** has a steroid structure and other steroids are synthesised from it. In the plasma, cholesterol occurs in two forms. About two-thirds is esterified with fatty acid (Fig. 16) to form *cholesterol ester*, and the remainder is free. For routine clinical purposes the estimation of total cholesterol is adequate.

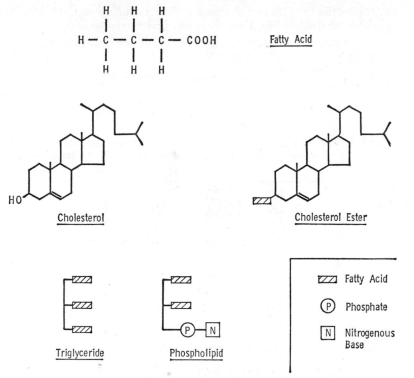

FIG. 16.—Plasma lipids.

## LIPOPROTEINS

Lipids are insoluble in water. Solution in plasma is achieved by combining lipid with a protein carrier to form a water-soluble *lipoprotein*. Several classes of lipoprotein occur, each containing protein, cholesterol, phospholipid and triglyceride, albeit in widely differing proportions. It is probable that the lipoprotein molecule is so constituted that the protein and water-miscible groups of the phospholipids are on the outside, in contact with the aqueous plasma, while the insoluble lipids are in the centre.

### Classification of Lipoproteins

The major lipoprotein classes may be separated and classified in a number of different ways.

(*a*) **Electrophoresis** separates the molecules by means of the electrical charge on the protein carrier. The technique is similar to that of protein electrophoresis (p. 316) except that a different staining method is used to render the lipid visible. Four fractions may be recognised:

α lipoproteins—contain mostly cholesterol and phospholipids
β lipoproteins—contain mostly cholesterol
pre-β lipoproteins ⎱
chylomicrons ⎰ contain mostly triglycerides

Normal fasting plasma contains α and β lipoproteins with small amounts of pre-β lipoproteins.

(*b*) **Ultracentrifugation** separates the molecules on the basis of their density. After centrifugation at high speeds some of the lipoproteins

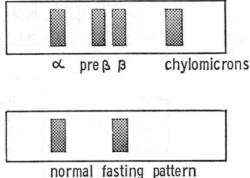

Fig. 17.—Lipoprotein electrophoretic strips (diagrammatic).

sediment with the other plasma proteins and are called *high density lipoproteins (HDL)*. The remainder, of lower density, tend to float, the rate of flotation being expressed in $S_f$ (Svedberg flotation) units. The greater the content of triglycerides, the lower the density of the lipoprotein class and the higher its $S_f$ number. Besides HDL, three other classes are recognised—*low density lipoproteins (LDL)* of $S_f$ 0–20; *very low density lipoproteins (VLDL)* of $S_f$ 20–400 and *chylomicrons* of $S_f$ $10^3$–$10^5$.

(*c*) **Filtration** separates lipoproteins by *molecular size*. Three groups are recognised—small (S), medium (M) and large (L) particles (SML classification).

These nomenclatures are correlated in Fig. 18. Note that the HDL or α lipoprotein class has no counterpart in the SML classification. Most laboratories identify lipoproteins by electrophoresis or filtration. Ultracentrifugal analysis is the reference method but is only rarely necessary in routine diagnostic work, and is not generally available.

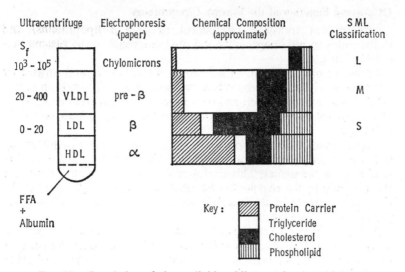

FIG. 18.—Correlation of plasma lipid and lipoprotein nomenclature.

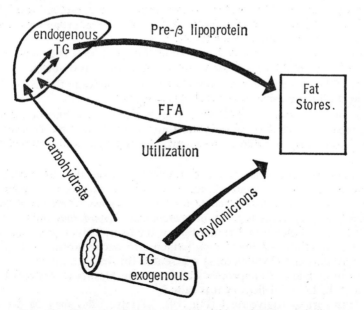

FIG. 19.—Normal and abnormal triglyceride and FFA metabolism.

## Origin and Function of the Plasma Lipoproteins

Lipoproteins transport cholesterol ($\alpha$ and $\beta$ lipoproteins) and triglycerides (pre-$\beta$ lipoproteins and chylomicrons) in the plasma in a soluble form.

**Transport of cholesterol.**—Cholesterol is synthesised and utilised by most tissues of the body and it is a component of cell membranes. It is catabolised only by the liver and consequently any excess of cholesterol, or cholesterol derived from cell breakdown, must be transported to that organ.

$\alpha$ *lipoproteins* (HDL) are probably synthesised and released from the liver with their phospholipid component. They accumulate free cholesterol from other tissues. This cholesterol is esterified with a fatty acid from lecithin by the enzyme *lecithin cholesterol acyltransferase (LCAT)*. Esterified cholesterol is then transferred to the other lipoproteins, particularly $\beta$ lipoprotein.

$\beta$ *lipoproteins* (LDL) are the major carriers of *cholesterol* in the plasma. They are possibly synthesised in the liver but there is evidence to suggest that much of the plasma $\beta$ lipoprotein is derived from the metabolism of pre-$\beta$ lipoprotein (see below).

Plasma levels of $\alpha$ lipoproteins are slightly higher in women than in men. They remain constant throughout life and are not influenced by dietary changes. $\beta$ lipoprotein concentrations, on the other hand, are affected by diet and by a number of factors which are further considered under cholesterol metabolism (p. 225).

Several rare *genetic disorders* of lipid transport have been described. They illustrate the role of the various factors described above.

1. In *LCAT deficiency* free cholesterol accumulates in the tissues and almost all the plasma cholesterol is unesterified.

2. In $\alpha$ *lipoprotein deficiency* (*Tangier disease*) esterification seems unimpaired as cholesterol esters accumulate in the tissues. This results in characteristic tonsillar enlargement, hepatomegaly and lymphadenopathy.

3. *Abetalipoproteinaemia* is characterised clinically by intestinal malabsorption with steatorrhoea, progressive ataxia, retinitis pigmentosa and acanthocytosis (crenation of erythrocytes). There is complete *absence of $\beta$ lipoprotein, pre-$\beta$ lipoprotein and chylomicrons* and inability to transport lipids from the intestine or the liver. This suggests that the protein fraction of $\beta$ lipoprotein (which also occurs normally in pre-$\beta$ lipoproteins and chylomicrons) is essential for lipid transport.

In both $\alpha$ and $\beta$ lipoprotein deficiency the plasma cholesterol level is low. In LCAT deficiency it is raised.

**Transport of triglycerides** (Fig. 19).—Triglycerides may be derived from the diet (exogenous) or be synthesised in the liver (endogenous).

The interrelationship of these two forms is best explained by following the fate of ingested fat.

(a) *Digestion and absorption.*—This is considered more fully on p. 266. Hydrolysed fat (a mixture of glycerol, fatty acids and monoglycerides) is absorbed into the intestinal mucosa, where FFA and glycerol are resynthesised into triglycerides and combined with small amounts of carrier protein, cholesterol and phospholipid to form *chylomicrons.* These large aggregates enter the lymphatics and pass via the thoracic duct into the blood stream. Because of their large size they scatter light and account for the turbidity of plasma seen after a fatty meal.

Some short and medium chain fatty acids enter the portal system and pass directly to the liver. They are, however, a minor component of the usual diet.

The fatty acid composition of chylomicrons reflects that of the diet.

(b) *Metabolism of chylomicrons.*—Although several tissues are capable of handling chylomicrons the major site of removal is adipose tissue. Here the enzyme, *lipoprotein lipase,* hydrolyses the chylomicron triglycerides and releases FFA which are taken up by the fat cells. This process reduces the size of the molecule and the plasma clears. The remaining particles, now containing mainly cholesterol, are further metabolised in the liver.

Lipoprotein lipase occurs in capillary endothelial cells, particularly of adipose tissue. The enzyme activity is not normally detectable in the blood stream except after an injection of heparin; consequently heparin can clear lipaemic plasma *in vivo* but not *in vitro.* Lipoprotein lipase appears to be activated physiologically by peptides of the α lipoprotein carrier protein. *Insulin* is also necessary for its normal activity. Lipoprotein lipase may be measured as post-heparin lipolytic activity (PHLA), although the estimation is not generally available. There is a rare condition in which lipoprotein lipase activity is congenitally absent (p. 231). This is characterised by the presence of circulating chylomicrons in the fasting state because they cannot be broken down.

(c) *Metabolism of adipose tissue.*—Adipose tissue is metabolically highly active. It consists almost entirely of triglyceride which is subject to continual breakdown and resynthesis. As this process is central to many metabolic problems it will be considered in some detail. The essential features are outlined in Fig. 20.

Triglycerides are hydrolysed by *triglyceride lipase* to FFA and glycerol. Adipose tissue lacks the necessary enzymes for resynthesis of triglyceride from glycerol; instead, the FFA are combined with α glycerophosphate. This in turn is derived by intracellular glycolysis, so that the most crucial factor in this reaction is the availability of intracellular glucose. If intracellular glucose is not available, as in fasting, starvation, or insulin lack (diabetes mellitus), breakdown exceeds synthesis and

FFA accumulate and pass into the blood stream. Plasma FFA levels are raised in these conditions.

(*d*) *Formation and metabolism of pre-β lipoproteins.*—Pre-β *lipoproteins* carry *endogenous triglycerides* in the plasma and are synthesised mainly in the liver with some contribution from the intestinal mucosa. There are two main sources of endogenous triglycerides:

(i) Excess FFA reaching the liver (see above) are re-esterified to form triglycerides and incorporated into pre-β lipoproteins. The fatty acid composition reflects that of the adipose tissue.

(ii) FFA, synthesised *de novo* in the liver from dietary carbohydrate, may be esterified to form triglycerides which are then incorporated into pre-β lipoproteins.

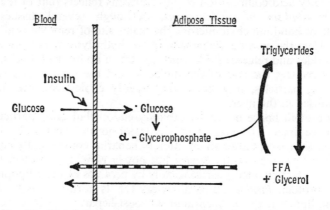

Fig. 20.—Metabolism of adipose tissue with (■■) and without (■ ■) adequate intracellular glucose.

Pre-β lipoproteins are broken down peripherally, mainly in adipose tissue, in the same way as chylomicrons. The removal of triglycerides results in molecules identical to α and β lipoproteins. This process may be the major source of plasma β lipoprotein.

The different origins of chylomicrons and pre-β lipoproteins have a bearing on the therapy of pathological increases. Hyperchylomicron-aemia (fat-induced hypertriglyceridaemia) is treated by reducing dietary fat, while raised levels of pre-β lipoproteins (carbohydrate-induced hypertriglyceridaemia) frequently respond to dietary carbohydrate reduction.

**Interrelationship of lipoproteins.**—The foregoing account of the lipoproteins is a simplification. All the classes contain many fractions and have features in common. We have mentioned the probable formation of one ($\beta$ lipoprotein) from another (pre-$\beta$ lipoprotein) and the role of the $\beta$ lipoprotein protein moiety in the transport of lipids by all classes. It is possible that the sub-fractions of the lipoprotein classes will form the basis of a future classification.

To recapitulate:

1. $\alpha$ *lipoproteins* carry mainly *cholesterol and phospholipids*.

2. $\beta$ *lipoproteins* carry mainly *cholesterol*.

3. *Pre-$\beta$ lipoproteins* carry mainly *endogenous triglycerides* (synthesised from FFA and dietary carbohydrate).

4. *Chylomicrons* carry mainly *exogenous triglycerides* (from dietary fat).

## METABOLISM OF CHOLESTEROL

Cholesterol is both absorbed from the gut and synthesised in the body. *Absorption* is largely proportional to intake in the range commonly found in the usual diet. In affluent societies this is about 1·5 to 2·0 mmol (600 to 800 mg) per day, with meat, dairy products and particularly egg yolk as the main sources. One egg contains about 0·65 mmol (250 mg) of cholesterol. To a certain extent high intakes are compensated by reduced endogenous synthesis, but this is not complete and plasma cholesterol values are usually higher with a high cholesterol intake than with a low one. The *nature of fat* in the diet influences plasma cholesterol: *saturated fatty acids*, which are found chiefly in animal fats, as is cholesterol, increase plasma levels while *polyunsaturated fatty acids*, in vegetable oils, tend to lower them.

Cholesterol in the gut is absorbed after incorporation in mixed micelles (p. 266). Bile salts are necessary for this process. In the intestinal mucosa the absorbed cholesterol is esterified and transported in the lymph in chylomicrons and pre-$\beta$ lipoproteins synthesised in the intestinal wall.

Most tissues of the body, particularly the liver and small intestine, synthesise cholesterol (about 2·6 mmol [1 g] a day). Cholesterol released from the peripheral tissues is probably taken up and esterified in $\alpha$ lipoproteins and transferred to $\beta$ lipoproteins where most of the plasma cholesterol is found (p. 222). About two-thirds is esterified.

The amount of cholesterol added daily to the body pool is balanced by an equivalent excretion in the bile. Part is degraded by the liver to bile acids and bile salts (p. 308), while the remainder is excreted as cholesterol and, together with dietary cholesterol, is available for reabsorption. *Cholestyramine*, which binds bile salts and interferes with their reabsorption, can be used to reduce plasma cholesterol and $\beta$ lipoprotein levels.

Many hormones, particularly those of the thyroid, influence cholesterol metabolism (p. 173). Hypothyroidism is associated with raised cholesterol levels due to reduced catabolism and d-thyroxine has been used to treat hypercholesterolaemia. High plasma cholesterol levels are found during pregnancy.

## "Normal" Plasma Cholesterol

It is accepted that certain types of hyperlipoproteinaemia are associated with an increased risk of cardiovascular disease. The best single chemical indicator of this risk is the plasma cholesterol level. It therefore becomes important to know the "normal" levels of cholesterol at different ages. It is also important to note that "normal" (i.e. 95 per cent limits) are not necessarily desirable levels. The following is offered as a guide in a very controversial field.

Plasma cholesterol at birth (cord blood) is usually below 2·6 mmol/l (100 mg/dl). Levels increase slowly throughout childhood, mostly during the first year, but do not usually exceed 4·1 mmol/l (160 mg/dl). In most affluent populations a further progressive increase occurs after the second decade. This increase is greater in men than in women during the reproductive years. The "normal" (95 per cent) upper limit in many societies is as high as 8·4 mmol/l (330 mg/dl) in the fifth and sixth decades. This rise is not seen in less affluent societies where the incidence of ischaemic heart disease is much lower. Most authorities, however, consider the desirable upper limit of cholesterol in the adult to be about 6·5 mmol/l (250 mg/dl).

There is little difference between the fasting cholesterol level and that of a random sample. A haemolysed or jaundiced specimen gives artefactually high results by certain methods. A raised value should always be confirmed by a repeat estimation by a reliable laboratory. Results obtained by simple "office" procedures should never be accepted uncritically. As mentioned on p. 227, values are influenced by posture.

## METABOLISM OF PHOSPHOLIPIDS

Phospholipids are widely distributed in all tissues and are major constituents of biological membranes. In bile phospholipids are important in keeping cholesterol in solution, and in the lung lecithin is an essential component of surfactant (p. 433).

Dietary phospholipids may be absorbed as such because of their relative solubility but the phospholipids of plasma are derived mainly from synthesis in the liver.

The role of plasma phospholipids is uncertain. They are concerned with blood coagulation (the lipid portion of thromboplastin) and their role in lipoproteins has been mentioned (p. 222). Plasma lecithin is the source of fatty acids for esterification of cholesterol in α lipoproteins.

# HYPERLIPOPROTEINAEMIA
## (*Syn.* Hyperlipaemia; hyperlipidaemia)

Hyperlipoproteinaemia should be suspected in patients with *premature cardiovascular disease* or *xanthomatosis*. It is often first detected by finding a raised cholesterol level or a turbid plasma in a patient without specific symptoms.

Before describing the classification of hyperlipoproteinaemia, three general points will be discussed.

### Correct Blood Sample

Plasma lipoprotein patterns are labile and are affected by eating, by changes in diet and by stress. For the *initial typing* of a hyperlipoproteinaemia it is *essential* that the sample be taken under *standard conditions*. It may, in some cases, be necessary to repeat the analysis to establish the most consistent pattern. The following points are mportant.

1. The patient must have fasted for 14 to 16 hours.

2. For the two weeks preceding the test the patient should have been on his "normal" diet and his weight should have remained constant.

3. The patient must not be on any treatment that affects his plasma lipids.

4. As in the case of all large molecules, lipoprotein concentrations are affected by venous stasis (p. 461) and posture. Plasma cholesterol concentration may be up to 10 per cent higher in the upright than in the recumbent position; triglyceride concentration may change slightly more than this. A standardised collection procedure, e.g. sitting for 30 minutes before blood is taken, is important in serial estimations to assess the effect of treatment.

5. Diagnosis and typing of hyperlipoproteinaemia is preferably deferred for three months after a myocardial infarction or major operation.

6. The blood sample must be separated as soon as possible.

### Primary and Secondary Hyperlipoproteinaemia

The types of hyperlipoproteinaemia described below refer to the *patterns found in plasma*, not to the cause of the disorder. In some cases no cause will be found. These are referred to as *primary* hyperlipoprotein-aemias and may or may not be familial. In many cases, however, the plasma lipid abnormality is *secondary* to other disease and may disappear on successful treatment of the cause.

The main secondary causes of each pattern will be mentioned, but in general *diabetes mellitus, hypothyroidism* and the *nephrotic syndrome* should be sought in all instances of hyperlipoproteinaemia. *Dysglobulinaemia*, either paraproteinaemia or a marked increase in $\gamma$ globulin as found in systemic lupus erythematosus, should also be considered. Abuse of *alcohol* is a common cause of secondary hyperlipoproteinaemia.

### Xanthomatosis

Xanthomata are yellowish deposits of lipids in the tissues of the body. Several different types exist and usually seem to be correlated with elevations of different lipid fractions in the plasma, although this correlation is not a constant one.

(*a*) *Eruptive xanthomata* are crops of small itchy yellow nodules that occur in association with raised triglycerides, either pre-$\beta$ lipoproteins or chylomicrons. They disappear rapidly if plasma lipid levels are restored to normal.

(*b*) *Tuberous xanthomata* are yellow plaques, in some cases large and disfiguring, found mainly over the elbows and knees. They occur with elevated $\beta$ lipoproteins, including the abnormal $\beta$ lipoprotein of Type III (see below) and, less commonly, pre-$\beta$ lipoproteins.

(*c*) Deposits in tendons (*xanthoma tendinosum*), eyelids (*xanthelasma*) and cornea (*corneal arcus*) at an early age occur typically with elevated $\beta$ lipoproteins.

(*d*) *Planar xanthomata*, raised yellow linear deposits in the creases of the palm, are seen typically with the abnormal $\beta$ lipoprotein of Type III hyperlipoproteinaemia.

PATTERNS OF HYPERLIPOPROTEINAEMIA (Fig. 21)

Hyperlipoproteinaemia is usually typed by lipoprotein electrophoresis. The most generally used classification is that of Fredrickson and is the one recommended by the World Health Organisation. Six types are described and will be considered in approximate order of frequency.

**Types II (a and b)** are characterised by an *increase in $\beta$ lipoprotein* alone (Type IIa) or with an associated increase in *pre-$\beta$ lipoprotein* (Type IIb). Cholesterol levels are elevated in both and triglyceride levels in Type IIb.

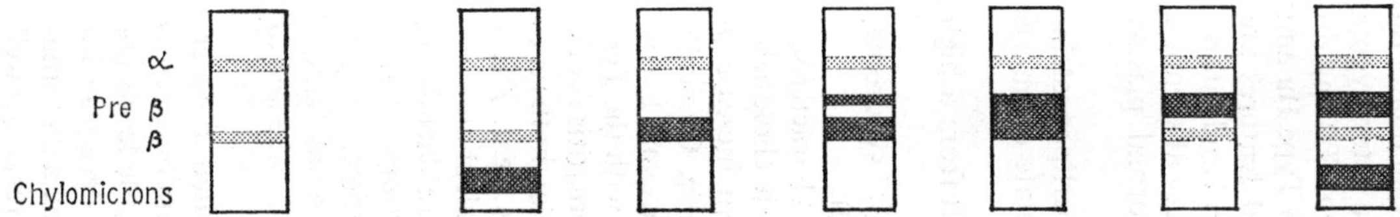

| Type | Normal | I | IIa | IIb | III | IV | V |
|---|---|---|---|---|---|---|---|
| Main chemical abnormality | | T | C | C & T | C & T | T | T |
| Appearance of fasting plasma (after 12 hours at 4°C) | | Creamy surface layer | Clear | Slightly turbid | Turbid, small creamy layer | Turbid | Turbid with creamy layer |
| Usual age of presentation | | Under 10 | Any age; often young | | Over 30 | Over 20 | Over 20 |
| Associated with :— | | | | | | | |
| Cardiovascular disease | | No | Yes | | Yes | Probably | Probably not |
| Abdominal pain or pancreatitis | | Yes | No | | No | Rarely | Yes |
| Impaired carbohydrate tolerance | | No | No | | Yes | Yes | Yes |
| Hyperuricaemia | | No | No | | Yes | Yes | Yes |
| Xanthomata – eruptive | | + | − | | − | + | + |
| – tuberous | | − | + | | + | Rarely | − |
| – tendinous | | − | + | | + | − | − |
| – planar | | − | Occasionally | | + | − | − |

(Electrophoresis strip labels: α, Pre β, β, Chylomicrons)

FIG. 21.—Features of the main types of hyperlipoproteinaemia (Fredrickson classification). T = raised plasma triglyceride; C = raised plasma cholesterol.

The familial form is the most serious of the inherited hyperlipo-proteinaemias. Xanthomata and cardiovascular disease are common and frequently severe. In the homozygote particularly, myocardial infarction may occur at a young age. Plasma cholesterol levels in the heterozygote range from approximately 8·0 to 13·0 mmol/l (300 to 500 mg/dl) and in the homozygote may be 2 or 3 times as high. Both Type IIa and IIb patterns occur in the same families and the clinical features are indistinguishable. *Treatment* is to restrict dietary cholesterol and satura-ted fat, as well as administration of cholestyramine to reduce absorption of bile salts (p. 267). These measures rarely achieve normal plasma cholesterol levels.

There is some controversy as to whether Type II hyperlipoprotein-aemia can be diagnosed at birth on cord blood. A definite diagnosis should probably be made only after the first year of life.

Non-familial Type II hyperlipoproteinaemia may result from a high cholesterol and saturated fatty acid intake.

*Secondary* causes are hypothyroidism, the nephrotic syndrome, dysglobulinaemia and biliary obstruction (see also p. 232).

**Types IV and V** may be considered together. Both are characterised by a turbid plasma due to raised triglyceride levels with cholesterol levels normal or only slightly raised. In both there is an increase in *pre-β lipoproteins*, and, in Type V, *hyperchylomicronaemia*. Clinically they are associated with obesity, impaired glucose tolerance and hyper-uricaemia. Eruptive xanthomata may be present. Patients with the Type V pattern may suffer attacks of abdominal pain and pancreatitis (as do patients with Type I hyperchylomicronaemia). When familial, both types occur in a single family and the distinction is not always clear. *Treat-ment* is by weight reduction and diet, with avoidance of alcohol.

These two patterns, with endogenous hypertriglyceridaemia in common, are frequently *secondary*. Associated conditions include diabetes mellitus, hypothyroidism, the nephrotic syndrome, dysglo-bulinaemia, glycogen storage disease, pancreatitis and renal failure. They are often found in subjects with an excessive alcohol intake and in women taking oral contraceptives.

**Type III** is a rare pattern, usually occurring as an inherited disorder. It is due to an *abnormal β lipoprotein* which usually appears on electro-phoresis as a broad β band. Both cholesterol and triglyceride levels are raised. It is diagnosed most often in the fourth and fifth decades with xanthomata and peripheral vascular disease. *Treatment* is by a diet simi-lar to that used in Type II patients but, unlike the latter, it responds well to clofibrate.

Rarely the pattern occurs in diabetes mellitus, hypothyroidism and dysglobulinaemia.

Although the diagnosis of Type III hyperlipoproteinaemia may be suggested by the electrophoretic pattern, special techniques are necessary to confirm it.

Type I is very rare and is characterised by a milky fasting plasma, due to *hyperchylomicronaemia*. Levels of triglyceride are very high and those of cholesterol usually normal. It is usually an autosomal recessive condition presenting in the first decade. It presents clinically with eruptive xanthomata, lipaemia retinalis, hepatomegaly and attacks of abdominal pain. The basic abnormality is a deficiency of lipoprotein lipase (demonstrated as reduced PHLA, p. 223). If fat is removed from the diet the plasma clears in a few days (fat-induced hyperlipoproteinaemia) and this is usually the only therapy required to prevent abdominal pain. There is no increased incidence of cardiovascular disease.

Rarely the Type I pattern is *secondary* to diabetes mellitus (insulin is necessary for normal lipoprotein lipase action—p. 223), hypothyroidism or dysglobulinaemia.

## Mechanisms of Hyperlipoproteinaemia

Hyperlipoproteinaemia may theoretically arise as a result of overproduction or of reduced breakdown of lipoproteins. The abnormality in Type I is clearly due to reduced removal of chylomicrons from plasma. There is evidence that deficient or abnormal catabolism is also an important factor in the other types of hyperlipoproteinaemia.

## Approach to the Differential Diagnosis of Hyperlipoproteinaemia

*The diagnosis of hyperlipoproteinaemia is made* by demonstrating an increase in plasma lipids by *chemical estimation*. For this the measurement of plasma *cholesterol* and fasting *triglyceride* levels is adequate. If the latter estimation is not available, inspection of the separated plasma suffices in most cases (see below). Once diagnosed, hyperlipoproteinaemia is *typed*, usually by lipoprotein electrophoresis. Again, if this is not possible, many cases can be typed using the following approach.

The vast majority of hyperliproteinaemias encountered will be of Types II, IV or V, that is, they will have as the predominant abnormality a raised cholesterol (Type II) or raised triglycerides (Types IV and V). Turbidity in a fresh, fasting plasma indicates raised triglycerides. The following combinations may be seen:

(*a*) **Raised cholesterol, normal triglycerides** (clear plasma) almost always indicates raised $\beta$ lipoproteins (Type IIa).

(*b*) **Raised triglycerides** (turbid plasma) with **normal or mildly raised cholesterol** usually indicates an excess of pre-$\beta$ lipoproteins and/or chylomicrons. These may be distinguished by standing the separated plasma in the refrigerator overnight. Pre-$\beta$ lipoproteins remain dispersed while chylomicrons float to the surface, forming a creamy layer. In Type IV there will therefore be a diffusely turbid plasma while in Type V there will in addition be a creamy layer. Very rarely a creamy layer with a clear infranatant will be seen, indicating a Type I pattern.

(c) **Raised cholesterol and raised triglycerides** (turbid plasma) usually indicates Type IIb, but may be Type III or even Types IV or V. Electrophoresis is required to distinguish between them.

It must be remembered that all the lipoprotein groups contain cholesterol and triglycerides (as well as protein and phospholipid), and elevation of triglyceride is usually accompanied by some elevation of cholesterol levels and *vice versa*.

This simple scheme is only a guide and it must be stressed that proper typing, by electrophoresis, is highly desirable before treatment is started.

Once the hyperlipoproteinaemia has been typed, the next step is to establish if it is primary or secondary. If primary, relatives should be investigated to establish a familial incidence.

As a summary, the main causes of raised plasma cholesterol and fasting triglycerides are listed in Table XX.

### TABLE XX

THE CAUSES OF RAISED PLASMA CHOLESTEROL AND TRIGLYCERIDE LEVELS

(Where both are indicated it does not imply that they always occur together)

|  | Cholesterol | Triglycerides |
|---|---|---|
| *Primary:* (Usual findings) | | |
| Type I | | + |
| Type IIa | + | |
| Type IIb | + | + |
| Type III | + | + |
| Type IV | | + |
| Type V | | + |
| *Secondary:* | | |
| Diabetes mellitus | + | + |
| Nephrotic syndrome | + | + |
| Hypothyroidism | + | + |
| Dysglobulinaemia | + | + |
| Pancreatitis | | + |
| Renal failure | | + |
| Excessive alcohol intake | | + |
| Pregnancy | + | |
| Oral contraceptives | | + |
| Biliary obstruction | + | |
| Glycogen storage disease | | + |
| Acute porphyria | + | |

## SUMMARY

1. Lipids in plasma are carried in the form of lipoproteins.

2. Plasma lipids may be classified according to their chemical structure, or as lipoproteins by electrophoresis or ultracentrifugation. Chemical estimations are most frequently available but lipoprotein analysis offers the best understanding of abnormalities.

3. The chemical lipid fractions are cholesterol, triglycerides, phospholipids and free fatty acids.

4. The lipoprotein classes separated by electrophoresis are $\alpha$ lipoproteins, $\beta$ lipoproteins, pre-$\beta$ lipoproteins and chylomicrons.

5. Lipoproteins transport lipids in plasma. $\alpha$ and $\beta$ lipoproteins carry mainly cholesterol while pre-$\beta$ lipoproteins and chylomicrons carry triglycerides.

6. Hyperlipoproteinaemias may be classified according to electrophoretic patterns. The clinical associations of these conditions include cardiovascular disease, xanthomatosis, carbohydrate intolerance and obesity.

7. Hyperlipoproteinaemias may be primary or secondary in origin. Primary hyperlipoproteinaemias often have a familial incidence.

8. The most serious hyperlipoproteinaemia is Type II (hyperbeta-lipoproteinaemia), characterised by a raised plasma cholesterol concentration. It may be primary, or secondary to other disease.

9. If lipoprotein electrophoresis is not available, many cases of hyperlipaemia may be classified on the basis of chemical estimation and inspection of fasting plasma.

## FURTHER READING

LEWIS, B. (1972). Metabolism of the plasma lipoproteins. *Scientific Basis of Medicine Annual Reviews*, p. 118. London: Athlone Press.

STONE, N. J., and LEVY, R. I. (1972). The hyperlipidemias and coronary artery disease. *Disease a Month* (August). Chicago: Year Book Medical Publishers.

LEVY, R. I. (1972). Dietary and drug treatment of primary hyperlipoproteinemia. *Ann. intern. Med.*, **77**, 267.

McGOWAN, G. K., and WALTERS, G. (Eds.) (1973). Disorders of lipid metabolism. *J. clin. Path.*, **26**, Suppl. 5.

**Reference**

Classification of Hyperlipidaemias and Hyperlipoproteinaemias (1970). *Bull. Wld. Hlth. Org.*, **43**, 891.

# Chapter X

# CALCIUM, PHOSPHATE AND MAGNESIUM METABOLISM

MOST of the body calcium is in bone, and a significant deficiency causes bone disease. However, the extraosseous fraction, although amounting to only 1 per cent of the whole, is of great importance because of its effect on neuromuscular excitability and on cardiac muscle. Most calcium salts in the body contain phosphate. This radical is very important in its own right (for instance in the formation of "high energy" phosphate bonds), but clinically, apart from its buffering power, it is mainly of interest because of its association with calcium.

More than half the body magnesium is in bone: most of the remainder is intracellular, but, as in the case of calcium, the small extracellular fraction (about 1 per cent) is important because of its neuromuscular action. Because of this similarity in distribution and physiological action, calcium and magnesium metabolism will be discussed in one chapter.

## CALCIUM METABOLISM

### TOTAL BODY CALCIUM

The total amount of calcium in the body depends on the balance between intake and loss.

### Factors Affecting Intake

The amount of calcium entering the body depends on the amount in the *diet*, which normally amounts to about 25 mmol (1 g) a day. Between 25 and 50 per cent of this is absorbed. *Vitamin D*, in the form of 1: 25-dihydroxycholecalciferol (p. 237), is required for adequate calcium absorption.

### Factors Affecting Loss

Calcium is lost in urine and faeces. *Faecal calcium* consists of that which has not been absorbed, together with that which has been secreted into the intestine. Calcium in the intestine, whether endogenous or exogenous in origin, may be rendered insoluble by large amounts of *phosphate, fatty acids* or *phytate* (in cereals). Oral phosphate may be used therapeutically to control calcium absorption and reabsorption (p. 255): an excess of fatty acids in the intestinal lumen in steatorrhoea contributes to calcium malabsorption, and high phytate diets may account for the relative frequency of nutritional osteomalacia in subjects in whom chapattis form an important part of the diet, especially in those living in countries with a low sunshine record.

*Urinary calcium.*—This depends on the *amount of calcium circulating* through the glomerulus, on *renal function, parathyroid hormone levels* and to a lesser extent on *urinary phosphate excretion.* The amount of calcium circulating through the glomerulus is increased after a calcium load or during decalcification of bone not due to calcium deficiency (e.g. osteoporosis or acidosis). Such decalcification rarely causes hypercalcaemia because of the rapid renal clearance of calcium. Conversely, hypercalcaemia, whatever the aetiology, causes hypercalcuria if renal function is normal. Glomerular insufficiency reduces calcium loss in the urine, even in the presence of hypercalcaemia.

## PLASMA CALCIUM

Estimation of plasma calcium should be accurate to about 0·05 mmol/l (0·2 mg/dl) and changes of less than this are probably insignificant. Calcium circulates in the plasma in two main states. The albumin bound fraction accounts for a little less than half the total calcium as measured by routine analytical methods: it is physiologically inactive and is probably purely a transport form comparable to iron bound to transferrin (p. 392). Most of the remaining plasma calcium is ionised ($Ca^{++}$), and this is the physiologically important fraction (compare with thyroxine, p. 163, and cortisol, p. 136). Ideally, ionised rather than total calcium should be measured, but techniques for this are not available in most routine laboratories. Whenever a plasma calcium value is assessed an effort should be made to decide whether the ionised fraction is normal by comparing the value for total calcium with that for protein (preferably albumin) on the same specimen. The total calcium concentration is lower in the supine than in the erect position, because of the effect of posture on plasma protein concentration.

The ionisation of calcium salts is greater in acid than in alkaline solution. In alkalosis tetany may occur in the presence of normal levels of total calcium because of the reduction in the ionised fraction. In clinical states associated with prolonged acidosis (for instance after transplantation of the ureters into the colon or in renal tubular acidosis, p. 95), the increased solubility of bone salts removes calcium from bone and leads to osteomalacia, even in the presence of an adequate supply of the ion to bone and of normal parathyroid function.

### Control of Plasma Calcium

The level of circulating ionised calcium is controlled within narrow limits by parathyroid hormone (PTH), secreted by the parathyroid glands. Calcitonin produced in the thyroid has an opposite action to that of PTH on calcium levels: its importance in calcium homeostasis is less certain than that of parathormone. Both these hormones control ionised calcium by using bone as a reservoir from which plasma levels can be controlled. Vitamin D increases calcium absorption from the intestine, and in physiological amounts appears to be necessary for the action of PTH.

**Action and control of parathyroid hormone.**—Parathyroid hormone raises the plasma ionised calcium concentration. It has two direct actions on plasma calcium and phosphate levels:

1. By acting directly on osteoclasts it releases bone salts into the extracellular fluid. This action tends to increase both calcium *and* phosphate in the plasma.

2. By acting on the renal tubular cells it decreases the reabsorption of phosphate from the glomerular filtrate, causing phosphaturia. This action tends to *decrease* the plasma phosphate: this in turn, by a mass action effect, increases release of phosphate salts (and therefore calcium) from bone.

In addition to these direct effects of PTH on plasma calcium and phosphate levels, it may function as a trophic hormone, stimulating the renal and hepatic production of active vitamin D.

The secretion of PTH, like that of insulin from the pancreas, is not controlled by any other endocrine gland. It is influenced only by the concentration of calcium ions circulating through it. Reduction of this concentration increases the rate of release of hormone, and this increase is maintained until the ionised calcium level returns to normal, when hormone production stops.

Bearing in mind the actions of PTH on bone and on the renal tubules, and the fact that its secretion is controlled by the level of ionised calcium, the consequences of most disturbances in calcium metabolism on plasma calcium and phosphate levels can be predicted. A low ionised calcium from any cause (other than hypoparathyroidism) by stimulating the parathyroid, causes phosphaturia: this loss of phosphate over-rides the tendency to hyperphosphataemia due to the direct action of PTH on bone, and plasma phosphate levels are normal, or even low. Conversely, a high calcium concentration due to any cause (except excess of PTH) cuts off production of the hormone and causes a high phosphate level. As a general rule, therefore, calcium and phosphate levels tend to vary in the same direction unless there is an inappropriate excess or deficiency of PTH (as in primary hyperparathyroidism or hypopara-thyroidism), or in the presence of renal failure when the phosphaturic effect is absent (see, however, "Action of vitamin D", below).

**Action and metabolism of vitamin D** (Fig. 22).—Vitamin D (D$_3$: cholecalciferol) may be formed in the skin from 7-dehydrocholesterol by the action of ultra-violet light; this is the form in animal tissues, particularly liver, and is therefore the most important dietary "vitamin D". This term is also applied to ergocalciferol (D$_2$), which can be obtained from plants after artificial ultraviolet irradiation. These vitamins D are transported in the blood bound to specific carrier proteins: they are inactive until metabolised.

In the *liver* the molecule is hydroxylated to *25-hydroxycholecalciferol* (*25-HCC*), the main circulating form of the vitamin. Any excess of vitamin D is stored in the organ, or is converted to inactive forms and excreted. Inactivation may be stimulated by *anticonvulsant drugs*; hypocalcaemia and bone disease may be found after long-term therapy with such drugs. Deficiency of 25-HCC has not been reported in liver disease.

Further hydroxylation of 25-HCC by the *kidney* to *1:25-dihy-droxycholecalciferol* (*1:25-DHCC*) is necessary to confer biological activity on the vitamin. 1:25-DHCC acts on the intestine, increasing calcium absorption; in conjunction with PTH it releases calcium from bone. The kidney is thus an endocrine organ, manufacturing and releasing the hormone 1:25-DHCC; failure of the final hydroxylation may, at least partially, explain the hypocalcaemia of renal disease (p. 245).

The lag between administration of therapeutic vitamin D and its action (p. 261) may be due to the time taken for the two-phase hydroxy-lation.

Vitamin D has a phosphaturic effect. Usually in the hypercalcaemia of vitamin D excess this effect is over-ridden by the reduction in phosphate excretion following reduction of parathormone secretion. Occasionally, however, phosphate levels are low or low normal when hypercalcaemia is due to this cause.

**Interrelationship of parathyroid hormone and vitamin D.**—The rate of production of 1:25-DHCC is probably controlled by the circulating level of PTH: the action of PTH on bone is impaired in the absence of 1:25 DHCC.

A reduction of extracellular ionised calcium levels stimulates PTH production by the parathyroid glands. PTH stimulates 1:25-DHCC synthesis, and the two hormones act synergistically on the bone "reser-voir" releasing calcium into the circulation; 1:25-DHCC also increases calcium absorption. In short-term homeostasis the bone effect is more important: after prolonged hypocalcaemia more efficient absorption

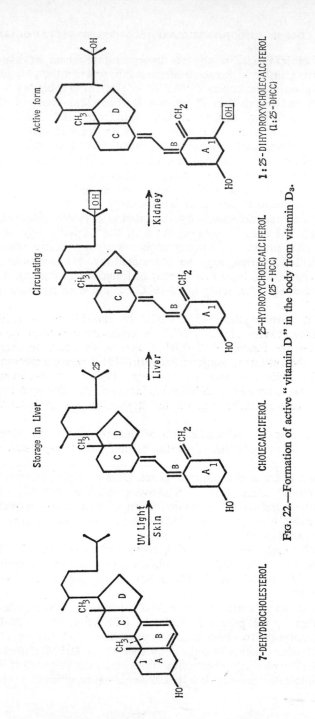

7-DEHYDROCHOLESTEROL

CHOLECALCIFEROL

25-HYDROXYCHOLECALCIFEROL
(25 - HCC)

1:25-DIHYDROXYCHOLECALCIFEROL
(1:25-DHCC)

Storage in Liver

Circulating

Active form

UV Light
Skin

Liver

Kidney

Fig. 22.—Formation of active "vitamin D" in the body from vitamin D₃.

becomes important. As soon as the ionised calcium concentration is corrected PTH production ceases, and that of 1:25-DHCC is reduced.

**Action and control of calcitonin.**—Calcitonin reduces plasma calcium by decreasing osteoclastic activity, and therefore bone resorption. It also has a phosphaturic effect. It is produced in the C cells of the thyroid and its secretion is stimulated by high ionised calcium levels. Its clinical importance is doubtful. Circulating concentrations may be very high in patients with medullary carcinoma of the thyroid, but abnormalities of plasma calcium concentration have very rarely been reported associated with these high levels.

**Action of thyroid hormone on bone.**—Thyroid hormone excess may be associated with the histological lesion of osteoporosis, and with increased faecal and urinary excretion of calcium, probably due to release from bone. Unless there is very gross excess of thyroid hormone, effects on plasma calcium levels are over-ridden by the homeostatic reduction of PTH secretion and hypercalcaemia rarely occurs. Some cases of post-thyroidectomy hypocalcaemia may be due, not to parathyroid damage, but to re-entry of calcium into depleted bone when the circulating thyroid hormone levels fall (p. 246).

Calcium homeostasis follows the general rule that body content is regulated by control of plasma levels. In the presence of normal parathyroid, kidney and intestinal function, and of sufficient supply of calcium and vitamin D, this is usually adequate. In the absence of one or more of these, control of plasma levels may be effected at the expense of the total amount of calcium in the body (compare sodium and water, p. 49).

## CLINICAL EFFECTS OF DISORDERS OF CALCIUM METABOLISM

### Clinical Effects of Excess Ionised Calcium

1. *On the kidney.*—The most dangerous consequence of prolonged, mild hypercalcaemia is *renal failure*. The high ionised calcium causes the solubility product of calcium phosphate in the plasma to be exceeded, and this salt is precipitated in extraosseous sites. The most important of these sites is the kidney, where it may cause renal damage. At an earlier stage calcification of the renal tubules may reduce their ability to reabsorb water, so that the presenting symptom of hypercalcaemia may be *polyuria*. It is because of this danger of renal damage that every attempt should be made to diagnose the cause of even mild hypercalcaemia and treat it at an early stage.

The high ionised calcium concentration in the glomerular filtrate may, if circumstances favour it, result in precipitation of calcium salts in the urine: the patient will present with *renal calculi*, and these may occur without significant renal parenchymal damage. All patients with renal stones should have plasma calcium estimated on more than one occasion.

2. *On neuromuscular excitability.*—High ionised calcium levels depress neuromuscular excitability in both voluntary and involuntary muscle. The patient may complain of *constipation* and *abdominal pain.* Muscular hypotonia may also be present. The *anorexia, nausea* and *vomiting* of hypercalcaemia are more probably due to a central effect.

3. *On the heart.*—Hypercalcaemia causes changes in the electro-cardiogram. If severe (as a rough guide, more than 3·75 mmol/l (15 mg/dl)) it carries the risk of sudden *cardiac arrest,* and for this reason should be treated as a matter of urgency.

### Clinical Effects of Reduced Ionised Calcium Levels

Low ionised calcium levels (including those associated with a normal total calcium in alkalosis, p. 235) cause increased neuromuscular activity leading to *tetany.*

Prolonged hypocalcaemia, even when mild, interferes with the metabolism of the lens and causes *cataracts*: it may also cause mental depression and other *psychiatric symptoms.* Because of the danger of cataracts, asymptomatic hypocalcaemia should be sought when there has been a known risk of damage to the parathyroid glands (e.g. after partial or total thyroidectomy).

### Effects of High Parathyroid Hormone Levels on Bone

1. **High PTH levels with a normal supply of vitamin D and calcium** (for example, primary hyperparathyroidism).—Parathyroid hormone acts on the osteoclasts of bone, increasing their activity. In cases in which the supply of vitamin D and calcium is normal and in which PTH has been circulating in excess over a long period of time the effects are as follows:

*Clinical:*
Bone pains.
Bony swellings.

*Radiological:*
Generalised decalcification.
Subperiosteal erosions.
Cysts in the bone.

*Histological:*
Increased number of osteoclasts.

A secondary osteoblastic reaction may occur: osteoblasts manufacture the enzyme alkaline phosphatase which, when these cells are increased in number, is released into the extracellular fluid in abnormal amounts and leads to a *rise in plasma alkaline phosphatase activity.*

The effects of excess parathyroid hormone on bone are only evident in long-standing cases. In disease of short duration (for instance, when the excess of PTH is due to malignant disease, or to early primary hyperparathyroidism) they are absent. Osteoblastic reaction is a late manifestation and plasma alkaline phosphatase levels are usually normal or only moderately raised in primary hyperparathyroidism.

2. **High PTH levels in vitamin D and calcium deficiency** (causing *rickets and osteomalacia*).—Vitamin D must be present in adequate amounts for the complete action of PTH on bone. In cases of vitamin D deficiency PTH levels may be very high, but the effects described above are seen to only a minor degree. Calcium deficiency means that osteoblastic reaction is ineffective, and uncalcified osteoid is a characteristic histological finding: this osteoblastic activity is marked and increases plasma alkaline phosphatase levels.

The effects are therefore as follows:

*Clinical:*

Bone pains.
*Rarely* bony swelling.

*Radiological:*

Generalised decalcification.
Pseudofractures.
*Rarely* subperiosteal erosions or cysts in bone.

*Histological:*

Uncalcified osteoid: wide osteoid seams.
*Occasionally* increased numbers of osteoclasts.

*In the plasma:*

Raised alkaline phosphatase activity.

These changes in the bone are also seen in certain inborn errors of renal tubular reabsorption of phosphate (p. 248).

## DISORDERS OF CALCIUM METABOLISM

In this section disorders of calcium metabolism will be considered in relation to circulating parathyroid hormone levels. They can be summarised as follows:

Diseases associated with high PTH levels.
Inappropriate secretion of PTH (*with raised plasma calcium concentration*).
Appropriate secretion of PTH (*with normal or low plasma calcium concentration*).

Diseases associated with low PTH levels.
Disease of the parathyroid gland (*with low plasma calcium concentration*).
Appropriate suppression of PTH secretion by *high plasma calcium levels* due to causes other than inappropriate PTH levels.

## Diseases Associated with High Circulating Parathyroid Hormone Levels

**Diseases due to inappropriate secretion of parathyroid hormone.**—The term "inappropriate secretion of parathyroid hormone" is used here to mean the production of the hormone under circumstances in which it should normally be absent (i.e. in the presence of normal or high ionised calcium concentrations). This syndrome should be compared with that of "inappropriate" ADH secretion (p. 446).

Such inappropriate secretion occurs in three circumstances:

1. Primary hyperparathyroidism.
2. Tertiary hyperparathyroidism.
(In these two cases the hormone is produced by the parathyroid gland.)
3. Ectopic production of PTH ("pseudohyperparathyroidism").

The findings, if renal function is normal, are those of excess PTH production, that is a raised plasma calcium concentration, with a normal or low plasma phosphate level. With the onset of renal damage, usually due to hypercalcaemia, both these tend to return to normal, the phosphate because of the inability of the kidney to respond to the phosphaturic effect of PTH and the calcium because of the lowering of plasma calcium in renal failure (p. 245). Diagnosis at this stage may be very difficult.

The *clinical features* of these cases are due to:

1. Excess of circulating ionised calcium (p. 239).
2. The effect of PTH on bone (p. 240).

Differences between these syndromes depend on the duration of the disease more than any other factor.

*Primary hyperparathyroidism.*—Primary overaction of the parathyroid glands is usually due to an adenoma, which may occasionally be present in an ectopic gland. Parathyroid tumours are almost always benign, but very rarely primary hyperparathyroidism is due to parathyroid carcinoma. Occasionally it results from diffuse hyperplasia of the glands.

Primary hyperparathyroidism presents most commonly with signs and symptoms associated with high ionised calcium levels, such as nausea, anorexia and abdominal pain, or with renal changes (calculi, polyuria or even renal failure). In many of these cases bone changes are absent and the plasma alkaline phosphatase concentration is normal or only slightly elevated. Patients presenting with overt bony lesions, on the other hand, often show no renal changes. It may be that when conditions in the urine are favourable for calcium precipitation (for instance an alkaline pH), or when a patient complains of symptoms early, the disease is usually diagnosed before bony changes occur: if these warning signs are absent the patient presents with the overt picture of bone disease. However, it should be pointed out that some of the renal cases have not developed bone disease after remaining untreated for many years: the reason for the two types of presentation is not fully understood.

Occasionally, as with hypercalcaemia due to any cause, the patient is admitted as an emergency with abdominal pain, vomiting and constipation due to severe hypercalcaemia.

*Tertiary hyperparathyroidism.*—Tertiary hyperparathyroidism is the name given to the condition in which the parathyroid has been under prolonged feed-back stimulation by low ionised calcium levels, and the resulting hypersecretion has become autonomous. Except for the history of previous hypocalcaemia, the disease is identical with primary hyperparathyroidism. It is usually diagnosed when the cause of the original hypocalcaemia is removed (e.g. by renal transplantation or correction of calcium or vitamin D deficiency). The typical biochemical findings of excess PTH are superimposed on those of osteomalacia, and because the condition is of long duration the plasma alkaline phosphatase level is usually very high. In some cases calcium levels later return to normal.

*Ectopic production of parathyroid hormone.*—Sometimes malignant tumours of non-endocrine tissues manufacture peptides foreign to them: PTH can be one of these. The production of hormones at these sites is not subject to feed-back control, and the initial findings are those of inappropriate PTH excess. Because of the nature of the underlying disease the condition is rarely of long standing, and the bony lesions due to excess circulating PTH are not evident: the alkaline phosphatase, however, may be raised because of bony or hepatic metastases, or both.

The subject of ectopic hormone production is discussed more fully in Chapter XXI.

**Secondary hyperparathyroidism ("appropriate" secretion of parathyroid hormone).**—In secondary hyperparathyroidism, the parathyroids are stimulated to produce hormone in response to low plasma ionised calcium levels. This secretion is therefore "appropriate" in that it is necessary to restore ionised calcium levels to normal. If it is effective in doing so the stimulus to production is removed, and the glands are "switched off". In secondary hyperparathyroidism, therefore, plasma calcium concentration is low (if the increased production of hormone is inadequate to correct the hypocalcaemia) or normal: it is *never* high. Hypercalcaemia, or a plasma calcium level in the upper normal range, suggests either a diagnosis of primary hyperparathyroidism (and this may be a *cause* of renal failure), or that prolonged calcium deficiency has led to tertiary hyperparathyroidism.

If renal function is normal the high PTH levels cause phosphaturia with hypophosphataemia. In all cases of secondary hyperparathyroidism, except that due to renal failure, both the plasma calcium and phosphate levels tend to be low.

If the PTH feed-back mechanism is normal any factor tending to lower the ionised calcium concentration causes secondary hyperparathyroidism. In long-standing cases, if the bone cells can respond to PTH, decalcification of bone occurs, leading to *osteomalacia* in adults, or *rickets* in children (before fusion of the epiphyses). The findings are described on p. 241. In very prolonged or severe depletion the high concentration of PTH may fail to maintain plasma calcium levels and the effects of reduced ionised calcium concentration may be seen (p. 240).

*Secondary hyperparathyroidism with osteomalacia or rickets* occurs in associated chronic calcium and vitamin D deficiency. This may be due to:

1. Reduced dietary intake of vitamin D and calcium in *malnutrition*.

2. Reduced absorption of vitamin D in *steatorrhoea*.

3. Reduced absorption of calcium in subjects with *high phytate intakes*.

4. Deficient metabolism of vitamin D to 1:25 DHCC due to
   ? *renal disease*
   ? *anticonvulsant therapy*.

*Secondary hyperparathyroidism without osteomalacia or rickets* occurs in:

1. Acute calcium and vitamin D depletion.

2. Most cases of pseudohypoparathyroidism (very rare).

*Osteomalacia and rickets without secondary hyperparathyroidism* may rarely occur in phosphate depletion due to:

*Renal tubular disorders of phosphate reabsorption* (p. 248).

*Calcium and vitamin D deficiency.*—Vitamin D is essential for normal absorption of calcium from the intestinal tract and calcium depletion is more commonly the result of deficiency of intake of this factor than of calcium itself. The commonest cause of calcium and vitamin D deficiency in Britain is the malabsorption syndrome: dietary deficiency is more important in the world as a whole. In the presence of steatorrhoea, vitamin D, which is fat soluble, is poorly absorbed. In addition calcium combines with the excess of fatty acids in the intestine to form insoluble soaps. Some cases of calcium deficiency in chapatti-eating communities may be due to a *high phytate diet*. Patients on prolonged *anticonvulsant therapy* sometimes develop hypocalcaemia, and even osteomalacia, possibly due to interference with vitamin D metabolism in the liver (p. 237).

Malnutrition is rarely selective, and protein deficiency may lead to a reduction of the protein bound fraction of the calcium. The ionised calcium level may not, therefore, be as low as that of total calcium would indicate.

Patients in renal failure are relatively resistant to the action of vitamin D and the resultant malabsorption of calcium is undoubtedly an important factor in the development of hypocalcaemia. The discovery that vitamin D (cholecalciferol) is inactive until finally hydroxylated in the kidney may explain this "resistance". The previously puzzling fact that the onset of hypocalcaemia is very rapid in renal failure in spite of adequate hepatic stores of vitamin D might now be explained: if active vitamin is formed in the kidney as required, failure of this mechanism would produce rapid effects. By contrast, dietary deficiency of the vitamin, with normally functioning kidneys, would not be evident until all these stores had been utilised. Low plasma protein levels sometimes contribute to the reduction of total calcium. Tetany rarely occurs, possibly because of the accompanying acidosis.

The resulting low plasma ionised calcium concentration stimulates PTH production, with the consequences described above. However, in renal failure, there is hyperphosphataemia. In chronic cases a rising alkaline phosphatase level heralds the onset of the bone changes of osteomalacia or rickets.

*Pseudohypoparathyroidism.*—In this rare inborn error circulating levels of parathormone are high. However, because in most cases neither the kidney nor bone can respond to the hormone, the biochemical changes of hypoparathyroidism, rather than hyperparathyroidism, are present.

*Acute pancreatitis.*—Acute pancreatitis causes temporary hypocalcaemia because of the precipitation of calcium soaps (calcium salts of fatty acids) following the local release of excess fatty acids from fats by the liberated lipase. Stimulation of PTH production rapidly restores levels to normal. Although this disease rarely presents as a problem of differential diagnosis of hypocalcaemia, the finding may be of use in the diagnosis of acute pancreatitis after serum amylase levels have returned to normal (p. 277).

## Diseases Associated with Low Circulating Parathyroid Hormone Levels

A low level or absence of circulating PTH leads to the changes described on p. 236.

**Primary parathyroid hormone deficiency (hypoparathyroidism).**— Primary hypoparathyroidism may be due to primary atrophy of the parathyroid glands (probably of autoimmune origin—compare Hashimoto's disease, p. 171), or more commonly to surgical damage, either directly to the glands or to their blood supply during partial thyroidectomy (during total thyroidectomy removal of the glands is almost inevitable). Evidence for asymptomatic hypocalcaemia should always be sought after partial thyroidectomy because of the danger of cataracts (p. 240): early post-operative parathyroid insufficiency may recover, but a low calcium level persisting more than a few weeks should be treated. Even with highly skilled surgery there is a danger of parathyroid damage during this operation. Post-thyroidectomy hypocalcaemia may not always be due to parathyroid damage, but may be due to the rapid entry of calcium into depleted bone when the thyroid hormone levels fall to normal (see p. 239): it may be these cases that appear to recover from hypoparathyroidism.

The clinical symptoms are those due to a low ionised calcium concentration (p. 240).

**Secondary suppression of parathyroid hormone secretion.**—Just as hypocalcaemia from any cause leads to secondary hyperparathyroidism, so hypercalcaemia (unless due to inappropriate PTH secretion) suppresses PTH secretion, with the consequences described on p. 236.

The causes of hypercalcaemia with suppression of PTH secretion are:

1. Vitamin D excess.
2. Idiopathic hypercalcaemia of infancy.
3. Sarcoidosis.
4. Thyrotoxicosis.
5. Malignant disease of bone and myeloma (possibly).
6. Milk-alkali syndrome.

The *clinical picture* in these cases is due to the high ionised calcium level (p. 239). Decalcification of bone only occurs in this group in the hypercalcaemia of thyrotoxicosis and in sarcoidosis. The alkaline phosphatase level is usually normal in conditions in which bone is not involved.

*Hypercalcaemia due to vitamin D excess.*—Overdosage with vitamin D increases calcium absorption and may cause dangerous hypercalcaemia. PTH secretion is suppressed, and phosphate levels are usually normal. However, as vitamin D has a phosphaturic effect (p. 237) plasma phosphate levels may occasionally be in the low normal range.

Such overdosage may be caused by over-vigorous treatment of hypocalcaemia, and this therapy should always be controlled by frequent plasma calcium and alkaline phosphatase estimations.

Occasionally patients overdose themselves with vitamin D. In any obscure case of hypercalcaemia a drug history should be taken.

The hypercalcaemia associated with two diseases is thought to be due to oversensitivity to the action of vitamin D. These diseases are:

1. Idiopathic hypercalcaemia of infancy.
2. Hypercalcaemia of sarcoidosis.

*Idiopathic hypercalcaemia of infancy.*—In these days of milk and vitamin supplements rickets is a rare disease in Great Britain. By contrast, the number of infants presenting with hypercalcaemia of obscure origin increased during the 1950s. The incidence of this syndrome has declined since the recommended dosage of vitamin D for infants was reduced in 1957, although the correlation between these two variables is uncertain. This is still a rare disease. An even rarer and more severe form—an inborn error—is associated with, amongst other stigmata, mental deficiency.

*Hypercalcaemia of sarcoidosis.*—Hypercalcaemia as a complication of sarcoidosis may also be due to vitamin D sensitivity. *Chronic beryllium poisoning* produces a granulomatous reaction very similar to that of sarcoidosis, and may also be associated with hypercalcaemia: beryllium is used in the manufacture of fluorescent lamps and in several other industrial processes.

Other diseases associated with hypercalcaemia are:

*Hypercalcaemia of thyrotoxicosis.*—Thyroid hormone probably has a direct action on bone (p. 239). In rare cases of very severe thyrotoxicosis significant hypercalcaemia may occur. If the hypercalcaemia fails to respond when hyperthyroidism is controlled, the possibility of co-existent hyperparathyroidism should be borne in mind, since some patients show multiple endocrine abnormalities.

*Malignant disease of bone.*—The hypercalcaemia due to ectopic PTH production by malignant tumours has been described on p. 243. A different syndrome has been reported in cases with multiple bony metastases, or with myeloma: in these the parallel rise of phosphate is said to indicate that the hypercalcaemia is caused by direct breakdown of bone by the local action of malignant deposits. Critical examination of the findings in most of these cases shows a rise of plasma urea accompanying that of phosphate, and such findings are compatible with the action of excess PTH with renal failure. Moreover, there is little correlation between the incidence of hypercalcaemia and the rate of extension of the bony lesions. Probably most, if not all, such cases are due to ectopic PTH production.

The raised alkaline phosphatase level associated with bony secondary deposits is due to the resultant stimulation of local osteoblasts (p. 241). This osteoblastic reaction does not occur in myelomatosis, in which the bone alkaline phosphatase level is therefore normal. This fact may be a useful pointer to the diagnosis of myeloma in the presence of extensive bony deposits of unknown origin.

The paraproteins of myeloma do not bind calcium to any significant extent, and are unlikely to account for hypercalcaemia in these cases.

*Milk-alkali syndrome.*—Hypercalcaemia has been reported due to massive intake of calcium in milk, when alkalis are also being taken, during treatment of peptic ulcer. The hypercalcaemia is probably the consequence of the very high calcium intake, and the alkalis, by causing glomerular damage, may temporarily aggravate the hypercalcaemia by delaying excretion of the excess ion.

This is an exceedingly rare cause of hypercalcaemia nowadays. Most modern antacids are acid adsorbents rather than alkalis, and do not cause alkalosis. The syndrome should not be diagnosed in the absence of alkalosis, and until other causes have been excluded. In true milk-alkali syndrome the calcium level falls rapidly when therapy is stopped.

There is a well-documented association between peptic ulceration and parathyroid adenomata: the association of hypercalcaemia with medication for dyspepsia should be assumed to be due to primary hyperparathyroidism until proved otherwise.

### Diseases of Bone Not Affecting Plasma Calcium Levels

Diseases producing biochemical abnormalities, or important in the differential diagnosis from those already mentioned, are:

*Osteoporosis.*—In this disease the primary lesion is a reduction in the mass of bone matrix with a secondary loss of calcium. Clinically it may resemble osteomalacia, but normal plasma calcium, phosphate and alkaline phosphatase concentrations (especially the latter) are more in favour of osteoporosis.

*Paget's disease of bone.*—Calcium and phosphate levels are rarely affected in Paget's disease. The alkaline phosphatase level is typically very high.

*Renal tubular disorders of phosphate reabsorption.*—This group of diseases comprises a number of inborn errors of renal tubular function in which *phosphate is not reabsorbed normally* from the glomerular filtrate. These usually present as cases of *rickets* or *osteomalacia*, but, unlike the usual form of this disease, response to vitamin D therapy is poor: very large doses may be needed to produce an effect. The syndrome has, therefore, been called *"resistant rickets"*. The defect in phosphate reabsorption may be an isolated one, as in *familial hypophosphataemia*, or part of a more general reabsorption defect as seen in the *Fanconi syndrome* (p. 14). In these cases phosphate deficiency is the primary lesion, and failure to calcify bone may be due to this. Plasma levels of phosphate are usually very low and fail to rise when vitamin D is given: there is a relative phosphaturia. The osteomalacia is reflected in the high plasma alkaline phosphatase levels. Plasma calcium concentration is usually normal, and in this respect the findings differ from those of classical osteomalacia; because of the normocalcaemia there is rarely evidence in the bone of hyperparathyroidism.

## DIFFERENTIAL DIAGNOSIS

In the preceding sections diseases of calcium metabolism have been discussed according to their aetiology, in an attempt to explain biochemical and clinical findings. Clinically they more often present as a problem of differential diagnosis of the causes of an abnormal plasma calcium concentration.

### THE DIAGNOSIS OF PRIMARY HYPERPARATHYROIDISM AND TESTS USED IN THE DIFFERENTIAL DIAGNOSIS OF HYPERCALCAEMIA

Primary hyperparathyroidism must, in its early stages, exist with plasma calcium levels within the normal range, but diagnosis at this time is almost impossible. A plasma-calcium concentration in the upper normal range, with that of phosphate in the low normal range in a patient with recurrent, calcium containing renal calculi is suggestive of primary hyperparathyroidism: in such a case plasma calcium levels should be estimated at three-monthly intervals and if primary hyperparathyroidism is present, these will eventually become unequivocally raised.

In the presence of hypercalcaemia the causes must be differentiated. These fall into three groups:

Raised protein bound, with normal ionised calcium.

Raised ionised calcium due to inappropriate PTH excess.

Raised ionised calcium due to other causes, and associated with appropriately low PTH levels.

#### Plasma Levels

**Raised protein bound calcium.**—Mild hypercalcaemia may be due to a raised albumin concentration, with no change in that of ionised calcium (p. 235). The only *in vivo* cause of this is dehydration with haemoconcentration (p. 41), and in such cases the clinical state of the patient, and the raised plasma *protein* and *haemoglobin* levels, should provide the clue. Such hypercalcaemia will disappear when the patient is rehydrated.

A more common cause of a raised protein bound calcium is the artefactual one caused by prolonged stasis during venesection (p. 461). When hypercalcaemia is found, especially if mild, the analysis should be repeated on a specimen taken without stasis.

In both the *in vivo* and artefactual causes of a raised protein bound calcium, plasma phosphate tends to rise because of leakage from cells (compare potassium).

Calcium levels, especially serial ones, should never be interpreted without an accompanying protein level. If the cause of a change in protein concentration is uncertain, the serum albumin should be estimated, as this is the calcium binding fraction. In most cases, however, total protein estimation is adequate.

**Raised ionised calcium.**—Hypercalcaemia found in a specimen taken without stasis from a well-hydrated patient can be assumed to be due to a rise in the physiologically active ionised calcium. It should be noted that a normal total plasma calcium in the presence of significant hypoproteinaemia also suggests an increase in the ionised fraction.

An accompanying low plasma *phosphate* is suggestive of inappropriate PTH secretion due to primary or tertiary hyperparathyroidism, or to ectopic hormone production. In the presence of a high plasma urea level the presence of hyperphosphataemia does not exclude these causes, and because renal disease also tends to lower calcium levels the plasma urea, as well as phosphate and protein, should always be estimated when calcium metabolism is being investigated. If this estimation is normal an increase of both calcium and phosphate suggests that these three causes are unlikely.

In primary hyperparathyroidism the plasma *alkaline phosphatase* concentrations rise significantly only at a relatively late stage of the disease. A finding of a very high level with no clinical evidence of severe bone disease makes the diagnosis improbable. In the hypercalcaemia of malignant disease the level of this enzyme may rise because of bony or hepatic metastases, and other signs of these should be sought. The alkaline phosphatase concentration is normal in myelomatosis (p. 248) unless there is hepatic infiltration by the malignant plasma cells: in such a rare case the enzyme can be shown to be of hepatic origin (p. 350). In sarcoidosis, as in malignant disease, the plasma alkaline phosphatase level may increase because of bony or hepatic infiltration. In the rare hypercalcaemia of severe thyrotoxicosis there may be some elevation of plasma concentrations of the enzyme, but in all other uncomplicated causes of hypercalcaemia it is normal.

A marked increase in plasma *protein* level, in the absence of clinical dehydration or of stasis in taking the blood, suggests myelomatosis as a cause, and this will be evident in the electrophoretic pattern. If this cause is considered bone marrow aspiration must be performed, and the presence of Bence Jones protein sought (p. 334), even in the presence of normal serum protein fractions.

## Steroid Suppression Test

This test is by far the most useful one in the differential diagnosis of hypercalcaemia of obscure origin. In most cases it differentiates primary or tertiary hyperparathyroidism from any other cause. Large doses of hydrocortisone (or cortisone) result in a fall of high plasma calcium levels to within the normal range in almost all cases *except primary* (or tertiary) *hyperparathyroidism.* Exceptions do occasionally occur, but in most of them the clinical diagnosis is clear and the test unnecessary (for instance in very advanced malignant disease of bone or very severe hyperthyroidism). Details of the test are given in the Appendix (p. 260).

The reason for this difference in reaction to steroids is not clear. It is particularly interesting that the calcium level in many cases of ectopic parathormone production does suppress, whereas it fails to do so in those in which the hormone is produced in the parathyroid glands.

## Urinary Calcium

This frequently requested estimation is of little diagnostic value in the differential diagnosis of hypercalcaemia. In the presence of normal renal function, hypercalcaemia from any cause, by increasing the load on the glomerulus, causes hypercalcuria. Causes of a high urinary calcium excretion without hypercalcaemia are the so-called "*idiopathic hyper-calcuria*" and some cases of *osteoporosis* in which calcium cannot be deposited in normal amounts in bone because of a deficient matrix.

Calcium excretion may be normal or low if renal function is diminished, even in the presence of hypercalcaemia (whether due to hyperparathyroidism or to other causes).

## Plasma Parathyroid Hormone Levels

Parathyroid hormone levels may be measured by radioimmunoassay, although the estimation is not routinely available. The specificity of the assay is still uncertain, and not all cases of primary hyperparathyroidism can be shown to have raised PTH levels, even when hypercalcaemia is present. In our experience careful attention to the plasma calcium levels, together with the clinical history and the other findings discussed in this chapter, still provide the most useful information.

TABLE XXI

DIFFERENTIAL DIAGNOSIS OF HYPERCALCAEMIA

| Diagnosis | Plasma | | | Steroid suppression | Comments |
|---|---|---|---|---|---|
| | Phosphate | Proteins | Alk. phos. | | |
| *Due to Raised Ionised Calcium* | | | | | |
| *Group 1. Inappropriate Parathyroid Hormone* | | | | | |
| Primary or tertiary hyperparathyroidism | N or ↓ | N | N or ↑ | No | ⎫ Calcium and phosphate changes masked in uraemia. |
| Malignancy with ectopic hormone production | N or ↓ | N (↑ myeloma) | N or ↑ (N in myeloma) | Yes (usually) | ⎭ |
| *Group 2. Appropriately low PTH* | | | | | |
| *Vitamin D* | | | | | |
| Vitamin D overdosage | Variable | N | N | Yes | |
| Sarcoidosis | Variable | γ glob. ↑ | N or ↑ | Yes | |
| Idiopathic hypercalcaemia of infancy | Variable | N | N | Yes | |
| *Milk-Alkali Syndrome* | N or ↑ | N | N | Yes | Rare |
| *Thyrotoxicosis* | N or ↑ | N | N or ↑ | Yes (usually) | Very rare. $HCO_3$ ↑ Severe thyrotoxicosis. Thyroid function tests abnormal. |
| *Due to Raised Protein Bound Calcium* | | | | | |
| Dehydration | N or ↑ | ↑ | N | | Clinical signs of dehydration. Urea ↑ Corrected by rehydration. |
| Artefactual (excess stasis) | N or ↑ | ↑ | N | | Specimen taken without stasis gives normal values. |

Hypercalcaemia due to raised protein bound calcium should *NOT* be treated.

### Tests Used in the Differential Diagnosis of Hypocalcaemia

As in the case of hypercalcaemia, the causes of hypocalcaemia fall into three groups:

Reduced protein bound, with normal ionised calcium.

Reduced ionised calcium due to primary PTH deficiency.

Reduced ionised calcium due to other causes and associated with appropriately high PTH levels.

### Plasma Levels

**Reduced protein bound calcium.**—As in the case of hypercalcaemia, hypocalcaemia due to a low protein bound fraction must be distinguished from that due to a low ionised fraction.

Just as protein bound calcium concentrations may be raised in haemoconcentration due to dehydration, so they may be reduced in overhydration. In such cases the protein and calcium levels will return to normal as hydration is corrected.

In any condition associated with significant hypoalbuminaemia (p. 320) the protein bound calcium may be reduced. In malnutrition, whether due to malabsorption or to a deficient diet, there may be an accompanying reduction of ionised calcium due to associated vitamin D and calcium deficiency. In such cases treatment should not aim to restore the total calcium level to normal, but to maintain that of alkaline phosphatase within the normal range (i.e. to prevent osteomalacia).

**Reduced ionised calcium levels.**—If plasma albumin levels are normal, or only slightly reduced, a significant reduction in total calcium concentration can be assumed to be due to that of ionised calcium.

Accompanying high *phosphate* levels suggest either reduced circulating PTH, or the failure of bone and kidney to respond to it in the very rare pseudohypoparathyroidism. Hypophosphataemia with hypocalcaemia is associated with calcium and vitamin D deficiency.

High plasma *alkaline phosphatase* levels also suggest that calcium deficiency is the primary abnormality, which has led to increased PTH secretion and resultant decalcification of bone (secondary hyperparathyroidism). This is most commonly due to malabsorption or to dietary deficiency. A dietary history should be taken and tests for malabsorption (p. 285) performed if indicated by the bowel history.

A high plasma *urea* suggests that the hypocalcaemia is due to renal failure, and caution should be exercised in basing treatment on plasma calcium levels alone.

### Plasma Parathyroid Hormone Levels

In most cases estimation of PTH levels is unnecessary: the present sensitivity of the method is such that low levels are less significant than

TABLE XXII

DIFFERENTIAL DIAGNOSIS OF HYPOCALCAEMIA

| Diagnosis | Plasma | | | | Comments |
|---|---|---|---|---|---|
| | Phosphate | Proteins | Urea | Alk. phos. | |
| *Due to Low Ionised Calcium* | | | | | |
| 1. Hypoparathyroidism | ↑ | N | N | N | |
| 2. Calcium and vitamin D deficiency | ↓ | N or ↓ | N | ↑ | May be accompanied by low protein bound calcium. |
| 3. Renal failure | ↑ | N or ↓ | ↑ | N or ↑ | Should be treated with caution in absence of bone disease. |
| 4. Pseudohypoparathyroidism | ↑ | N | N | N | Very rare. |
| *Due to Low Protein Bound Calcium* | | | | | |
| Overhydration | N | ↓ | N or ↓ | N | Responds to fluid restriction. |
| Hypoalbuminaemia | N | ↓ | N or ↓ | N | |

Hypocalcaemia due to lowered protein bound calcium should *NOT* be treated.

high ones. In the very rare cases in which pseudohypoparathyroidism is suspected this estimation is useful. Very high levels of PTH together with hypocalcaemia are incompatible with the diagnosis of primary hypoparathyroidism. If secondary hyperparathyroidism has been excluded and if hyperphosphataemia is present, they are suggestive of end organ unresponsiveness (p. 245).

## BIOCHEMICAL BASIS OF TREATMENT

### Hypercalcaemia

Mild hypercalcaemia.—If raised calcium levels of less than about 3·75 mmol/l (15 mg/dl) are present in a patient without serious symptoms of hypercalcaemia, there is no *immediate* need for urgent therapy. However, treatment should be instituted as soon as a diagnosis is made because of the danger of renal damage.

If possible the cause should be removed (e.g. a parathyroid adenoma, a primary malignant lesion, hyperthyroidism). If this is not possible hydrocortisone will reduce the levels within a day or two in most cases not due to primary (or tertiary) hyperparathyroidism (see "Steroid Suppression Test", p. 251). It may sometimes be relatively ineffective in malignant hypercalcaemia.

Oral sodium phosphate will tend to precipitate calcium phosphate in the intestine, preventing reabsorption of secreted calcium and thus removing calcium from the body. This treatment may be preferable to steroid therapy in many cases. It may be used, in conjunction with steriods, to treat intractable hypercalcaemia of late malignancy. The effect of oral phosphate, like that of steroids, is only manifested after 24 hours.

Severe hypercalcaemia.—If plasma calcium levels exceed about 3·75 mmol/l (15 mg/dl) treatment is indicated as a matter of urgency, because of the danger of cardiac arrest.

Most measures available for lowering plasma calcium acutely (i.e. within a few hours) depend, partially at least, on precipitation of insoluble calcium salts. Since some of this precipitation may occur in the kidney, the treatment carries a slight risk of initiating or aggravating renal failure, and this risk should always be weighed against the danger of cardiac arrest. It is small in subjects with normal or only slightly raised plasma urea levels: although many cases show a transient rise of plasma urea at the start of treatment, this usually subsides rapidly.

Solutions which have been used for intravenous administration are either a mixture of sodium and potassium phosphates (see Appendix, p. 260) or sodium sulphate. The latter, though somewhat less effective, increases calcium loss in the urine with less theoretical danger of ectopic calcification: an effective dose of sulphate must be dissolved in relatively large volumes of water, and it should not be used in patients in whom there is a danger of fluid overloading. In practice intravenous phosphate is usually satisfactory (Appendix, p. 260).

Ethylene diamine tetraacetate (EDTA, sequestrene), a substance which chelates calcium, has been used, but has been shown to be nephrotoxic.

As with most other extracellular constituents, rapid changes in calcium level may be dangerous because time is not allowed for equilibration across cell membranes (see e.g. urea, p. 23 and sodium, p. 43). The aim of emergency treatment should be to lower calcium temporarily to safe levels, while initiating treatment for mild hypercalcaemia. A too rapid reduction of calcium concentration may induce tetany, or, more seriously, hypotension, even though the calcium is within or above normal levels. (Tetany in the presence of normocalcaemia may also occur during the rapid fall of serum calcium concentration after removal of a parathyroid adenoma.) A slow reduction of calcium level also reduces the risk of renal calcification.

### Hypocalcaemia

**Asymptomatic hypocalcaemia.**—Hypocalcaemia, whatever the cause, if asymptomatic or accompanied by only mild clinical symptoms, is usually treated with large doses of vitamin D by mouth. It is difficult to give enough oral calcium, by itself, to make a significant difference to plasma calcium levels and vitamin D, by increasing absorption of calcium, is usually adequate without calcium supplementation.

The hypocalcaemia of renal failure should be treated with caution in the absence of signs of osteomalacia (raised alkaline phosphatase, etc.), because of the danger of ectopic calcification in the presence of hyper-phosphataemia. Hypocalcaemia associated with low protein levels should not be treated.

**Hypocalcaemia with severe tetany.**—In the presence of severe tetany hypocalcaemia should be treated, as an emergency, with intravenous calcium (usually as the gluconate).

## MAGNESIUM METABOLISM

Magnesium is present with calcium in bone salts, and tends to move in and out of bone with calcium. It is also present in all cells of the body in much higher concentrations than those in the extracellular fluid, and therefore tends to enter and leave cells under the same conditions as do potassium and phosphate.

Magnesium can be lost in large quantities in the faeces in diarrhoea.

## PLASMA MAGNESIUM AND ITS CONTROL

Probably some of the plasma magnesium, like calcium, is protein bound. However, less is known about the importance of this than in the case of calcium.

The mechanism of control of magnesium levels is poorly understood, but may involve the action of aldosterone and PTH.

## CLINICAL EFFECT OF ABNORMAL PLASMA MAGNESIUM LEVELS

### Hypomagnesaemia

This causes symptoms very similar to those of hypocalcaemia. If a patient with tetany has normal calcium, protein and bicarbonate (or, more accurately, blood pH) levels, blood should be taken for magnesium estimation. If there is good reason to suspect magnesium deficiency (e.g. in the presence of severe diarrhoea), and if it is not possible to have the result of the estimation reasonably quickly, intravenous magnesium should be administered as a therapeutic test (see Appendix, p. 262). Less severe magnesium deficiency should be treated orally (Appendix, p. 262).

### Hypermagnesaemia

This causes muscular hypotonia, but as this condition is rarely seen in isolation, symptoms are difficult to distinguish from those of co-existent abnormalities (e.g. hypercalcaemia).

## CAUSES OF ABNORMAL PLASMA MAGNESIUM LEVELS

### Hypomagnesaemia

1. **Excessive loss of magnesium.**—Excessive loss of magnesium occurs in severe diarrhoea, and this is by far the most important cause of a clinical disturbance of magnesium metabolism requiring treatment.

2. **Hypomagnesaemia accompanied by hypocalcaemia.**—Magnesium tends to move in and out of bones in association with calcium, and hypocalcaemia is often accompanied by hypomagnesaemia. This may occur in hypoparathyroidism, or during the fall of calcium after the removal of a parathyroid adenoma, especially if severe bone disease is present.

3. **Hypomagnesaemia accompanied by hypokalaemia.**—Since magnesium moves in and out of cells with potassium, hypomagnesaemia tends to occur in association with hypokalaemia. These conditions include diuretic therapy and primary aldosteronism (p. 51). Such hypomagnesaemia is rarely of clinical importance.

## Hypermagnesaemia

The commonest cause of hypermagnesaemia is probably renal failure (when plasma potassium is also high). It rarely, if ever, requires treatment on its own, and it responds to measures to treat the underlying condition (e.g. haemodialysis). Magnesium salts should never be administered in renal failure.

## SUMMARY

### Calcium Metabolism

1. About half the calcium in plasma is bound to protein, and half is in the ionised form.

2. The ionised calcium is the physiologically important fraction and calcium levels should be interpreted together with protein levels.

3. Plasma calcium levels are controlled by parathyroid hormone. Parathyroid hormone secretion is increased if ionised calcium concentrations are reduced.

4. Parathyroid hormone acts:

(a) On bone, releasing calcium and phosphate into the plasma, which, if prolonged, increases osteoblastic activity and increases plasma alkaline phosphatase levels.

(b) On kidneys, causing phosphaturia, which lowers the plasma phosphate.

5. Symptoms and findings in diseases of calcium metabolism can be related to circulating levels of ionised calcium, to levels of parathyroid hormone, and to renal function.

6. Excessive "inappropriate" levels of parathyroid hormone are present in primary parathyroid disease, or if the hormone is produced at ectopic sites. In these circumstances plasma calcium is high.

7. Increased "appropriate" levels of parathyroid hormone are present whenever plasma calcium levels fall.

8. Parathyroid hormone levels are low in hypoparathyroidism (associated with a low serum calcium), or in the presence of hypercalcaemia other than that of inappropriate parathyroid hormone production. The causes and differential diagnosis of hypercalcaemia and hypocalcaemia are summarised in Tables XXI and XXII.

### Magnesium Metabolism

1. Hypomagnesaemia may cause tetany in the absence of hypocalcaemia.

2. The commonest cause of significant hypomagnesaemia is severe diarrhoea.

3. Magnesium levels tend to follow those of calcium (in and out of bone) and potassium (in and out of cells).

## FURTHER READING

FOURMAN, P., and ROYER, P. (1968). *Calcium Metabolism and the Bone*, 2nd edit. Oxford: Blackwell Scientific Publications.

WILLS, M. R. (1971). Calcium homeostasis in health and disease. In *Recent Advances in Medicine*, **16**, 57. Edinburgh: Churchill-Livingstone.

WILLS, M. R. (1974). Hypercalcaemia. *Brit. J. hosp. Med.*, **11** 279.

STAMP, T. C. B. (1973). Vitamin D metabolism—recent advances. *Arch. Dis. Childh.*, **48**, 2.

KODICEK, E. (1974). The story of vitamin D. *Lancet*, **1**, 325.

# APPENDIX

## DIAGNOSIS

<small>STEROID SUPPRESSION TEST IN DIFFERENTIAL DIAGNOSIS OF HYPERCALCAEMIA</small>
(Dent, C. E., and Watson, L. *Lancet*, 1968, **2**, 662)

All specimens for calcium estimation should be taken without venous stasis (p. 250).

### Procedure

1. At least two specimens are taken on two different days before the test starts for estimation of plasma calcium, protein and urea concentrations.

2. The patient takes 120 mg of oral hydrocortisone a day in divided doses (40 mg, 8-hourly), for 10 days.

3. Blood is taken for estimation of plasma calcium, protein and urea concentration on the 5th, 8th and 10th day.

4. After the 10th day the dose of hydrocortisone is gradually reduced, as usual after steroid therapy.

### Calculations

During administration of steroids there may be significant fluid retention. This will dilute the protein bound calcium and may lead to an apparent fall of calcium. To correct this it is desirable to "correct" total calcium results for changes in protein concentration, and a "standard" total protein of 72 g/l is used in the calculation.

For every 3·7 g/l that the protein concentration is *below* 72 g/l, 0·06 mmol/l (0·25 mg/dl) is *added* to the calcium level.

For every 3·7 g/l that the protein concentration is *above* 72 g/l, 0·06 mmol/l (0·25 mg/dl) is *subtracted* from the calcium level.

This correction is a rough one, and is inaccurate in the presence of very abnormal plasma protein concentrations.

### Interpretation

Failure of plasma calcium concentration to fall to within the normal range by the end of the test is strongly suggestive of *primary hyperparathyroidism*. For exceptions see p. 251.

## TREATMENT

<small>HYPERCALCAEMIA</small>

### Emergency Treatment of Hypercalcaemia

Solution for intravenous infusion contains

$Na_2HPO_4$ (anhydrous)—11·50 g
$KH_2PO_4$ (anhydrous)— 2·58 g $\Big\}$ made up to 1 litre with water.

500 ml of this should be infused over 4–6 hours. This 500 ml will contain a total of 81 mmol of sodium and 9·5 mmol of potassium.

## Long Term Oral Phosphate Treatment for Hypercalcaemia

Oral phosphate is given as the disodium or dipotassium salt. The choice depends on the serum potassium level.

1. The solution should contain 32 mmol in 100 ml.

This is        46 g/litre of $Na_2HPO_4$ (anhydrous)
or           56 g/litre of $K_2HPO_4$ (anhydrous)

The dose is 100–300 ml per day in divided doses. 100 ml contains approximately 65 mmol of sodium or potassium respectively.

2. Phosphate Sandoz Effervescent tablets contain:

| | | |
|---|---|---|
| Phosphorus | 16 mmol (500 mg) | ⎤ |
| Sodium | 21 mmol | ⎬ per tablet |
| Potassium | 3 mmol | ⎦ |

The dose is 1–6 tablets daily.

3. Alternatively the phosphate can be given as sodium cellulose phosphate (Whatman) 5 g (the contents of one sachet), three times a day.

*Note* that hypokalaemia is a common accompaniment of hypercalcaemia. When this is present the phosphate preparation of choice is the $K_2HPO_4$.

HYPOCALCAEMIA

## Emergency Treatment of Hypocalcaemia

Calcium Gluconate Injection (B.P.)—0·23 mmol (9 mg) calcium/ml.
Dose—10 ml intravenously in the first instance.

## Long Term Treatment of Hypocalcaemia

*Vitamin D Therapy*

**Warning.**—There may be several day's lag in response to either starting or stopping therapy with vitamin D: plasma calcium levels may continue to rise for weeks after stopping therapy. Treatment should be carried out with caution, using intelligent anticipation based on laboratory assessment of plasma calcium and, in osteomalacia, alkaline phosphatase levels. *Patients on maintenance doses should be seen at regular intervals, because requirements may change, with the danger of the development of hypercalcaemia.* Calcium levels must be assessed with those of plasma protein.

**Osteomalacia with raised alkaline phosphatase levels.**—1. *Initially*—500 000 IU (12·50 mg) daily orally or I.M. Rarely 750 000 to 1 000 000 IU (18·75–25·00 mg) may be needed.

2. When the calcium level reaches about 1·75 mmol/l (7·0 mg/1 dl), and the alkaline phosphatase is near normal, reduce to 100 000 to 250 000 (2·50–6·25 mg) daily. At this stage bone calcification is nearing normal, and hypercalcaemia may develop rapidly if high dosage is continued.

3. *Maintenance doses* vary between 25 000 and 100 000 IU (0·63–2·50 mg),

and are determined by trial and error. The aim should be to keep plasma calcium, phosphate, and alkaline phosphatase levels normal.

**Hypocalcaemia of hypoparathyroidism with normal alkaline phosphatase levels.**—The lower dosage of 100 000 IU (2·50 mg) daily should be used from the outset, monitoring being based on plasma calcium levels; again, the exact dose is found by trial and error.

*Oral Calcium Tablets*

Calcium Gluconate (B.P.C.) (600 mg) = 1·35 mmol (54 mg) of calcium per tablet.

Calcium Gluconate Effervescent (B.P.C.) (1000 mg) = 2·25 mmol (90 mg) of calcium per tablet.

"Sandocal" Effervescent (4·5 g calcium gluconate) = 10 mmol (400 mg) of calcium per tablet.

Calcium Sandoz Chocolate (1·5 g calcium gluconate) = 3·38 mmol (135 mg) of calcium per tablet.

Calcium Lactate (B.P.) (300 mg) = 1.4 mmol (55 mg) of calcium per tablet.

Calcium Lactate (B.P.) (600 mg) = 2.8 mmol (110 mg) of calcium per tablet.

## MAGNESIUM

**Emergency Treatment of Magnesium Deficiency**

Magnesium Chloride ($MgCl_2.6H_2O$)—20 g/100 ml.

This contains 1 mmol of magnesium/ml and may be added to other intravenous fluid. If renal function is normal, up to 40 mmol (40 ml) may be infused in 24 hours.

**Oral Magnesium Therapy**

Magnesium Chloride ($MgCl_2.6H_2O$)—20 g/100 ml.

The dose is 5–10 ml, q.d.s. Each 5 ml contains 5 mmol of magnesium. Note, however, that magnesium is poorly absorbed.

# Chapter XI

# INTESTINAL ABSORPTION: PANCREATIC AND GASTRIC FUNCTION

THE most important function of the gastro-intestinal tract is the digestion and absorption of nutrients. Many substances which are absorbed from the food also enter the intestinal tract, so that net absorption may be less than true absorption (insorption): for instance, dietary fat is nearly completely absorbed under physiological circumstances, and almost all that found in normal faeces is derived from intestinal cells—a fact of importance in interpretation of faecal fat values (p. 286).

Digestion of larger molecules involves the action of intestinal enzymes: if these are to act under optimal conditions the nutrient molecules must be dispersed as much as possible, firstly by mechanical action such as chewing, and secondly by mixture of the food with fluid. Enzymes act best in the presence of certain electrolytes and at certain pH values: the fluid secreted contains these ions and varying amounts of hydrogen ion. Before the smaller water-soluble molecules can pass through the intestinal cells they must be in solution and some of these, too, require fluid at the correct pH. The very large amounts of electrolyte and water secreted into the gastro-intestinal tract have already been mentioned, and it has been pointed out that normally about 99 per cent of the sodium and water are reabsorbed (p. 31). In this context the gastro-intestinal tract acts in much the same way as the kidney (if we consider that gastro-intestinal secretions are in some ways analogous with the glomerular filtrate): water, sodium and potassium are reabsorbed throughout the small intestine, as they are in the proximal renal tubule. Final adjustment is made in the colon where, for instance, aldosterone stimulates sodium reabsorption and potassium secretion as it does in the distal renal tubule, and where final water reabsorption occurs. Disturbances of water and electrolyte metabolism and of hydrogen ion homeostasis are therefore common in diarrhoea due to extensive small intestinal or colonic disease, or when there is direct loss of fluid and electrolyte from the upper intestinal tract through vomiting or fistulae. Such disturbances also occur if absorption from the upper intestinal cells is so grossly impaired that the amounts of fluid and electrolyte entering distal parts exceed the reabsorptive capacity: this effect is similar to that of an increased glomerular filtration rate and of osmotic diuretics on the volume and composition of urine (p. 38 and p. 9). These disturbances of electrolyte, water and hydrogen ion homeostasis have been discussed in Chapters II, III and IV.

In this chapter we are concerned with disturbances of the absorptive mechanisms, either due to abnormality of absorptive cells of the small intestine, or to failure of normal digestion of food. Unless malabsorption is gross, causing severe intestinal hurry, water, electrolyte and hydrogen ion disturbances are relatively unimportant in malabsorption syndromes. The effects are largely those of disturbed nutrition, since nutrient cannot pass through the intestinal cells.

## NORMAL DIGESTION AND CONVERSION OF NUTRIENT TO AN ABSORBABLE FORM

Complex molecules such as those of protein, polysaccharide and fat are usually broken down by digestive enzymes. This process starts in the mouth, where food is mechanically broken down by chewing and is mixed with saliva containing *amylase*. In the stomach further fluid is added and the low pH initiates protein digestion by *pepsin*. The stomach also secretes *intrinsic factor* necessary for absorption of vitamin $B_{12}$. However, quantitatively by far the most important digestion takes place in the duodenum and upper jejunum, where large volumes of alkaline fluid are added to the already liquid food. Pancreatic enzymes in this fluid convert protein to amino acids and small peptides, polysaccharides to mono- and disaccharides and oligosaccharides (consisting of a small number of monosaccharide units), and fat to monoglycerides and fatty acids. From the pathological point of view, severe generalised malabsorption due to failure of *digestion* is most commonly due to pancreatic disease.

## NORMAL ABSORPTION

Normal absorption depends on:

1. The integrity and normal surface area of absorptive cells.

2. The presence of the substance to be absorbed in an absorbable form (and therefore on normal digestion).

3. A normal ratio of speed of absorption to speed of passage of contents through the intestinal tract.

The *absorptive area* of the intestine is normally very large. Macroscopically the mucosa forms *folds*, increasing the area considerably. Microscopically, these folds are covered with *villi*, lined with absorptive cells: this further increases the area about eight-fold.

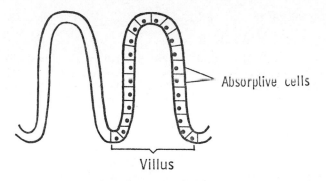

Villus

If these villi are flattened, as they are in, amongst other conditions, gluten sensitive enteropathy, the absorptive area is much reduced.

Each intestinal absorptive cell has on its surface a large number of minute projections (*microvilli*) detectable with the electron microscope, further increasing absorptive area about 20-fold.

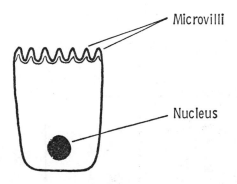

Minute spaces exist between the microvilli (*microvillous spaces*).

To be absorbable, the substances must be in the form of relatively *small molecules*, such as result from normal digestion. Vitamin $B_{12}$ absorption requires the formation of a complex with intrinsic factor, which can attach to the intestinal cells and bring the substance in a spatially advantageous position for absorption. The method of absorption depends on whether the molecule is *water soluble* or *lipid soluble*. Lipid soluble nutrients frequently share the mechanisms for fat absorption. Absorption may be *active*, in which case it can occur against a physicochemical gradient, or *passive* when it passes along physicochemical gradients (for a more complete definition of these processes see p. 2).

LIPID ABSORPTION

## Digestion of Neutral Fats

Neutral fats (triglycerides) contain glycerol esterified with three fatty acids. These fatty acids are usually three different ones ($R_1$, $R_2$, $R_3$).

$$CH_2OR_1$$
$$|$$
$$CHOR_2$$
$$|$$
$$CH_2OR_3$$

In shorthand notation this can be written

The *lipase* of pancreatic juice releases the fatty acids from glycerol by hydrolysis and has a predilection for the fatty acids in positions 1 and 3. The end result is production of some diglycerides, but mainly the 2-monoglycerides together with free fatty acids.

*Bile salts* emulsify the fat.

## Micelle Formation

The resultant monoglyceride and fatty acids aggregate with bile salts to form a *micelle*: the micelle also contains free *cholesterol* (liberated by hydrolysis in the lumen from cholesterol esters) and *phospholipids*, as well as *fat soluble vitamins* (A, D and K). The diameter of the micelle is between 100 and 1000 times as small as that of the emulsion particle and is small enough to pass through the microvillous spaces: it is also negatively charged, which is necessary if it is to pass through these spaces.

### Lipids in the Intestinal Cell

In the intestinal cell triglycerides are resynthesised from monoglycerides and fatty acids. Cholesterol is re-esterified. The triglycerides, cholesterol esters and phospholipids, together with the fat soluble vitamins, are coated with a layer of lipoprotein (manufactured in the cell) to form *chylomicrons* (p. 223). These are readily suspended in water and pass into the lymphatic circulation, probably through the cell wall.

### Absorption of Free Fatty Acids

Some free fatty acids pass through the intestinal cell into the portal blood stream.

From this account it will be seen that absorption of the following constituents of food,

> Neutral fat
> Cholesterol
> Phospholipids
> Fat soluble vitamins,

depends on:

1. Normal emulsification of fats, and therefore the presence of *bile salts*.

2. Normal digestion of neutral fat by *lipase* and therefore normal pancreatic function.

3. Normal *intestinal mucosa* for formation of the chylomicron and a normal absorptive area.

*Bile salts* are synthesised in the liver from cholesterol. They enter the intestinal tract in the bile and are actively reabsorbed in the distal ileum. They are recirculated to the liver and resecreted into the bile (the "enterohepatic circulation"). Resection of the distal ileum may prevent reabsorption and re-use of bile salts, and stimulate their resynthesis from cholesterol. The plasma cholesterol concentration may fall as a result, but this is probably of little clinical importance.

Some intestinal bacteria convert bile salts to bile acids and thus inactivate them: some of the bile acids formed may be toxic.

### CARBOHYDRATE ABSORPTION

**Polysaccharides** such as starch and glycogen are hydrolysed by amylase (salivary and pancreatic, the latter being of the greater importance) mainly to 1:4 disaccharides such as maltose (glucose + glucose): a few larger branch chain saccharides remain.

**Disaccharides** (maltose, sucrose (glucose + fructose) and lactose (glucose + galactose)) are hydrolysed to their constituent mono-saccharides by the appropriate disaccharidase (maltase, sucrase or lactase): isomaltase hydrolyses isomaltose (two glucose molecules joined by 1:6 linkages). These enzymes are not present in significant amount in intestinal secretions, but are on the surface of the intestinal cell where the hydrolysis takes place.

**Monosaccharides** are absorbed in the duodenum. Glucose and galac-tose are probably absorbed by a common active process, while fructose is absorbed by a different mechanism.

Thus carbohydrate absorption depends on:

1. The presence of *amylase*, and therefore on normal pancreatic function (polysaccharides only).

2. The presence of *disaccharidases* on the intestinal cell (disac-charides).

3. Normal *intestinal mucosa* with normal active transport mechan-isms (monosaccharides).

Note that absorption of polysaccharides requires normal function of all three mechanisms.

### PROTEIN ABSORPTION

The diet is not the only source of protein in the intestinal lumen: a significant proportion of that absorbed originates from intestinal secre-tions and desquamated mucosal cells.

Protein is broken down by pepsin, followed by trypsin and the other proteolytic enzymes of pancreatic juice. The products are amino acids and small peptides.

**Amino acids** are actively absorbed in the small intestine.

**Small peptides** are probably absorbed into the cell intact by special transport mechanisms, and hydrolysed intracellularly.

Protein absorption therefore depends on:

1. The presence of *pancreatic proteolytic enzymes* and therefore on normal pancreatic function.

2. Normal *intestinal cells* with normal active transport mechan-isms.

### VITAMIN $B_{12}$ ABSORPTION

Vitamin $B_{12}$ can be absorbed only when it has formed a complex with *intrinsic factor*, a mucopolysaccharide secreted by the stomach: in this form it can bind to the intestinal cell where it is absorbed, mainly in the lower ileum. Its absorption therefore depends on:

1. Normal *gastric secretion*.

2. Normal *intestinal cells* in the *lower ileum*.

Some intestinal bacteria require vitamin $B_{12}$ for growth and prevent its absorption by mechanisms probably involving competition with the intestinal cells for it. Normal absorption therefore also depends on:
3. Normal *intestinal flora.*

## ABSORPTION OF OTHER WATER SOLUBLE VITAMINS

Most of the water soluble vitamins (C and all those of the B group except $B_{12}$) are absorbed, probably by special mechanisms, mainly in the upper small intestine. Since their absorption is not dependent on fat absorption, clinical deficiencies of these (with the exception of folate) are relatively uncommon in malabsorption syndromes.

## ELECTROLYTE AND WATER ABSORPTION

Much electrolyte and water absorption is really reabsorption of the contents of intestinal secretions.

**Sodium and potassium** are probably absorbed by an active process throughout the small intestine, much as they are in the renal tubule: many other active transport mechanisms depend on metabolic energy provided by the sodium pump. In the colon sodium-potassium exchange is stimulated by aldosterone.

**Chloride** absorption in the small intestine probably follows the electrochemical gradient created by absorption of sodium and other cations. In the colon chloride is absorbed in exchange for bicarbonate (see p. 91).

**Water** is probably absorbed passively, as in the proximal renal tubule, along an osmotic gradient created by absorption of sodium, sugars, amino acids and other solutes. If these solutes cannot be absorbed normally water is "held" in the intestinal lumen (see "dumping syndrome" p. 275 and disaccharidase deficiency p. 279). Final water absorption occurs in the colon.

## CALCIUM AND MAGNESIUM ABSORPTION

**Calcium** is actively absorbed in the upper small intestine, particularly the duodenum and upper jejunum. Normal calcium absorption depends on its presence in an ionised form (it is inhibited by the formation of insoluble salts with phosphate, fatty acids and phytate), and on the presence of vitamin D in the form of 1:25-dihydroxycholecalciferol (p. 237). Much of the faecal calcium is endogenous and is derived from calcium in intestinal secretions.

**Magnesium** is also absorbed by an active process and may share in the calcium pathway.

Normal calcium and magnesium absorption depends on:

1. A *low concentration of fatty acids*, phosphate and phytate in the intestine.

2. The absorption and metabolism of *vitamin D* and therefore on normal fat absorption.

3. Normal *intestinal mucosa*.


## IRON ABSORPTION

Iron is absorbed by an active process in the duodenum and upper jejunum, and absorption is stimulated by anaemia (p. 390). Some of the iron absorbed into the intestinal cell enters the blood but some stays in the cell and is lost into the intestine when this is desquamated (p. 390).

We are now in a position to understand the consequences of disorders of digestion and absorption. From the above account it should be noted that:

1. *Pancreatic failure* affects absorption of large molecules only (fats, polysaccharides and protein).

2. *Disease of the intestinal mucosa* affects absorption of small molecules and products of digestion of large molecules.

3. Absence of *bile salts* causes malabsorption of fats and of those substances sharing mechanisms for fat absorption.


## MALABSORPTION SYNDROMES

From the clinical point of view the malabsorption syndromes can be divided into those associated with generalised malabsorption of fat, protein, carbohydrate and other nutrients, and those associated with failure to absorb one or more specific substances.


### GENERALISED MALABSORPTION (ASSOCIATED WITH STEATORRHOEA)

**Intestinal Disease**

1. Reduction of absorptive surface or general impairment of transport mechanisms.

> Gluten sensitivity causing coeliac disease.
> Tropical sprue.
> Idiopathic steatorrhoea.
> Extensive surgical resection of small intestine.

2. Extensive infiltration or inflammation of the small intestinal wall (for example, Crohn's disease, amyloidosis, scleroderma).

3. Increased rate of passage through the small intestine:
Post-gastrectomy
Carcinoid syndrome.

## Pancreatic Dysfunction (Failure of Digestion)

Chronic pancreatitis
Fibrocystic disease of the pancreas.
Inactivation of lipase by low pH in the Zollinger-Ellison syndrome (p. 281).

### FAILURE OF ABSORPTION OF SPECIFIC SUBSTANCES

1. Altered bacterial flora (mainly fat, p. 267; and vitamin $B_{12}$, p. 269):

Blind loop syndrome
Diverticula
Surgery
Neomycin therapy.

2. Biliary obstruction (fat and substances dependent on fat absorption, p. 267).

3. Local disease or surgery (for instance, disease of the terminal ileum affects vitamin $B_{12}$ absorption).

4. Pernicious anaemia (vitamin $B_{12}$).

5. Disaccharidase deficiency:
Congenital
Acquired.

6. Protein-losing enteropathy.

7. Isolated transport defects resulting from inborn metabolic errors (p. 365).

### GENERALISED MALABSORPTION

**Gluten sensitive enteropathy (coeliac disease)** occurs at any age. Sensitivity to gluten (in wheat germ) causes flattening of intestinal villi and considerable reduction in absorptive area. The flattening of the villi may be demonstrated on intestinal biopsy specimens. These cases respond to treatment with a gluten-free diet, but this improvement may not be evident for some months.

In **tropical sprue** there is also flattening of the villi, but these cases do not respond to a gluten-free diet. They do respond to broad spectrum antibiotics and folate, suggesting that a bacterial factor is important in the aetiology of the disease.

The term *idiopathic steatorrhoea* should be reserved for those cases of steatorrhoea which do not respond, either to a gluten-free diet, or to broad-spectrum antibiotics, and for which no cause can be found.

**Extensive surgical resection of the small intestine** may so reduce the absorptive area as to cause malabsorption.

**Extensive infiltration and inflammation of small intestinal mucosa** may damage its ability to carry out absorptive processes. This may be aggravated by the presence of altered bacterial flora in these conditions.

**After gastrectomy** normal mixing of food with fluid, acid and pepsin does not occur in the stomach, so that the activity of enzymes in the small intestine is less efficient than usual. Passage of intestinal contents through the duodenum is also more rapid than usual. In spite of this, post-gastrectomy malabsorption is rarely severe, and clinically "dumping" and hypoglycaemic attacks are more troublesome (p. 275).

**The carcinoid syndrome** (p. 438) is due to the excessive production of 5-hydroxytryptamine (5HT) by tumours of argentaffin cells usually arising in the small intestine, and by their metastases, usually in the liver. This is a rare disease which even more rarely presents a problem of differential diagnosis of malabsorption. The malabsorption is probably the result of increased intestinal motility induced by 5HT. Diagnosis depends on the presence of the other clinical features of the disease, and on an increased excretion of 5-hydroxyindole acetic acid in the urine.

**In pancreatic disease** (due most commonly to *chronic pancreatitis*) failure of absorption is due to failure of digestion and predominantly affects large molecules. *Fibrocystic disease of the pancreas (cystic fibrosis: mucoviscidosis)* is an inherited disease, usually presenting in early childhood but, more rarely, first recognised in the adult patient. Pancreatic and bronchial secretions are viscid, and by blocking pancreatic ducts and bronchi cause obstructive disease of these organs. Sweat glands are also affected and the diagnosis depends on the demonstration of an *increase in concentration of sweat sodium*, often to about twice normal (p. 44). The patient may present with pulmonary disease, or with malabsorption.

### Results of Generalised Malabsorption

The most obvious finding common to generalised intestinal and pancreatic malabsorption is *steatorrhoea* (more than 18 mmol (5 g) of fat in the stools per day—see Appendix, p. 285). In gross steatorrhoea the stools are usually pale, greasy and bulky and, if the condition is severe, there may be diarrhoea. In intestinal malabsorption fat can be acted on by lipase, but the products cannot be absorbed normally. In pancreatic steatorrhoea the intestinal wall is normal but fat cannot be digested. The collections of specimens for faecal fat estimation must be carefully controlled if a reliable answer is to be obtained (Appendix, p. 285).

Malabsorption of fat is always accompanied by malabsorption of the substances listed on p. 267. *Vitamin D deficiency* impairs calcium absorption and this is one of the causes of a low level of circulating ionised calcium and of osteomalacia in malabsorption syndromes. *Vitamin K* is required for hepatic synthesis of prothrombin and other clotting factors, and severe cases of malabsorption may develop a haemorrhagic diathesis associated with a prolonged prothrombin time: this prothrombin deficiency, unlike that of liver disease, can be reversed by parenteral administration of vitamin K. *Vitamin A* deficiency is rarely clinically evident, although malabsorption of an oral dose of vitamin A can be demonstrated. Malabsorption of lipids causes low plasma *cholesterol* levels—an incidental finding of no diagnostic importance.

In intestinal malabsorption the *low ionised calcium level and osteo-malacia* due to vitamin D malabsorption is aggravated by the formation of insoluble calcium soaps with unabsorbed fatty acids. If ionised calcium levels have been low for a sufficiently long time to cause osteomalacia the *plasma alkaline phosphatase* concentration will rise. Secondary hyperparathyroidism due to the low ionised calcium level causes phosphaturia with a low plasma phosphate (p. 236). If the ionised calcium levels fall very low *tetany* may rarely result.

*Protein malabsorption* occurs in intestinal disease because intestinal transport mechanisms are impaired: in pancreatic disease the digestion of protein is severely impaired. In either case prolonged disease causes generalised *muscle and tissue wasting* and *osteoporosis* (p. 248), and a lowering of all protein fractions in the blood. The low albumin may cause oedema (p. 52) and results in a reduction of *protein bound calcium* (p. 235). The total calcium level may therefore give a false idea of the severity of the hypocalcaemia. Reduced *antibody formation* (immuno-globulins) predisposes to infection.

Thus in generalised malabsorption, whether pancreatic or intestinal, the *clinical picture* may be as follows:

Bulky, fatty stools, with or without diarrhoea.
General wasting and malnutrition (protein deficiency),
Osteoporosis and osteomalacia (protein and calcium deficiency).
Oedema (albumin deficiency).
Haemorrhages (vitamin K deficiency).
Tetany (calcium deficiency).
Recurrent infections (immunoglobulin deficiency).

The *laboratory findings* are:

Increased excretion of fat in the stools.
Hypocalcaemia (ionised and protein bound)—with hypophos-phataemia.
Raised alkaline phosphatase levels in cases with osteomalacia.
Generalised lowering of the concentration of all protein fractions.

Possibly a low urea (due to decreased production from amino acids).

Hypocholesterolaemia.

Prolonged prothrombin time.

The complete picture is only seen in advanced cases of the syndrome. It is important to realise that generalised malabsorption may present as osteomalacia, with bone pains, or with anaemia.

### Differential Diagnosis of Generalised Intestinal and Pancreatic Malabsorption

Tests of pancreatic function are generally unsatisfactory (see below), and the differential diagnosis of steatorrhoea is usually made on indirect evidence. Remember that pancreatic malabsorption affects large molecules predominantly.

Malabsorption of fat occurs in both types of disease. Theoretically, in pancreatic malabsorption fat should be predominantly present as triglyceride (neutral fat), while in intestinal disease free fatty acids should be present. However, the estimation of so-called "split" and "unsplit" fat in the stools has been abandoned as it provides no useful information because of the action of fat-splitting bacteria in the lower intestinal tract: if a small amount of stool is emulsified on a microscopic slide the presence of many fat globules (triglyceride) suggests a pancreatic cause. Similarly, the presence of many undigested meat fibres (striated muscle) in the stool is suggestive of a pancreatic rather than intestinal origin. This examination requires experience, and is rarely helpful if other estimations are available.

**Differences in carbohydrate metabolism.**—Polysaccharide absorption is impaired in both conditions. However, some of the carbohydrate in the diet is in the form of mono- and disaccharides and these can be absorbed in pancreatic disease when neither intestinal disaccharidase activity nor active monosaccharide absorption is affected. Hypoglycaemia is rare in either condition, but is probably more common in intestinal malabsorption: however, in this type of malabsorption impaired absorption of a glucose load is often evident in a "flat" *glucose tolerance curve* (but see p. 198). In pancreatic disease this curve may be normal, but as insulin is of pancreatic origin the curve may even be diabetic in type.

Because glucose is rapidly metabolised in the body, interpretation of the glucose tolerance curve as an index of the type of malabsorption may be difficult. *Xylose* is a pentose, not rapidly metabolised in mammalian tissues and filtered by the renal glomerulus. There is an active transport mechanism for xylose in the upper small intestinal mucosa, although this is relatively inefficient (p. 286). If an oral dose of xylose is given, and if the intestinal mucosa is normal, it will be absorbed and appear in the urine: if, however, there is disease involving the upper small intestinal mucosa it will not be absorbed normally and less will appear in the blood and urine. This is the basis of the *xylose absorption test* which is usually normal when malabsorption is due to disease of the pancreas and abnormal in intestinal disease: however, disease involving only the ileum (such as Crohn's disease) may be associated with normal xylose absorption.

*Anaemia* is more common in intestinal than in pancreatic malabsorption since iron, vitamin $B_{12}$ and folate are not digested by pancreatic enzymes before absorption. In intestinal malabsorption the blood and bone marrow films typically show a mixed iron deficiency and megaloblastic picture. Malabsorption of iron, vitamin $B_{12}$ and folate may be demonstrable and the absorption of vitamin $B_{12}$ is not increased by the simultaneous administration of intrinsic factor, as it is in pernicious anaemia (see p. 278). Anaemia may be aggravated by protein deficiency.

The differential diagnosis is summarised in Table XXIII (p. 278).

## The Post-Gastrectomy Syndrome

Malabsorption after gastrectomy is usually mild. However, rapid passage of the contents of the small gastric remnant into the duodenum may have two clinical consequences:

1. **The "dumping syndrome".**—Soon after a meal the patient may experience abdominal discomfort and feel faint and sick. The syndrome is thought to be due to the sudden passage of fluid of high osmotic content into the duodenum. Before this abnormally large load can be absorbed, water passes along the osmotic gradient from the extracellular fluid into the lumen of the intestine. The reduction in plasma volume causes faintness and the large volume of duodenal fluid causes abdominal discomfort.

2. **Post-gastrectomy hypoglycaemia.**—If a meal containing much glucose passes more rapidly than normal into the duodenum glucose absorption is very rapid. The blood glucose level rises suddenly and causes an outpouring of insulin. The resultant "over-swing" of blood glucose concentration may cause hypoglycaemic symptoms which typically occur at about two hours after a meal. The glucose tolerance curve in this type of case is "lag storage" in type.

Both these disabilities can be mitigated if meals low in carbohydrate content are taken "little and often".

## Tests of Exocrine Pancreatic Function

It is convenient to digress here to discuss tests of the ability of the pancreas to secrete digestive juices. Unfortunately such tests of pancreatic function are very unsatisfactory. In chronic pancreatic hypofunction the diagnosis is usually made by the indirect methods described above. Carcinoma of the pancreas is very difficult to diagnose unless the lesion is in the head of the organ and causes obstructive jaundice: extensive gland destruction may cause late onset diabetes. Acute pancreatitis, however, may usually be diagnosed by estimation of plasma amylase levels.

**Plasma enzymes.**—Measurements of plasma *amylase* levels usually show little change from normal in pancreatic disease except in acute pancreatitis (see below), when levels may be raised. In chronic pancreatic hypofunction low levels can rarely be demonstrated, possibly because of the presence of amylase of hepatic and salivary gland origin, and the estimation is useless. Plasma *lipase* estimation is technically not very suitable for routine use.

**Faecal trypsin** levels are extremely variable, probably because of bacterial action. The estimation is of no value in diagnosis of pancreatic hypofunction in adults. In infants with diarrhoea its absence is suggestive of fibrocystic disease of the pancreas.

**Duodenal enzymes.**—Measurement of pancreatic enzymes and bicarbonate in duodenal aspirate before and after stimulation of the pancreas with secretin or pancreozymin (hormones stimulating the pancreas which are normally secreted during digestion) is not very suitable for routine use because of difficulty in positioning the duodenal tube correctly and in quantitative sampling of the secretions. In the *Lundh test* a mixture of corn oil, milk powder and glucose is given, and samples of duodenal secretion analysed for trypsin: this test, too, is only useful if performed by an expert.

## Acute Pancreatitis

In acute pancreatitis necrosis of the cells of the organ results in release of their enzymes into the peritoneal cavity and blood stream. The presence of pancreatic juice in the peritoneal cavity causes *severe abdominal pain* and *shock*: this picture is common to many acute abdominal emergencies. A vicious circle is set up as more pancreatic cells are digested by the released enzymes.

Acute pancreatitis is most commonly the result of obstruction of the pancreatic duct, or of regurgitation of bile along this duct. The most important predisposing factors are *alcoholism* and *biliary tract disease*. *Trauma* to the pancreas by damaging the cells may also initiate the vicious circle. There is an association between acute pancreatitis and *hypercalcaemia*: evidence for the latter should be sought; the finding is the more significant in this condition, since hypocalcaemia is the usual result of an attack.

Typically plasma amylase values increase five-fold or more. However, it is important to realise that concentrations of up to, and even above, this value may be reached in any acute abdominal emergency, but especially after gastric perforation into the lesser sac: very high levels may occur in renal failure when the enzyme cannot be excreted normally, and these are usually asymptomatic. Conversely, levels in acute pancreatitis may not reach very high levels, and usually fall very rapidly as the enzyme is lost in the urine. High plasma amylase levels are therefore only a rough guide to the presence of acute pancreatitis and normal or only slightly raised values do not exclude the diagnosis.

As the released fatty acids form soaps with calcium, plasma calcium levels may fall. This may be a useful sign when amylase concentrations have fallen to normal (a few days after the acute attack).

Malabsorption probably does occur during acute pancreatitis but is of little importance in the acute phase. Chronic pancreatic failure may follow a severe attack and is especially probable following repeated attacks.

### FAILURE OF ABSORPTION OF SPECIFIC SUBSTANCES

### 1. Altered Bacterial Flora (Malabsorption of Vitamin $B_{12}$ and Fat)

The "Blind Loop syndrome".—This syndrome is associated with stagnation of intestinal contents with a consequent alteration of bacterial flora, and occurs when the loops are the result of surgery, or in the presence of *diverticula*.

Treatment with neomycin, by altering bacterial flora and possibly by combining with bile salts and so disrupting micelles, can cause a similar syndrome.

Many intestinal bacteria require vitamin $B_{12}$ for metabolism, and *megaloblastic anaemia* is common in the syndrome. Bile salts may be metabolised to bile acids with resultant *steatorrhoea* and malabsorption of all those substances associated with fat absorption (p. 267). Protein and carbohydrate are usually normally absorbed.

### 2. Biliary Obstruction (Malabsorption of Fat)

Bile salts are synthesised and secreted by the liver. In biliary obstruction these cannot reach the intestinal lumen in normal amounts, and *steatorrhoea* results with the usual consequences (p. 272). Because of extreme jaundice there is rarely any difficulty in the differential diagnosis.

### Differential Diagnosis of Steatorrhoea

The malabsorptive conditions described so far are all associated with steatorrhoea. As has been mentioned, biliary obstruction is rarely a problem of diagnosis. The important points in laboratory findings in differential diagnosis of the other conditions are summarised in Table XXIII.

TABLE XXIII

DIFFERENTIAL DIAGNOSIS OF STEATORRHOEA

|  | Upper Small Intestinal Disease | Pancreatic Disease | Blind Loop Syndrome |
|---|---|---|---|
| Xylose absorption | Reduced | Normal | Usually normal |
| Anaemia | Mixed megalo-blastic and iron deficiency common | Rare | Megaloblastic common |
| Intestinal Biopsy | May show flattened villi or other cause | Normal | Normal |
| Glucose Tolerance Test | May be flat (except post-gastrectomy) | Normal or diabetic | Normal |

## 3. Local Disease or Surgery

As a few substances are absorbed in significant amounts only in certain areas of the small intestine, local disease or resection of such an area will cause selective malabsorption. The lower ileum is concerned with vitamin $B_{12}$ absorption and with reabsorption of bile salts, and resection or disease of this area (for example, Crohn's disease or tuberculosis), may cause macrocytic anaemia.

## 4. Pernicious Anaemia

Vitamin $B_{12}$ cannot be absorbed in the absence of intrinsic factor (p. 268). In pernicious anaemia the stomach cannot secrete this substance and significant deficiency may occur after total gastrectomy or with extensive malignant infiltration of the stomach.

The Schilling test.—In such subjects malabsorption of vitamin $B_{12}$ can be demonstrated if a small dose of the radioactive vitamin is given and its excretion in the urine measured (compare xylose absorption test, p. 274). If the malabsorption is due to pernicious anaemia, administration of the labelled vitamin together with intrinsic factor results in normal absorption: if it is due to intestinal disease malabsorption persists. A "flushing" dose of non-radioactive vitamin $B_{12}$ is given parenterally at the same time as, or just after, the labelled dose, to ensure quantitative urinary excretion. Haematological tests such as examination of blood and marrow films should have been completed before the vitamin $B_{12}$ is given.

## 5. Disaccharidase Deficiency

Generalised disease of the intestinal wall usually causes a non-selective disaccharidase deficiency, because these enzymes are located on the surface of the absorptive cells: this is relatively unimportant compared with general malabsorption and tests for this syndrome are therefore useful only in the absence of steatorrhoea when a selective rather than a generalised malabsorption of carbohydrate may be present.

The *symptoms* of disaccharidase deficiency are those of the effects of unabsorbed, osmotically active sugars in the intestinal tract, and include faintness, abdominal discomfort and severe diarrhoea after ingestion of the offending disaccharide. (Compare the "dumping syndrome", p. 275.)

### Lactase deficiency

1. *Acquired lactase deficiency* is much more common than the congenital form and is probably the commonest type of disaccharidase deficiency: it may first present in adults. It may be due to a sensitivity reaction, perhaps in subjects with a genetic predisposition.

2. *Lactase deficiency associated with prematurity.*—In premature infants lactase may not be present in normal amounts in intestinal cells. Initially these cases resemble the congenital ones. However, the sensitivity to milk usually disappears within a few days of birth.

3. *Congenital lactase deficiency.*—This is very rare. Infants present soon after birth with severe diarrhoea. Stools are typically acid because of bacterial production of lactic acid from lactose. The syndrome is cured by removal of milk and milk products from the diet.

**Sucrase and isomaltase deficiency** usually co-exist.

*Congenital sucrase-isomaltase deficiency* is more common than congenital lactase deficiency.

*Acquired sucrase-isomaltase deficiency* and *maltase deficiency* of any kind are very rare.

The *diagnosis* of disaccharidase deficiency is most reliably made by estimation of the relevant enzymes in intestinal biopsy tissue.

The relevant disaccharide may be given orally and blood glucose estimated as in the glucose tolerance test. If the disaccharide cannot be hydrolysed the constituent monosaccharides (glucose, or those converted to glucose in the intestinal cell), cannot be absorbed, and the curve is flat. The result should usually be compared with a glucose tolerance curve. If the patient experiences typical symptoms, or if a child excretes disaccharides in the urine when the offending sugar is given, the comparison need not be made.

Radiological examination may assist in the diagnosis of disaccharidase deficiency. Barium is administered without, and then with, the disaccharide. When the sugar is not absorbed the flocculation pattern of barium is altered by its osmotic effect.

## 6. Protein-losing Enteropathy

This is a very rare syndrome in which the intestinal wall is abnormally permeable to large molecules (as the glomerulus is in the nephrotic syndrome). This is not strictly speaking a malabsorption syndrome, but is due to excessive loss of protein from the body into the gut. It occurs in a variety of conditions in which there is ulceration of the bowel, lymphatic obstruction and intestinal lymphangiectasis, or hypertrophic lesions of the bowel. Typically it is only rarely associated with steator-rhoea, and the clinical picture, like that of the nephrotic syndrome (p. 338), is due to hypoalbuminaemia: protein is not usually found in the urine.

In this syndrome the hypoproteinaemia is probably not entirely due to loss from the body, because some of the protein which passes into the intestinal tract is digested and the amino acids and small peptides reabsorbed. However, the rate of protein breakdown probably exceeds the rate at which the reabsorbed amino acids can be resynthesised into protein.

Diagnosis can be made by measuring the loss into the bowel after intravenous injection of a substance of a molecular weight approximating that of albumin: substances that have been used are radioactive polyvinylpyrrolidone (PVP), dextran or albumin. Typically the electrophoretic pattern is similar to that found in the nephrotic syndrome: as in that syndrome the relatively high molecular weight $\alpha_2$ fraction (on cellulose acetate, p. 322) is retained in the blood stream while all other fractions are lost from the body.

## GASTRIC FUNCTION

The important components of gastric secretion are *hydrochloric acid*, *pepsin* and *intrinsic factor*. All these factors have already been mentioned as being of importance in digestion and absorption, and loss of hydrochloric acid in pyloric stenosis has been discussed as a cause of metabolic alkalosis (p. 100).

Stimulation of gastric secretion occurs by two main pathways:

1. Through the *vagus nerve*, which in turn responds to stimuli from the cerebral cortex, normally resulting from the sight, smell and taste of food. Hypoglycaemia can stimulate this pathway and this fact can be used to assess the completeness of vagotomy.

2. By *gastrin*, a hormone normally produced by G cells in the gastric antrum in response to the presence of food; it is carried by the blood stream to the parietal area of the stomach where it stimulates

secretion. Acid in the pylorus, in turn, inhibits gastrin secretion, providing feed-back control.

## HYPERSECRETION

Hypersecretion of gastric juice may be associated with *duodenal ulceration*, when it may be neurogenic in origin. However, there is overlap between acid secretion in normal subjects and in those with duodenal ulceration, and the estimation is of very limited diagnostic value in this condition.

In the *Zollinger-Ellison syndrome* G cells, usually in the pancreatic islets (most commonly as tumours), produce large amounts of gastrin ectopically: of these tumours about 60 per cent are malignant; of the remaining 40 per cent two-thirds are multiple and only one-third (13 per cent of the total) are single, resectable adenomata. More rarely the syndrome is due to hyperplasia of G cells in the gastric antrum. Acid secretion by the stomach in this condition is very high. The consequent ulceration of the stomach and upper small intestine may cause severe diarrhoea: the low pH, by inhibiting lipase activity, may cause steatorrhoea. Associated benign adenomata may be found in other endocrine glands, such as the parathyroid, pituitary, thyroid and adrenal (multiple endocrine adenomatosis): these are only occasionally functional.

## HYPOSECRETION

Hyposecretion of gastric juice occurs in *pernicious anaemia* (p. 425) and is probably the result of antibodies: this is by far the commonest cause, and the achlorhydria is usually pentagastrin and histamine "fast" (p. 287). In extensive *carcinoma of the stomach* and in *chronic gastritis* there may also be gastric hyposecretion. However, in none of these conditions is estimation of gastric acidity of help in diagnosis: the diagnosis of pernicious anaemia should be made on haematological grounds, and on the result of the Schilling test, and results in the other two conditions are too variable to be useful.

## TESTS OF GASTRIC FUNCTION

Most routine tests of gastric function involve measurement of the acid secretion by the stomach, either at rest or in response to stimuli.

### "Tubeless" Gastric Analysis

This depends on oral administration of a substance in which hydrogen ions can replace another cation in the molecule. The released cation (usually a dye) is excreted in the urine and measured. Caffeine, which can act as a gastric stimulant, is usually included in the tablets. Unfortunately results of this test have been disappointing and this is, at best, a screening procedure. Excretion of any dye during the test probably excludes complete achlorhydria.

All other tests involve the passage of a tube into the stomach. This is unpleasant for the patient: moreover, valuable results will only be obtained when the operator is skilled and able to recognise when the specimens obtained are incomplete, either because of blockage or malpositioning of the tube. For these reasons the tests should only be carried out when really necessary for diagnosis and management, and only by a skilled operator.

### Resting Juice or Overnight Secretion

Gastric contents may be aspirated overnight and the total night secretion measured, or a timed specimen may be obtained without a stimulus and the hourly secretion determined. Both these secretions will usually be high in volume and acid content in duodenal ulceration. In the Zollinger-Ellison syndrome diagnosis depends on finding a very high rate of acid secretion in a one-hour basal collection (more than 15 mmol/hour) (see Appendix).

### Stimulation of Gastric Secretion

1. **Direct stimulation of parietal cells** is used to demonstrate achlorhydria, which is typically pentagastrin and histamine "fast" in pernicious anaemia: as already pointed out, the test is of limited use in the diagnosis of this condition, and is unpleasant for the patient.

*Pentagastrin* is a pentapeptide consisting of the physiologically active part of the gastrin molecule. It acts directly on the parietal cells of the stomach and can be used to stimulate secretion. It has fewer undesirable side-effects than histamine and is the stimulant of choice.

*Histamine or "Histalog" (ametazole hydrochloride).*—Histamine also acts directly on parietal cells and stimulates secretion. Because of the undesirable side-effects of histamine it should be given with antihistamine cover: antihistamines do not block the gastric stimulatory effect of histamine. "Histalog" is a synthetic analogue of histamine, said to have fewer undesirable side-effects than the parent substance.

2. **Vagal stimulation of gastric secretion** is used before vagotomy to test the therapeutic prospects of this operation in treatment of peptic ulcer, and after vagotomy to test completeness of the section of the nerve. The stimulus used is insulin induced hypoglycaemia (compare the use of insulin to stimulate cortisol secretion). If vagotomy is complete there should be no acid secretion even when the blood glucose

level falls below 2·2 mmol/l (40 mg/dl), and when there is clinical evidence of hypoglycaemia.

## Plasma Gastrin Levels

Radioimmunoassay of plasma gastrin is available in some centres. In the Zollinger-Ellison syndrome fasting concentrations are from 5 to 30 times the upper limit of normal. Although gastric acid secretion may be diagnostic of this condition, the simultaneous finding of high gastrin levels indicates autonomous hormone secretion not under feed-back control.

## SUMMARY

### Intestinal Absorption

1. Normal intestinal absorption depends on adequate digestion of food (and therefore on normal pancreatic function), and on a normal area of functioning intestinal cells.
2. Normal digestion and absorption of fat depends on the presence of bile salts as well as of lipase.
3. Absorption of cholesterol, phospholipids and fat soluble vitamins depends on normal neutral fat absorption.
4. In *intestinal* malabsorption there is malabsorption of small molecules, usually due to a reduced absorptive area.
5. In *pancreatic* malabsorption there is malabsorption of fats, proteins and polysaccharides, but small molecules are usually absorbed normally.
6. An *abnormal bacterial flora* may cause steatorrhoea (because of competition for bile salts), and macrocytic anaemia (because of competition for vitamin $B_{12}$).
7. There may be steatorrhoea in biliary obstruction (because of lack of bile salts).
8. Selective malabsorption of vitamin $B_{12}$ occurs in pernicious anaemia (lack of intrinsic factor).
9. Selective disaccharidase deficiencies cause malabsorption of disaccharide. These deficiencies are more commonly acquired than congenital in origin.

### Pancreatic Function

1. Direct tests for pancreatic hypofunction are unsatisfactory except in special centres.
2. Acute pancreatitis is associated with a transient rise in plasma amylase levels. This enzyme can also reach high concentrations in many acute abdominal emergencies and in renal failure.

**Gastric Function**

1. Hypersecretion of acid occurs in:
   (*a*) Duodenal ulceration.
   (*b*) The Zollinger-Ellison syndrome.

2. Hyposecretion of acid (pentagastrin and histamine "fast") occurs in:
   (*a*) Pernicious anaemia.
   (*b*) Extensive gastric infiltration.

3. The parietal cells can be directly tested by stimulation with pentagastrin, or histamine (or its analogue).

4. The vagal stimulation of gastric secretion can be tested by producing insulin induced hypoglycaemia.

5. Estimation of plasma gastrin may be of value in the diagnosis of the Zollinger-Ellison syndrome.

## FURTHER READING

Course in Gastroenterology. (1973). *Proc. Mayo Clin.*, **48.**

LOSOWSKY, M. S., WALKER, B. E., and KELLEHER, J. (1974). *Malabsorption in Clinical Practice*. Edinburgh: Churchill-Livingstone.

DAWSON, A. M., Ed. (1971). Symposium—Intestinal Absorption and its Derangements. *J. clin. Path.*, **24**, *Suppl. 5*.

BARON, J. H., and LENNARD-JONES, J. E. (1971). Gastric secretion in health and disease. *Brit. J. hosp. Med.*, **6**, 303.

# APPENDIX

## TESTS OF MALABSORPTION

### COLLECTION OF SPECIMENS FOR FAECAL FAT ESTIMATION

Estimation of faecal fat output measures the difference between the fat absorbed and that entering the intestinal tract from the diet and from the body (p. 263). Absorption of fat and addition of fat to the intestinal contents occurs throughout the small intestine. A single 24-hour collection of faeces will usually give inaccurate results for two reasons.

1. The transit time from the duodenum to the rectum is variable.
2. Rectal emptying is variable and may not be complete.

It has been shown that consecutive 24-hour collections yield answers which may vary by several hundred per cent, and which are therefore useless as an estimate of daily excretion from the body into the gut. The longer the period of collection, the nearer does the calculated daily mean value approach the "true" one.

For obvious reasons patients cannot be kept in hospital indefinitely. A usual compromise is to collect *at least a 3-day* and preferably a 5-day speci·men of stools. Precision may be increased by collecting between "markers"—usually dyes which can be taken orally and which colour the stool.

### Procedure

*Day 0.*—The first "marker" (usually two capsules of carmine) is given.

As soon as the "marker" appears in the stool the collection is started. This "marker" will gradually disappear.

*Day 5.*—The second "marker" is given.

As soon as the second "marker" appears in the stool the collection is stopped.

The estimation is made on all the specimens passed between the appearance of the two "markers" and including one of the marked stools.

**Important.**—1. It is very important that *all* stools should be collected during this period. To ensure that none is missing it is best to label each specimen with the following information:

> Name of Patient
> Ward
> Date of Specimen
> *Time* of Specimen ⎫ A record of these should also be kept on the
> *Number* in the series ⎭ ward.

The patient should not be allowed to go to the toilet, and should be impressed with the importance of a complete collection.

2. The time for collection of specimens will be about a week (including the time for appearance of the marker). Before starting the test try to make sure

that during this time the patient is *not to be discharged*, is *not going to be operated on* (except in emergency), does *not receive enemas or aperients*: any patient requiring aperients to keep his bowels open is most unlikely to have significant steatorrhoea at that time. *Neither barium enemas nor barium meals* should be performed during this time as the barium interferes with the estimation.

The collection and estimation of faecal fat excretion is time–consuming and unpleasant for all concerned (including the patient). It is important that specimen collection is carefully controlled so that the answer may be meaningful.

### Interpretation

A *mean* daily fat excretion of more than 18 mmol (5g) indicates steatorrhoea. Since, in the normal person, almost all the faecal fat is of endogenous origin, it is not affected by diet within very wide limits.

### XYLOSE ABSORPTION TEST

Until recently it was customary to give an oral dose of 25 g of xylose for this test. However, the absorption of xylose is relatively inefficient and the presence of this large dose in the intestine may, by its osmotic effect, cause unpleasant symptoms and further interfere with absorption. For this reason a 5 g dose is preferable.

**Warning.** In the presence of poor renal function the test is invalid. Oedema, because the volume through which the xylose is distributed is increased, also invalidates the test.

### Procedure

The patient is fasted overnight.

8 a.m.—The bladder is emptied and the *specimen discarded*. 5 g of xylose dissolved in a glass of water is given orally.

All specimens passed between 8 a.m. and 10 a.m. are put into Bottle 1.

10 a.m.—The bladder is emptied and *the specimen put into Bottle 1* which is now complete.

All specimens passed between 10 a.m. and 1 p.m. are put into Bottle 2.

1 p.m.—The bladder is emptied and *the specimen is put into Bottle 2*, which is now complete.

Both bottles are sent to the laboratory for analysis.

### Interpretation

In the normal person more than 23 per cent of the dose (1·15 g) should be excreted during the five hours of the test. Fifty per cent or more of the total excretion should occur during the first two hours. In mild intestinal malabsorption the total 5-hour excretion may be normal, but delayed absorption is reflected in a 2- to 5-hour excretion ratio of less than 40 per cent.

In pancreatic malabsorption the result should be normal (p. 275).

## TESTS OF GASTRIC FUNCTION
### PENTAGASTRIN OR "HISTALOG"

**Procedure** (applies to all types of test of gastric secretion)

The patient fasts from 10 p.m. on the evening before the test. On the morning of the test a radio-opaque Levin tube is passed into the stomach (preferably under x-ray control) until it lies in the gastric antrum: the tube is attached to the side of the face with plaster.

8 a.m.—All the gastric juice is aspirated from the stomach and is put in a bottle marked "*Resting Juice*".

8 a.m. to 9 a.m.—Gastric juice is aspirated continuously for the next hour and put in a bottle marked "*Basal Secretion*".

9 a.m.—Pentagastrin 6 $\mu$g/kg body weight, or "Histalog" 100 mg, is injected intramuscularly.

9 a.m. to 9.15 a.m.  
9.15 a.m. to 9.30 a.m.  } Four 15-minute samples are aspirated and put  
9.30 a.m. to 9.45 a.m.  { into bottles marked with the appropriate times.  
9.45 a.m. to 10 a.m.

All specimens are sent to the laboratory for analysis.

**Interpretation**

1. In pentagastrin and histamine fast *achlorhydria* no specimen has a pH as low as 3·5.

2. A *basal* acid secretion of greater than 15 mmol of hydrogen ion in the hour, with no further response to stimulation, is suggestive of the *Zollinger-Ellison syndrome*.

3. If any specimen is of pH 3·5 or less the hydrogen ion secretion is checked by titration. In *hyperchlorhydria* the highest mean acid secretion (calculated by adding the results of the two consecutive specimens of highest acidity and by multiplying by two, to give an hourly secretion), is greater than 40 mmol/ hour.

The resting juice is inspected for *blood*. A large quantity of altered blood is suggestive of carcinoma. Small flecks may be due to trauma during aspiration.

### INSULIN STIMULATION OF GASTRIC SECRETION (p. 282)

This test is usually carried out at about six months after vagotomy. Insulin induced hypoglycaemia stimulates gastric secretion via the vagus nerve.

**Procedure**

The preparation of the patient and positioning of the tube is described under "'Histalog' Test Meal".

8 a.m.—The stomach is emptied and the specimen discarded.

8 a.m. to 9 a.m.—A 1-hour "basal secretion" is collected.

9 a.m.—Soluble insulin (0·15 units/kg) is injected intravenously.

9 a.m. to 11 a.m.—Specimens are collected for blood glucose estimation every 30 minutes for two hours. During this time gastric juice is aspirated every 15 minutes.

The specimens are sent to the laboratory for analysis.

**Warning.**—1. The test is potentially dangerous and should be done only under *direct medical supervision. Glucose* for intravenous administration should be *immediately available* in case severe hypoglycaemia develops. At the conclusion of the test the patient should be given something to eat.

2. If it is necessary to administer glucose, *continue with the sampling.* The stress has certainly been adequate.

## Interpretation

Analysis of gastric samples is only useful if the blood glucose has fallen to levels below 2·2 mmol/l (40 mg/dl) and if clinical hypoglycaemia is present.

If more than 2 mmol of hydrogen ion is present in any 60-minute sample the response is positive, suggesting the presence of intact vagal fibres.

# Chapter XII

# LIVER DISEASE AND GALL STONES

## LIVER DISEASE

### OUTLINE OF FUNCTIONS OF THE LIVER

THE liver plays an important role in many metabolic processes. It receives blood from the portal vein and so all nutrient from the gut, with the exception of fats (p. 267), reaches the liver before entering the systemic circulation.

1. **Carbohydrate metabolism.**—*Glycogen is synthesised* and stored in the liver during periods of carbohydrate availability. During fasting, blood glucose levels are maintained within normal limits by breakdown of stored glycogen (*glycogenolysis*). The formation of glucose from such substrates as amino acids (*gluconeogenesis*) occurs mainly in the liver.

2. **Lipid metabolism.**—*Synthesis* of lipoproteins, phospholipids, cholesterol and endogenous triglyceride occurs largely, but not exclusively, in the liver and the breakdown products of cholesterol are excreted in bile. *Fatty acids* reaching the liver from fat stores are *metabolised* in the tricarboxylic acid cycle. Any excess is incorporated into endogenous triglyceride, or is converted into ketones.

3. **Protein synthesis.**—Many of the plasma proteins, including special carrier proteins and most of the coagulation factors, but with the notable exception of the immunoglobulins, are synthesised in hepatic cells. Prothrombin and factors VII, IX and X require vitamin K for their synthesis.

4. **Storage functions.**—In addition to glycogen, many *vitamins* such as vitamins D and $B_{12}$ are *stored* in the liver (p. 419) and it is a major site of iron storage (p. 388).

5. **Excretion and detoxication.**—*Bile pigments* and *cholesterol* (p. 291) are excreted in the bile. Numerous *drugs* are detoxicated by the liver and some are excreted in bile. *Ammonia*, derived from protein and amino acid metabolism or produced in the bowel by bacteria, is converted to *urea* and so rendered non-toxic. *Steroid hormones* are inactivated by conjugation with glucuronic acid and sulphate in the liver and are excreted in the urine.

6. **Reticulo-endothelial function.**—The Kupffer cells lining the sinusoids of the liver form part of the reticulo-endothelial system.

## BILE PIGMENT METABOLISM

At the end of their life span circulating red cells are broken down in the reticulo-endothelial system, mainly in the spleen. The released haemoglobin is split into globin, which enters the general protein pool, and haem which is converted to bilirubin after removal of iron. The iron is reutilised.

Bilirubin formed by this process is carried to the liver and accounts for about 80 per cent of the bilirubin metabolised daily. Other sources include the breakdown of immature red cells in the bone marrow and of compounds chemically related to haemoglobin, such as myoglobin and cytochromes. In all about 500 $\mu$mol (300 mg) of bilirubin is carried daily to the liver. Healthy hepatic cells are capable of handling much greater loads than this.

Bilirubin in transit to the liver is bound to plasma albumin and is not soluble in water. It is referred to as *unconjugated bilirubin* (or simply as *bilirubin*) and, because of its binding to albumin and its insolubility in water, does not appear in the urine. Most of the "normal" plasma bilirubin is in this form; some *drugs* compete with bilirubin for binding sites on albumin.

At the hepatic cell membrane bilirubin is split off from albumin and passes into the cell where it is accepted by specific binding proteins (X and Y proteins). Many other organic anions (including bromsulphthalein [BSP] and certain drugs) are similarly bound and compete with bilirubin. Bilirubin is then transported within the cell to the smooth endoplasmic reticulum (microsomes). Here it is *conjugated* with glucuronic acid to form bilirubin glucuronide, or *conjugated bilirubin*, by the enzyme uridyl diphosphate (UDP) glucuronyl transferase. The activity of this enzyme may be increased (enzyme induction) by phenobarbitone. Conjugated bilirubin is transported out of the liver cell into the bile canaliculi and *excreted in the bile*. This is probably an active, energy-requiring process and seems to require normal bile salt metabolism.

Conjugated bilirubin is not present in normal plasma, or is present in very small quantities and routine methods of estimation are inaccurate at this level. At pathologically high concentration it may also be bound to plasma albumin. Conjugated bilirubin is water-soluble and appears in the urine.

The conjugated bilirubin enters the gut in bile. In the colon it is broken down by bacteria to a group of products known collectively as *stercobilinogen* (often referred to as *faecal urobilinogen*). These are oxidised to *stercobilin*, a pigment that contributes to the brown colour of the stool. A small amount is absorbed into the portal circulation and most of this is re-excreted in bile: a very small fraction appears in the urine as *urobilinogen* which in turn can be oxidised to *urobilin*. These amounts are not always detectable by routine urine tests (p. 297). The sequence of events is outlined in Fig. 23.

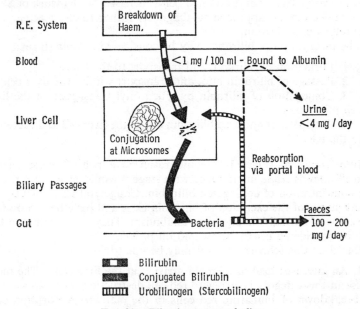

FIG. 23.—Bile pigment metabolism.

The term *bile pigments* is used to include bilirubin and its breakdown products. With the exception of urobilinogen and stercobilinogen all are coloured.

*Other terminology for bile pigments.*—An early test used in the investigation of jaundice was the van den Bergh reaction. In some cases of jaundice plasma bilirubin reacted directly with the colour reagent used (direct reaction) whereas in others the addition of methanol was required (indirect reaction). These terms are still sometimes used in expressing the results of quantitative tests and total plasma bilirubin may be subdivided into *direct* and *indirect* fractions. These correspond approximately to *conjugated bilirubin* and *bilirubin* respectively. The latter terms are preferable and will be used in this book.

*Bile salts* are one of the end products of cholesterol breakdown and are considered on p. 308.

### DISORDERS OF BILE PIGMENT METABOLISM AND THE CLASSIFICATION OF JAUNDICE

Numerous classifications of jaundice exist. Most are based on the mechanism or site of production of the excess bilirubin, but these are not of great diagnostic value. Many cases of jaundice overlap more than one category. The scheme outlined below is based on the work of Sherlock and offers an approach to diagnosis rather than an attempt at theoretical classification.

The metabolism of bilirubin may be considered in four stages:

1. Formation of bilirubin and transport to the liver.
2. Passage of bilirubin into, and transport within, the liver cell.
3. Conjugation of bilirubin by glucuronyl transferase in the liver cell microsomes.
4. Passage of conjugated bilirubin out of the liver cell and excretion in the bile.

Interference at stages 1–3 would result in a rise of unconjugated bilirubin levels while interference with stage 4 would result in a rise of the concentration of conjugated bilirubin. Congenital abnormalities of many stages of the excretion of bilirubin exist and have helped outline the steps of normal bile pigment metabolism. They are uncommon but will be considered briefly below and on p. 306.

Based on this scheme jaundice may be due to:

1. **An increased load of bilirubin arriving at the liver cell.**—The main cause in this category is increased red cell destruction, as in *haemolysis* or breakdown of immature red cells in the marrow. Absorption of a *large haematoma* has the same result and may, particularly in infants, produce jaundice. It is an important cause of jaundice in the newborn when plasma bilirubin levels may be very high. In adults, haemolytic jaundice is usually mild (30–50 $\mu$mol/l or 2–3 mg/dl) because of the very high excretory capacity of the adult liver.

2. **Defective uptake and transport by the liver cell.**—Reduced uptake by the liver cell may be a factor in the hyperbilirubinaemia of *Gilbert's disease*. Low levels of the Y protein in the newborn may be partially responsible for "physiological" neonatal jaundice.

3. **Disturbance of conjugation.**—This is a contributory factor in *neonatal jaundice*, especially in premature babies where the glucuronyl transferase enzyme system is immature. Congenital deficiency of this enzyme occurs in the *Crigler-Najjar* syndrome when jaundice may be severe. Novobiocin can produce jaundice at this stage, as can destruction of liver cells, such as in hepatitis. Reduced enzyme activity has also been found in some patients with Gilbert's disease.

TABLE XXIV

Causes of Jaundice

| Fraction of bilirubin elevated | Basic abnormality | Existing classifications | | |
|---|---|---|---|---|
| Unconjugated | 1. Excessive formation of bilirubin | Pre-hepatic | Haemolytic | *Retention jaundice* (a) Haemolytic |
| | 2. Impaired entry of bilirubin into, or transport within, the liver cell | | | (b) Non-haemolytic |
| | 3. Impaired conjugation of bilirubin | Hepatic | Hepatogenous | |
| Both, mainly conjugated | 4. Impaired excretion of conjugated bilirubin | Post-hepatic | Obstructive | *Regurgitation jaundice* (a) Parenchymal (b) Mechanical |

4. **Disturbances of excretion of conjugated bilirubin.**—This is the largest group and is characterised by a rise in conjugated as well as unconjugated bilirubin. *Liver cell destruction* and *intra- and extrahepatic cholestasis* fall into this group, as does the congenital *Dubin–Johnson syndrome* (p. 306). Most of the chemical tests of "liver function" are used to distinguish different causes of jaundice in this category.

Comparison with existing aetiological classifications is shown in Table XXIV.

BIOCHEMICAL TESTS IN LIVER DISEASE

Many tests of "liver function" exist. Discussion in this chapter is confined to those tests commonly performed and which the authors have found useful. Most suffer from two main disadvantages:

1. The liver, like the kidney, has considerable functional reserve and many tests only become positive when there is widespread involvement of the organ.

2. Most parameters, such as the changes in plasma proteins, and plasma enzyme activity, are also affected by diseases other than those of the liver. For this reason more than one test is usually performed, selection depending on the probable diagnosis.

**Basic Processes in Liver Disease**

There is no test or combination of tests completely specific for any particular disease. Most laboratories perform several tests of "liver function" which indicate the nature of the pathological process, but not necessarily the cause of it. There are two main pathological changes seen in most liver disorders and either or both may be present in a single disease. These are:

(*a*) Liver cell damage, with or without demonstrable liver dysfunction.

(*b*) Cholestasis.

(*a*) **Liver cell damage.**—This may vary from areas of focal necrosis to destruction of most of the liver leading to liver failure. Examples of the main causes are:

(i) *Acute hepatitis*, which is most commonly viral but may occur in the course of other diseases such as infectious mononucleosis, or septicaemia.

(ii) The action of *toxins* on the liver, especially chlorinated hydrocarbons such as carbon tetrachloride. Phosphorus is a hepatotoxin. A number of the hydrazine drugs such as iproniazid (a monoamine oxidase inhibitor) produce liver damage and therefore are no longer in clinical use.

(iii) *Chronic hepatitis* which may follow viral hepatitis (*chronic persistent hepatitis*), or which may be related to autoimmunity (*chronic active, or aggressive, hepatitis*).

(iv) Liver cell damage may occur secondarily to prolonged *biliary obstruction*.

(v) Cellular destruction of known or unknown aetiology may be followed by *cirrhosis*, and the process may persist in the cirrhotic liver.

(vi) *Hepatic congestion* and/or hypoxia such as occur in congestive cardiac failure or "shock".

The *consequences* of liver cell damage vary in degree. If mild, biochemical tests may provide the only evidence. With more severe damage there is usually jaundice and demonstrable failure of synthetic function. Severe destruction may be fatal.

(*b*) **Cholestasis.**—Cholestasis is "the syndrome associated with failure of bile to reach the duodenum" (Sherlock). The causes may be divided into intrahepatic and extrahepatic. The first are not amenable to surgery, the second may be. Distinction between the two is therefore important.

(i) *Intrahepatic cholestasis.*—This is frequently associated with liver cell destruction. The main causes are:

Viral hepatitis (cholangiolytic hepatitis).
Certain drugs such as methyl testosterone, chlorpromazine.
Pregnancy. Idiopathic cholestasis may develop in late pregnancy in some subjects, and may recur in subsequent pregnancies. These patients also tend to develop cholestasis with oral contraceptive therapy. The cholestasis is thought to be hormonal in origin.
Cholangitis.
Biliary cirrhosis.
Cirrhosis (some cases).
Infiltrations of the liver (e.g. Hodgkin's disease, malignancy).
Intrahepatic biliary atresia.

(ii) *Extrahepatic cholestasis.*—The main causes are:

Gall stone in the common bile duct.
Carcinoma of head of pancreas, ampulla of Vater or, rarely, bile duct.
Fibrosis of the bile duct.
External pressure from tumour; glands.
Extrahepatic atresia of bile duct.

The *consequences* of cholestasis depend largely on its duration. There is accumulation in the blood of substances normally excreted in the bile, including conjugated bilirubin, bile salts and cholesterol.

Retention of bilirubin leads to jaundice. The high levels of cholesterol found in some cases, especially in biliary cirrhosis, may cause xanthomatosis. In chronic cases, with prolonged absence of bile salts from the gut, absorption of fat and fat-soluble vitamins is impaired (p. 277).

Abnormalities in bile salt metabolism may be important in the initiation and maintenance of cholestasis—both intrahepatic and extrahepatic. A simplified version of a postulated sequence of events is as follows. Increased pressure in the biliary system (e.g. due to extrahepatic obstruction) or primary cellular damage may interfere with bile salt metabolism and decrease bile salt excretion. This leads to accumulation in the cell and damage, with perpetuation of the cholestasis. As bile salt excretion is important for normal bile flow, there is stagnation and formation of bile plugs in the canaliculi which damages them. These events are common to both intrahepatic and early extrahepatic cholestasis; it is therefore not surprising that distinction between the two by biochemical tests is rarely possible.

## Tests for Liver Disease

Tests may be classified as follows:

1. Tests for liver cell damage

   *Plasma enzymes* released by damaged cells (for example *transaminases*)

2. Tests for liver dysfunction

   (a) Tests of conjugating capacity
   *Bile pigment* metabolism
   *Bromsulphthalein* excretion

   (b) Tests of synthetic function

   Serum *albumin* (associated with a rise in $\gamma$ globulin due to increased production elsewhere)
   *Coagulation factors*

3. Tests for cholestasis

   *Plasma enzymes* arising from the biliary tract
   *Alkaline phosphatase* and *5′-nucleotidase*

4. Tests indicating aetiology

   *Hepatitis associated antigen* (HAA)
   Circulating *antibodies*
   $\alpha$ *fetoprotein*

Although all these tests are known collectively as "liver function tests" (LFTs), only those in group 2 truly test function.

## 1. Tests Indicating Liver Cell Damage

Damage to liver cells, with or without necrosis, causes the acute release of intracellular constituents into the blood stream. This is detected by measuring plasma enzymes.

*Transaminases.*—Raised levels of transaminases are found with liver cell damage due to any cause, most commonly viral hepatitis (p. 301). Particularly high levels are seen in cases of hepatic necrosis due to toxins. Mildly raised levels occur in cholestasis and some cases of cirrhosis. Both aspartate (AST: SGOT) and alanine (ALT: SGPT) transaminases are affected.

AST is present in both mitochondria and cytoplasm whereas ALT is found in the cytoplasm only. With mild cellular damage ALT levels are higher than those of AST whereas the latter predominates in cases with more severe cellular damage and necrosis. As ALT has a longer half-life than AST, raised levels often persist for a longer period of time. For the detection of cell damage either enzyme may be measured.

*Isocitrate dehydrogenase.*—The changes in isocitrate dehydrogenase parallel those of the transaminases. The enzyme is much more specific for liver disease (p. 353) than the transaminases, but its estimation offers no special advantage for the detection of liver cell damage. Levels in uncomplicated extrahepatic obstruction are often normal.

*Lactate dehydrogenase.*—A similar pattern is seen in levels of lactate dehydrogenase (LD). This test is less sensitive than those already discussed and offers little extra advantage in the investigation of liver disease.

## 2. Tests for Liver Dysfunction

**Tests of conjugating capacity.**—*Bile pigment metabolism.*—The physiology of bile pigments has been outlined on p. 290. The essential tests required to assess jaundice are estimation of plasma bilirubin (both fractions), qualitative tests for urinary bile products and visual inspection of the faeces for stercobilin.

1. Raised levels of *unconjugated bilirubin* only indicate excessive red cell destruction, defective entry of bilirubin into the liver cell or defective conjugation of bilirubin. The diagnosis is rarely a problem and the main causes have been outlined on p. 292.

2. In most cases of jaundice *both fractions* are raised. In cholestasis conjugated bilirubin predominates.

3. *Bilirubin in the urine* means that there is an increase in circulating conjugated bilirubin and is always pathological.

4. Increased *urobilin(ogen)* in the urine may indicate:

(i) Excessive amounts of bile pigment reaching the bowel and being reabsorbed, for example, due to haemolysis.

(ii) Normal amounts of bile pigment with liver damage and failure to re-excrete urobilinogen (Fig. 23), for example, due to hepatitis.

(iii) A positive result may occur in normal urine, if it is a concentrated specimen.

In fresh urine urobilinogen is present, but with time urobilin is produced. The significance of the finding remains unaltered. The usual test for urobilin (p. 313) converts any urobilinogen present into urobilin and so is probably the test of choice.

Quantitative estimation of urobilinogen in a timed collection of urine eliminates the variation due to urinary concentration. Excretion is, however, affected by other factors such as the pH of the urine, and the estimation is only rarely of value.

Complete absence of urobilinogen from the urine in a case of jaundice indicates complete cholestasis (no bilirubin reaches the bowel for conversion and reabsorption). Conversely the presence of urobilinogen is incompatible with the diagnosis of *complete* biliary obstruction. It is therefore important to report urinary urobilinogen as increased, "normal", or absent. The latter may, of course, be normal.

5. Complete absence of bile pigments in the faeces produces pale stools and indicates cholestasis. There are, however, other causes of pale stools such as steatorrhea (p. 272).

*Bromsulphthalein test of excretory function.*—As usually performed, less than 5 per cent of an injected dose of bromsulphthalein (BSP) remains in the circulation 45 minutes after administration.

Excretion depends not only on the integrity of the liver cell, but also on adequate circulation in the liver and patency of the biliary tract.

Increased retention of BSP is a sensitive index of hepatocellular dysfunction. *False positive* results occur where there is impaired circulation in the liver, for example, in congestive cardiac failure, or if biliary obstruction is present. The test is unnecessary if the patient is jaundiced and its main function is as a sensitive test of liver function when other tests are normal.

It has been pointed out that there are at least two phases in the disappearance of BSP from the blood, the first representing equilibration within the liver (and possibly uptake by the reticulo-endothelial system) while only the second represents hepatic excretion. To distinguish these two phases multiple blood samples at 10-minute intervals are required. It is not clear whether this procedure offers any diagnostic advantage over the single stage test in properly selected cases.

## Tests Mainly of Synthetic Function

**Plasma proteins and coagulation factors.**—*Albumin* is catabolised at a rate of about 4 per cent of the body pool daily; decreased synthesis is soon reflected in reduced plasma levels. Hypoalbuminaemia may occur in both acute and chronic liver disease and is an index of the severity and extent of liver damage.

Changes in the α *and β globulin* electrophoretic fractions are inconstant and non-specific. γ *globulin* levels increase in chronic liver disease and in cirrhosis a characteristic, but not specific, pattern often occurs: the β and increased γ fractions fuse due to an increase in fast-moving immunoglobulins.

Estimation of the separate *immunoglobulins* may occasionally be of value. IgA increases in *early cirrhosis* but in most established cases IgG and IgM are also elevated. Predominant elevation of IgG suggests *chronic active hepatitis. Primary biliary cirrhosis* often has a marked increase in IgM.

*Many of the coagulation factors* are synthesised by the hepatic parenchymal cells. Hepatocellular damage may give rise to a bleeding state, or laboratory tests of coagulation may be abnormal. Four factors, II (prothrombin), VII, IX and X are only synthesised in the liver in the presence of vitamin K. Their activities (with the exception of factor IX) are conveniently measured by the *one-stage prothrombin time.* Deficiencies may arise in two ways. In cholestasis the absorption of the fat-soluble vitamins, including vitamin K, is impaired (p. 273). In the presence of parenchymal damage synthesis is impaired despite an adequate supply of vitamin K. These two mechanisms offer a way of distinguishing cholestasis from cellular damage by the response in the former to parenteral vitamin K. Shortening of the prothrombin time after vitamin K injection suggests cholestasis whereas total lack of response indicates liver cell damage. As in many liver function tests there is considerable overlap of results.

In severe liver disease deficiencies of factors V and I (fibrinogen) may also be encountered. The latter is more often due to fibrinolysis than to deficient synthesis.

## 3. Tests Indicating Cholestasis

**Plasma alkaline phosphatase.**—As mentioned on p. 350, plasma alkaline phosphatase is derived from at least two sources—the osteoblasts and the cells lining bile canaliculi. The latter fraction is normally excreted in the bile and, if there is biliary obstruction, is regurgitated into the blood stream, as is conjugated bilirubin. Raised alkaline phosphatase levels in jaundice generally indicate cholestasis, although moderately raised levels, as in hepatitis, may be a result of liver cell damage.

The higher the level of alkaline phosphatase the greater the likelihood of extrahepatic cholestasis.

If obstruction involves one of the hepatic ducts or if it is due to local-ised lesions such as multiple secondary carcinomata, a characteristic picture develops. There is retention of both bilirubin and alkaline phosphatase. Bilirubin, unlike alkaline phosphatase, can be excreted by liver cells in the unaffected part of the biliary system, so plasma levels may be only minimally raised. In the absence of bone disease the finding of a raised alkaline phosphatase out of proportion to plasma bilirubin is suggestive of a space-occupying lesion(s) or obstruction of one of the radicles of the common bile duct.

**5'-nucleotidase** activity parallels that of liver alkaline phosphatase (p. 350) and may be used as an alternative test for cholestasis. Alter-natively, it may be reserved for distinguishing a raised alkaline phos-phatase due to liver disease from that due to bone disease if there is clinical uncertainty. It is particularly valuable in children (p. 350).

Raised alkaline phosphatase and 5'-nucleotidase may occur, without jaundice or raised transaminases, in cholangitis.

**γ Glutamyltransferase** (p. 352) is also an indicator of liver disease. The highest values occur with cholestasis but levels also rise in other liver disease such as hepatitis and cirrhosis. It is valuable both in detecting liver disease and in indicating the source of a raised alkaline phos-phatase level.

It must be emphasised that these tests *do not distinguish* between intra- and extrahepatic cholestasis.

### Tests Indicating Aetiology

Three newer tests are of value in the diagnosis of liver disease.

**Hepatitis-associated antigen** (HAA) (Australia antigen).—In some cases of viral hepatitis an antigen (HAA) appears in the blood; it is probably either the infectious agent or a marker of its presence. HAA has been found in the blood of:

(*a*) some patients with *acute hepatitis*, most commonly long-incubation hepatitis. It is detectable before the onset of symptoms and usually disappears within a few weeks. Epidemic viral hepatitis is usually HAA negative.

(*b*) some patients (particularly males) with *chronic active hepatitis*.

(*c*) patients with diseases associated with *immunological deficiency* such as leukaemia, leprosy and chronic renal failure. These patients may not develop clinical hepatitis and HAA persists in their blood for months or years. Such patients, particularly those with renal failure in dialysis units, should be considered infective.

(*d*) some apparently unaffected normal subjects (carriers). The test is therefore used to screen blood donors, and staff on dialysis units.

**Circulating antibodies.**—Circulating antibodies to mitochondria and smooth muscle, as well as antinuclear factor, often occur in three related conditions, *chronic active hepatitis, cryptogenic cirrhosis* and *primary biliary cirrhosis.* Positive results help to distinguish these conditions from clinically similar diseases. Mitochondrial (or M) antibody is especially helpful in differentiating between primary biliary cirrhosis (more than 80 per cent of cases positive) and extrahepatic biliary obstruction (usually negative).

α **Fetoprotein.**—α Fetoprotein, normally present in foetal serum, is undetectable by *routine immunodiffusion techniques* in children and adults. Detectable levels occur in most patients with *primary hepatocellular carcinoma* and some patients with germinal cell tumours of testis and ovary. Transient positive results may be found during acute viral hepatitis, particularly in children.

**Biochemical Changes in Individual Liver Diseases and the Selection of Tests**

**1. Acute hepatitis (viral or toxic).**—The basic pathological abnormality in this condition is cell damage: severity varies from that detectable only by plasma enzyme estimations to that which produces detectable disturbance in function. In some cases cholestasis is a dominant feature (cholestatic hepatitis).

*Bilirubinuria* is often the first biochemical evidence of the disease, and is present before the onset of clinically detectable jaundice in many cases. The patient may notice darkening of the urine. During the prodromal phase transaminase levels are raised, reach their peak at about the time jaundice is noticed, and remain elevated for about a further two weeks.

The plasma bilirubin level rises but rarely exceeds 340 μmol/l (20 mg/dl), unless there is impaired renal function. The jaundice is due both to cell damage and to cholestasis and the rise affects both fractions, predominantly the conjugated one. At this stage the urine contains *urobilin(ogen)* and bilirubin, and the *alkaline phosphatase* level is moderately increased. There is an element of cholestasis in most cases of hepatitis and this usually appears shortly after the onset of jaundice, when the stools become pale and urobilin(ogen) disappears from the urine. It reappears as the liver recovers only to disappear finally with complete recovery. Bilirubin disappears from the urine before urobilinogen. This pattern of urinary biliary constituents is seen in many, but not all cases of viral hepatitis. In some, cholestasis occurs from the onset and the patient notices pale stools as the first feature.

Tests for HAA may be positive in the early stages.

**In summary,** the sequence of events is:

TABLE XXV

|  | Prodromal phase | Jaundice (liver damage) | Cholestatic phase | Relief of cholestasis | Recovery phase |
|---|---|---|---|---|---|
| *Urine* |  |  |  |  |  |
| Bilirubin | + | + | + | + | − |
| Urobilin (ogen) | − | + | − | + | − |
| *Faecal* |  |  |  |  |  |
| colour | N | N | pale | N | N |
| *Blood* |  |  |  |  |  |
| Bilirubin | Usually N | + + | + + + | + + | + |
| AST | + + + | + + + + | + + | + | N |

Anicteric hepatitis is not uncommon and may be diagnosed by bilirubinuria and raised transaminase levels.

The *course* of hepatitis may be followed by plasma bilirubin estimations together with the changes in urine and faeces outlined in Table XXV. If relapse is suspected clinically, it may be confirmed by enzyme levels which remain high, or which show a second rise. The development of a chronic phase is suggested by an increase in $\gamma$ globulin levels.

The *severity and extent* of liver cell destruction are shown best by the prothrombin time, although this is also affected by bile salt deficiency (p. 299): occasionally, in very severe damage, albumin levels may be unequivocally low. Bilirubin levels are affected by both cholestasis and cell damage and the rise in transaminases reflects the rapidity rather than degree of destruction of liver cells.

Most cases of hepatitis recover, but a very small percentage die in liver failure and a small number develop cirrhosis.

2. **Chronic hepatitis** (whether chronic aggressive or chronic persistent). The only abnormal finding in these conditions may be raised transaminases. In this situation ALT is much more sensitive than AST (p. 297).

3. **Biliary obstruction.**—In complete extrahepatic biliary obstruction *plasma bilirubin* levels rise progressively for several weeks, when the curve tends to flatten out. Levels may fluctuate slightly but always tend to increase and may reach values of 850 $\mu$mol/l (50 mg/dl) or greater. The rise is predominantly in the conjugated fraction. The urine contains bilirubin but no urobilin(ogen) and the stools are pale.

*Plasma alkaline phosphatase* levels increase progressively as do those of cholesterol and phospholipids. After prolonged obstruction there may be secondary liver damage due to bile necrosis of liver cells or ascending cholangitis and tests of liver cell damage become positive.

**Differential diagnosis of prolonged cholestasis.**—The main causes to be considered are:

(i) Intrahepatic—cholestatic hepatitis
    drug-induced jaundice (for example, chlorpromazine)
    primary biliary cirrhosis

(ii) Extrahepatic obstruction

Surgery is usually indicated for extrahepatic obstruction: as anaesthesia may be detrimental in the presence of hepatocellular damage, it is important to distinguish between the types.

As outlined above not only does prolonged extrahepatic obstruction cause liver damage but cholestasis develops in many cases of hepatitis. *It is most important to perform the relevant tests as early as possible* if the primary process is to be diagnosed. This is second in importance only to an adequate clinical history (with emphasis on possible drug exposure) and examination.

In both conditions the rise of plasma bilirubin is predominantly in the conjugated fraction. Points helpful in differential diagnosis are the following:

(*a*) Transaminase levels which are very high, or are elevated early in the course of the disease are in favour of hepatitis. In extra-hepatic obstruction the levels of these enzymes may be normal, although they tend to increase as cell damage occurs.

(*b*) Raised alkaline phosphatase, 5′-nucleotidase or $\gamma$ glutamyltransferase levels indicate the presence of cholestasis but not its cause. The higher the alkaline phosphatase the greater the likelihood of extrahepatic obstruction, but there is no level clearly separating the two.

(*c*) A positive mitochondrial antibody (M) test is a strong point against extrahepatic obstruction (p. 301). A predominant increase in IgM also favours a diagnosis of primary biliary cirrhosis.

(*d*) In problem cases the *prednisolone test* (see Appendix, p. 312) has been used. An unequivocal fall of plasma bilirubin after prednisolone is strong evidence for hepatitis rather than extrahepatic obstruction. The mechanism of this lowering of plasma bilirubin is unknown.

These laboratory tests must be supplemented by other diagnostic procedures, such as radiological studies and liver biopsy.

3. **Cirrhosis.**—In cirrhosis during phases of active cellular destruction there are usually raised transaminase levels, sometimes with jaundice. *Urobilin(ogen)* is frequently present in the urine, with or without *bilirubin*. Many cases show the typical *plasma protein* changes (p. 299) of a low albumin and diffusely elevated $\gamma$ globulin with $\beta$–$\gamma$ fusion. If ascites is present there may be dilutional hyponatraemia (p. 53). Plasma $\gamma$ glutamyltransferase is a sensitive indicator of liver disease, and may be the only enzyme elevated, particularly in alcoholic cirrhosis (p. 352).

During quiescent periods, or in mild cases, the only detectable abnormality may be increased BSP retention.

4. **Hepatocellular failure.**—Most cases of hepatocellular failure occur during the course of severe hepatitis or decompensated cirrhosis, or following ingestion of liver toxins. Jaundice is usually present although fulminating cases may die before it develops. Any or all of the biochemical abnormalities of hepatitis may be found depending on the stage of the disease.

Other features may include:

(*a*) A low plasma urea. Normally ammonia from deamination of amino acids is utilised in the synthesis of urea in the liver. Impairment of this process results in deficient urea production and accumulation of amino acids in the blood with consequent overflow aminoaciduria (p. 364).

(*b*) A prolonged prothrombin time.

(*c*) Rarely hypoglycaemia (p. 205).

(*d*) Severe electrolyte disturbances, particularly hypokalaemia, due to secondary aldosteronism.

5. **Hepatic infiltration.**—Hepatomegaly may be a result of infiltration of the liver by carcinoma or lymphoma or of replacement by granulomata. Transaminase levels are often raised, AST being much more sensitive in this situation than ALT. If there is an element of cholestasis, alkaline phosphatase and 5'-nucleotidase levels may also rise, and may be the only abnormal findings. In many cases bilirubin levels are normal, and liver function is rarely demonstrably impaired.

6. **Haemolytic jaundice.**—Increased erythrocyte destruction of any aetiology produces an increase in the amount of bilirubin arriving at the liver for conjugation and excretion. The reserve capacity of the adult liver is such that jaundice, if present, is mild (usually below 70 $\mu$mol/l (4 mg/dl)). The rise in plasma bilirubin is confined almost entirely to the unconjugated fraction unless there is associated liver disease. The increased load of excreted bilirubin in the bile, however, results in increased stercobilinogen production in the gut and, after this has been

reabsorbed, an increased urinary urobilin(ogen). Bilirubin does not appear in the urine (acholuric jaundice). Other "liver function" tests are unaffected with the exception in some cases of raised AST and LD levels, due not to liver damage but to red cell destruction (p. 355).

In the newborn, however, the massive red cell destruction occurring in haemolytic disease of the newborn, coupled with the immature hepatic handling of bilirubin, can produce elevations of unconjugated bilirubin of 400–500 $\mu$mol/l (25–30 mg/dl) or greater. Such elevations are associated with the risk of developing *kernicterus* and may be reduced by exchange transfusion. In premature infants, too, the poorly developed conjugating mechanism may result in so-called "physiological" jaundice with markedly raised levels of unconjugated bilirubin, also necessitating exchange transfusion.

The effects of certain drugs on jaundice in the newborn are considered on p. 307.

In summary, the following are the most useful tests in individual conditions.

1. Suspected acute hepatitis
   Plasma bilirubin
   ALT *or* AST
   HAA
   Urinary bilirubin and urobilinogen.

2. Suspected chronic hepatitis
   ALT (AST may be normal)
   In some cases HAA and antibodies
   Plasma protein electrophoresis.

3. Suspected cirrhosis
   Plasma albumin and protein electrophoresis. In this condition any or none of the biochemical findings may be abnormal; in addition ALT, AST, alkaline phosphatase *or* 5′-nucleotidase, bilirubin and, *if all these are normal*, the BSP excretion test should be performed.

4. Suspected cholestasis or cholangitis
   Alkaline phosphatase *or* 5′-nucleotidase
   Bilirubin, with special reference to the conjugated fraction
   Transaminases are less useful, but may help to differentiate extra from intrahepatic cholestasis.

5. Suspected hepatic deposits or infiltration
   AST
   Alkaline phosphatase *or* 5′-nucleotidase.

6. Suspected primary hepatocellular carcinoma
   α fetoprotein
   AST
   Alkaline phosphatase *or* 5'-nucleotidase.

## CONGENITAL HYPERBILIRUBINAEMIA

A group of disorders, apparently of congenital origin, has been described in which there is defective handling of bilirubin. The three best defined entities will be described briefly.

1. **Gilbert's disease.**—In patients with this disease there is probably defective transport of bilirubin into the liver cell. In some, reduced glucuronyl transferase (p. 290) activity has been demonstrated. Plasma levels of *unconjugated bilirubin* are mildly raised (20–35 $\mu$mol/l or 1–2 mg/dl), and tend to fluctuate. The condition may be noted at any age. A common means of discovery is the failure of bilirubin levels to return to normal after an attack of hepatitis, or any mild illness which may be misdiagnosed as hepatitis. The only other abnormality detectable by routine tests is a slightly prolonged BSP retention in some cases. The condition is harmless but must be differentiated from haemolysis and from hepatitis.

2. **Crigler-Najjar syndrome.**—In contradistinction to the other two syndromes discussed the Crigler–Najjar syndrome is *not* harmless. It presents in the first few days of life as jaundice due to a rise in *unconjugated bilirubin* levels. They may often be high enough (350 $\mu$mol/l [20 mg/dl] or higher) to cause kernicterus. The probable cause is a deficiency of glucuronyl transferase. In infants who survive, the level of bilirubin tends to stabilise, suggesting the existence of alternative pathways of bilirubin excretion.

3. **Dubin-Johnson syndrome.**—This condition is characterised by mildly raised *conjugated bilirubin* levels that tend to fluctuate. There is defective excretion of conjugated bilirubin. Bilirubin is present in the urine. Alkaline phosphatase levels are normal. There may be hepatomegaly and the liver is dark brown due to the presence of a pigment with the staining properties of lipofuscin in the cells. The condition is harmless and the diagnosis may be confirmed by the characteristic staining in the liver biopsy.

There is overlap between these and related conditions.

## DRUGS AND THE LIVER

Many drugs are capable of producing jaundice with or without liver damage and a drug history is an essential part of the investigation of a patient with jaundice. This summary is based on the references given at the end of the chapter.

There are several mechanisms whereby drugs can produce jaundice.

1. Drug-induced *haemolysis* is usually not severe enough to cause jaundice. If it does the findings are those of haemolytic jaundice. The subject is dealt with in textbooks of haematology.

2. Novobiocin interferes with bilirubin *conjugation*.

3. Direct *hepatotoxins* such as carbon tetrachloride and *Amanita phalloides* poisoning produce liver cell necrosis of varying severity. Ingestion is usually accidental or with suicidal intent.

Tetracyclines by intravenous injection, especially during pregnancy, have produced acute fatty liver and death from liver failure.

Acute overdosage (usually accidental or suicidal) of paracetamol, and, in children, of ferrous sulphate, can produce severe hepatic necrosis.

4. *Hepatitis-like reaction.*—Several drugs may produce a clinical, biochemical and histopathological syndrome closely resembling viral hepatitis.

(*a*) Hydrazine derivatives, e.g. iproniazid, phenelzine and rarely isoniazid.

(*b*) Chlordiazepoxide (rare).

(*c*) Methyldopa (occasionally).

(*d*) Halothane anaesthetics rarely produce severe liver damage and death.

5. *Cholestatic reaction.*—Another group of drugs produces the picture of cholestatic jaundice:

(*a*) 17-α-alkylated steroids such as methyl testosterone, norethandrolone, norethisterone (components of certain oral contraceptives).

(*b*) Phenothiazines, such as chlorpromazine.

(*c*) Rarely thiouracil, chlorpropamide, phenylbutazone.

(*d*) Antibiotics such as erythromycin and oleandomycin.

(*e*) Para-aminosalicylic acid (PAS) (rare).

(*f*) Cholestasis may occur as part of a generalised sensitivity reaction (e.g. to penicillin).

**Neonatal jaundice and drugs.**—In the newborn, particularly if premature, the hepatic handling of bilirubin is immature (p. 292) and jaundice due to a rise in unconjugated bilirubin levels is common. This phase lasts two to three days in full term infants and five to six days in premature infants. The course and severity of unconjugated hyperbilirubinaemia may be influenced by drugs in three ways:

(a) Several drugs displace bilirubin from plasma albumin and increase the risk of kernicterus. These include salicylates and sulphonamides.

(b) Novobiocin inhibits the glucuronyl transferase system and aggravates unconjugated hyperbilirubinaemia.

(c) Any drug producing haemolysis aggravates the condition.

## BILE AND GALL STONES

### BILE ACIDS AND BILE SALTS

Four bile acids are produced in man. Two of these, *cholic acid* and *chenodeoxycholic acid*, are synthesised in the liver from cholesterol and are referred to as *primary bile acids*. These are excreted in the bile into the gut where bacterial action converts them to *secondary bile acids*, *deoxycholic acid* and *lithocholic acid*, respectively. In human bile the bile acids are in the form of sodium salts, all conjugated with the amino acids glycine or taurine (*bile salts*). Some of the secondary bile salts are absorbed and re-excreted by the liver (enterohepatic circulation of bile salts, p. 267). Bile therefore contains a mixture of primary and secondary bile salts.

Ths importance of the bile salts lies in their possession of both polar and non-polar chemical groups and their ability to coalesce into *micelles*. In these micelles the non-polar groups are orientated in the centre of the molecule and form a small pool of lipid solvent, while the polar (water-soluble) groups are on the outside.

*Deficiency of bile salts* in the intestinal lumen leads to impaired micelle formation, and malabsorption of fat (p. 277). Such deficiency may be caused by cholestatic liver disease (failure to reach the gut) or by ileal resection or disease (failure of reabsorption, with reduced bile-salt pool). The role of deficient bile salts in cholelithiasis is discussed below.

### FORMATION OF BILE

About 1 to 2 litres of bile is produced in the liver daily. This *hepatic bile* contains bilirubin, bile salts, phospholipids and cholesterol as well as electrolytes in similar concentrations to those in plasma. Small amounts of protein are also present. In the gall bladder there is active reabsorption of sodium, chloride and bicarbonate, together with an isosmotic amount of water. The end result is *gall bladder bile* which is ten times more concentrated than hepatic bile and in which sodium is the major cation and bile salts the major anion. The concentration of other non-absorbable molecules, conjugated bilirubin, cholesterol and phospholipids also increases.

## GALL STONES

Although most gall stones contain all constituents of bile, several types differing in the main constituent are recognised. Unlike renal calculi only about 10 per cent of gall stones contain sufficient calcium to be radio-opaque.

1. *Cholesterol gall stones* may be single or multiple and are not usually radio-opaque. They are white or yellowish and the cut surface has a crystalline appearance. They may be mulberry shaped.

2. *Pigment gall stones* consist largely of bile pigments with organic material and variable amounts of calcium. They are small multiple stones, dark green or black and are hard. Rarely they are radio-opaque.

3. *Mixed gall stones*, the commonest form, are, as the name suggests, composed of a mixture of cholesterol, bile pigments, protein and calcium. They are multiple and appear as faceted dark brown stones with a hard shell and softer centre. They may be radio-opaque.

## CAUSES OF GALL STONE FORMATION

*Pigment stones* are found in *chronic haemolytic states* such as hereditary spherocytosis where there is an increase in bilirubin formation and excretion. In all other forms of gall stones, however, the aetiology is uncertain. The role of infection as a precipitating factor in gall stone formation is at present controversial.

Current theories are based on the demonstration of differences between bile in which gall stones have formed and normal bile. Such pathological bile has a greater concentration of cholesterol and a lower concentration of bile salts than normal. As mentioned above bile salts are required to maintain cholesterol in solution. In addition, the salts of cholic acid (with three OH groups) form more stable micelles than do those of chenodeoxycholic acid and deoxycholic acid (with two OH groups). Stone-forming bile has been shown to have a higher proportion of the dihydroxy bile salts than normal. This combination of bile salts deficient in quantity and quality would favour the precipitation of cholesterol. The cause of such alterations is not clear. It may be due to primary hepatic or gall bladder dysfunction or it may be the result of reduced enterohepatic circulation of bile salts due to biliary stasis or obstruction. It has been suggested that inflammation of the gall bladder wall leads to greater absorption of the more water-soluble bile salts (those with more OH groups).

Routine chemical analysis of gall stones is of little value.

## CONSEQUENCES OF GALL STONES

Gall stones may remain silent for an indefinite length of time and be discovered only at laparotomy for an unrelated condition, or on x-ray of the abdomen. They may, however, lead to several clinical consequences:

1. Acute cholecystitis, due to obstruction of the cystic duct by a gall stone with chemical irritation of the gall bladder mucosa by trapped bile and secondary bacterial infection.

2. Chronic cholecystitis.

3. Common bile duct obstruction, if a stone lodges in the bile duct. This may present as biliary colic, obstructive jaundice or acute pancreatitis if the pancreatic duct is also occluded.

4. Extremely rarely, carcinoma of the gall bladder occurs.

## SUMMARY

### LIVER DISEASE

1. The liver has a central role in many metabolic processes.

2. Bilirubin derived from haemoglobin is conjugated in the liver and excreted in bile. Conversion to stercobilinogen (faecal urobilinogen) takes place in the bowel. Some reabsorbed faecal urobilinogen is excreted in the urine.

3. Bilirubin metabolism may be assessed by plasma levels of total and conjugated bilirubin, by qualitative tests for bilirubin and urobilin(ogen) in the urine and by visual inspection of the stool.

4. Jaundice is due to a raised plasma bilirubin. No satisfactory classification exists other than separating jaundice due to a raised unconjugated bilirubin only from that in which both fractions are raised. The majority of cases of jaundice fall into the latter group.

5. The two basic processes in liver disease are liver cell damage and cholestasis. Tests help to distinguish the underlying process, but not necessarily its cause.

6. Tests of liver function are considered in the following categories:

   (a) Tests for liver cell damage.

   (b) Tests for liver dysfunction.

   (c) Tests for cholestasis.

   (d) Tests indicating aetiology.

7. The selection of tests is governed by the particular problem. The main indications are:

(*a*) Differential diagnosis of jaundice.
(*b*) Assessment of the severity and progress of liver disease.
(*c*) Detection of liver disease.

8. A group of congenital conditions exist, characterised by hyper-bilirubinaemia. Most are relatively harmless, but the Crigler-Najjar syndrome may lead to kernicterus.
9. Drugs may produce jaundice in several ways and a drug history is important in the assessment of the jaundiced patient. In the newborn, drugs may increase the risk of kernicterus.

## Bile and Gall Stones

Bile secreted by the liver is concentrated in the gall bladder before passing into the gut. Cholesterol in bile is held in solution by bile salt micelles. A change in either the concentration or type of bile salts may lead to defective micelle formation and permit precipitation of cholesterol with gall stone formation.

## FURTHER READING

Sherlock, Sheila (1968). *Diseases of the Liver and Biliary System*, 4th edit. Oxford: Blackwell Scientific Publications.
Symposium on Liver Disease (1970). *Amer. J. Med.*, **49**, 573. (Articles on many aspects of liver disease including bilirubin metabolism, HAA and immunology.)
Gocke, D. J. (1973). New Faces of Viral Hepatitis. *Disease-a-Month*, (January). Chicago: Year Book Medical Publishers.
Maxwell, J. D., and Williams, R. (1973). Drug-induced jaundice. *Brit. J. hosp. Med.*, **9**, 193.
Boucher, I. A. D. (1972). Clinical aspects of alterations in bile acid metabolism. *Brit. J. hosp. Med.*, **7**, 107.
Symposium on Bile Salts (1971). *Amer. J. Med.*, **51**, 565.

# APPENDIX

## 1. Handling of Blood Samples from Patients with Possible Hepatitis

Samples from all patients with viral hepatitis, undiagnosed jaundice, or positive HAA tests, as well as at-risk patients from dialysis units, should be considered *infective*. Anyone handling such a sample (medical and nursing staff, porters and laboratory staff) is at risk. It is *the duty of the clinician sending the blood to identify it clearly as potentially dangerous*. The sample should be sent to the laboratory in a sealed plastic bag.

## 2. Bromsulphthalein (BSP) Retention Test

The test is preferably performed on the fasted patient to avoid turbidity of the plasma from recent fat ingestion.

(*a*) BSP solution is injected *slowly* intravenously. The recommended dose is 5 mg/kg body weight.

(*b*) 45 minutes after the injection blood is taken from a vein on the opposite arm to avoid contamination from the injection site.

The residual dye is estimated in the plasma by comparison with a standard solution. As different batches of BSP may vary, it is important to retain some of the solution injected for use as a standard. Normally less than 5 per cent of the dye is retained at 45 minutes. *Interpretation* is considered on p. 298.

**Notes.**—(*a*) Severe systemic reactions very rarely occur. *Local tissue necrosis*, however, is seen if the solution is injected outside the vein or if extravasation occurs.

(*b*) The test should not be performed immediately after cholecystography when falsely impaired retention has been described, presumably due to competition for excretion mechanisms.

(*c*) Urinary loss of BSP can normally be discounted. In patients with marked proteinuria, however, it may be significant and urine should be collected over the period of the test to assess the extent of the loss. As BSP is purple in alkaline solution it is advisable to reassure the patient that the urine may become coloured temporarily.

(*d*) If necessary the test may be repeated after 24 hours but it is then advisable to take a pre-injection sample of blood to exclude residual dye.

## 3. Prednisolone Test (p. 303)

This test may be used for the differential diagnosis of jaundice.

Plasma bilirubin is estimated before and after administering 30 mg prednisolone daily for 5 days. A fall of concentration of more than 40 per cent strongly suggests hepatitis rather than extra-hepatic obstruction. Lesser falls are equivocal.

#### 4. Rapid Tests for Urinary Bile Constituents

Urine containing bilirubin is usually dark yellow or brown, whereas fresh urine containing urobilinogen only is initially of normal colour.

Reagent strips (Ames) are available for the detection of these substances. *Fresh urine is essential.*

(*a*) **Ictostix** includes stabilised diazotised 2:4-dichloraniline which reacts with *bilirubin* to form azobilirubin.

The test will detect about 3 µmol/l (0·2 mg/dl) of bilirubin. Drugs (such as large doses of chlorpromazine) may give *false positive* reactions.

(*b*) **Urobilistix** includes paradimethylaminobenzaldehyde which reacts with *urobilinogen*. It does *not* react with porphobilinogen.

This test will detect urobilinogen in some normal urines. *False positive* results may occur with drugs such as *p*-aminosalicylic acid and certain sulphonamides. Positive results should be confirmed by a test for *urobilin* (below).

Both these reagent strips must be stored, and the test performed, strictly according to the instructions of the manufacturers.

(*c*) **Alternative test for urobilinogen**

1. Mix about 2 ml of *fresh* urine with an equal volume of Ehrlich's reagent and allow to stand for 5–10 minutes.

2. Add 4 ml of saturated sodium acetate solution and mix.

A red colour denotes the presence of urobilinogen. Distinction from porphobilinogen (p. 415) is made by shaking the coloured solution with *n*-butanol. The colour due to urobilinogen is extracted into butanol (upper layer) whereas that of porphobilinogen is not.

(*d*) **Urobilin.**—This test is preferable to the above if the urine is not fresh, as urobilinogen is converted to urobilin on standing.

1. Mix 5 ml of urine with 2 drops of alcoholic iodine solution (converts urobilinogen to urobilin).

2. Add 5 ml of zinc acetate suspension, mix and allow to settle.

3. A greenish fluorescence in the supernatant is due to a zinc-urobilin complex. This is best seen in darkened surroundings by shining the light from a pencil torch through the fluid.

The interpretation of these tests is discussed on p. 297.

# Chapter XIII

# PLASMA PROTEINS AND IMMUNOGLOBULINS: PROTEINURIA

## PLASMA PROTEINS

PLASMA contains a complex mixture of proteins at a concentration of about 70 g/l. These include simple proteins as well as those incorporating carbohydrates and lipids. These different proteins have different functions and originate from several different cell types. Changes in disease are frequently non-specific, but can be of diagnostic value in a limited number of conditions.

### Functions

The following is an outline of the main functions of the plasma proteins.

1. **Control of extracellular fluid distribution (ECF).**—Distribution of water between the intra- and extravascular compartments is affected by the concentration of plasma proteins. Because of its low molecular weight, albumin is the most important in this respect (p. 35).

2. **Antibodies** are proteins (p. 325).

3. **The complement system** consists of proteins (p. 322).

4. **Transport.**—Plasma proteins fulfil a transport function for many hormones (for example, cortisol, p. 135 and thyroxine, p. 162) and for vitamins, lipids (p. 219), calcium, trace metals and some drugs. Combination with proteins renders them soluble (for example, lipids), or physiologically inactive (for example, calcium).

5. **Blood clotting factors.**—Most of these are proteins.

6. **Nutrient.**—Circulating protein, particularly albumin, is a source of nutrient for tissues.

7. **Buffering.**—Proteins form a minor part of the plasma buffering system (p. 82),

8. **Enzymes** are proteins. Only a few enzymes are functional in the circulation, most of them arising from leakage from cells.

9. **Hormones.**—Many circulating hormones are peptides.

This list is by no means complete. The function of many of the proteins which have been identified in plasma is unknown.

## Source of Plasma Proteins

Many proteins, notably albumin, some coagulation factors, carrier proteins and lipoproteins are synthesised in the liver. Plasma protein concentrations alter in hepatic disease (p. 299). Immunoglobulins (antibodies) are synthesised by plasma cells and lymphocytes in the lymphoreticular (reticulo-endothelial) system. Some lipoproteins are synthesised in the intestinal wall.

### TOTAL PROTEIN

## Limitations of Total Protein Estimations

Changes in one protein fraction may be masked by opposite changes in another; for instance, in chronic inflammation a low albumin is frequently accompanied by a raised $\gamma$ globulin level. Examples of conditions in which total protein concentration may be abnormal are listed below, the fractions most affected being indicated in brackets.

## Causes of Raised Total Protein Concentration

1. **Changes due to relative water deficiency** (all fractions). These are concentration changes only and *do not indicate alterations in absolute amounts of protein.*

(*a*) *Dehydration.*

(*b*) *Artefactual*—stasis during venepuncture. Stasis induced by keeping the tourniquet on too long during venepuncture causes fluid to escape into the extravascular compartment, leading to haemoconcentration and a falsely raised protein concentration. *This is the commonest cause of a high total protein level.*

2. **Paraproteinaemia** (paraprotein, p. 331).

3. **Certain chronic diseases** ($\gamma$ globulin).

(*a*) *Chronic inflammatory conditions* (p. 323).

(*b*) *Cirrhosis* of the liver.

(*c*) *Autoimmune disease,* for example, systemic lupus erythematosus (SLE).

(*d*) *Sarcoidosis.*

## Causes of Low Total Protein Concentration

1. **Changes due to relative water excess** (all fractions). These are concentration changes only and *do not indicate alterations in absolute amounts of protein.*

(*a*) *Overhydration.*

(*b*) *Artefactual*—blood taken from the "drip" arm.

2. **Excessive loss of protein** (mainly albumin).

(a) *Through the kidney* in the nephrotic syndrome.

(b) *From the skin* after severe burns.

(c) *Through the intestine* in protein-losing enteropathy.

3. **Decreased synthesis of protein.**

(a) *Severe dietary protein deficiency*, for instance in kwashiorkor (all fractions but mainly albumin).

(b) *Severe liver disease* (mainly albumin).

In both of the above there may be no fall in total protein because of a rise in $\gamma$ globulin.

(c) *Severe malabsorption* (all fractions).

Total protein estimation may be helpful in the assessment of hydration. Extracellular proteins, because of their large size, are mostly confined to the vascular compartment: changes in plasma volume will cause a change in total protein *concentration* (and in haematocrit). Serial estimations may be useful in following hydration.

## PLASMA PROTEIN FRACTIONS

### Measurement of Plasma Protein Fractions

Plasma contains a large number of chemically different proteins, of which only a few are present in concentrations detectable by routine methods. This mixture may be separated analytically into groups.

(a) **Qualitative indices of altered ratios of plasma proteins.**—*The erythrocyte sedimentation rate* (ESR) is a frequently performed test It is influenced by plasma levels of fibrinogen and, to a lesser extent, by those of $\alpha_2$ and $\gamma$ globulins. The raised ESR found in infective, neoplastic and degenerative diseases reflects these changes. Very high ESR readings may be found in myelomatosis and macroglobulinaemia.

*Flocculation tests* are less commonly performed than previously. However, they may still be useful if electrophoresis is not readily available. They are based on the observation that flocculation occurs when thymol, or the salts of certain metals, are added to serum in which the various proteins are present in abnormal ratios. Abnormal turbidity or flocculation is seen most commonly when $\gamma$ globulin levels are increased, albumin levels are low, or both factors are present.

(b) **Zone electrophoresis.**—Electrophoresis uses the differences in electrical charge on different proteins to separate them. It is routinely performed by applying a small amount of serum to a strip of cellulose acetate and applying a current across it for a standard time. In this way five main groups of protein may be distinguished: these are albumin and the $\alpha_1$, $\alpha_2$, $\beta$ and $\gamma$ globulins. The "strip" is stained to render these fractions visible, and may then be compared with a normal one. Alternatively, it may be passed through a light path (*scanned*); the absorption or reflection of light by each fraction is graphically represented as the electrophoretic "scan" (Fig. 24). The scanned fractions may be quantitated, but this adds little to the diagnostic value of the technique: if fractions *are* quantitated the resultant values will differ according to the dye and scanning technique used. The student *must* consult the "normal" ranges of his own laboratory.

Zone electrophoresis in *starch or polyacrylamide gels* separates the proteins into a larger number of groups, but is rarely necessary for routine diagnostic purposes.

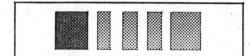

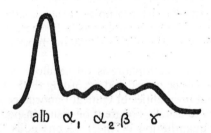

FIG. 24.—Normal electrophoretic strip and scan.

*Zone electrophoresis should always be performed on serum.* The fibrinogen present in plasma may be misinterpreted as, or mask, a paraprotein.

(*c*) **Immunochemical methods.**—Immunochemical methods depend on the precipitation which occurs when a protein reacts with its specific antiserum under defined conditions. Their main use is in the determination of single proteins, most commonly the individual *immunoglobulins*. When combined with electrophoresis (one- or two-dimensional) a large number of different specific proteins may be identified in a single sample. This method (*immunoelectrophoresis*) is rarely necessary for routine purposes. Proteins present in very low concentrations may be estimated by *radioimmunoassay*.

Recently an automated procedure has been developed which quantitates individual proteins, after mixture of the specimen with specific antiserum, by measuring the turbidity due to the antigen-antibody complex (*automated immunoprecipitation—AIP*).

(*d*) **Chemical methods.**—Some proteins can be measured by chemical methods. As an example, transferrin can be measured by its iron binding capacity (p. 394). These methods are not ideal, and are slowly being replaced by more specific immunochemical methods.

(*e*) **Ultracentrifugation** separates proteins on the basis of their different densities. It is a specialised technique requiring expensive apparatus, and is not routinely available.

## INVESTIGATION OF SUSPECTED PROTEIN ABNORMALITIES

For most routine purposes (for example, for following changes in hydration of the patient) total protein estimation is adequate. If an abnormality of plasma proteins is suspected this should, in the first instance, be combined with albumin estimation and inspection of a stained electrophoretic strip.

Assessment of the clinical findings in conjunction with the results of these simple tests may occasionally indicate the need for immunoglobulin or other specific protein estimations: examples of such conditions are discussed in the section on "Immunoglobulins". In these cases consultation with the laboratory is essential, so that the most useful tests may be chosen.

As most laboratories use zone electrophoresis on cellulose acetate the changes in serum proteins will be considered according to the findings by this method. The few abnormalities of clinical importance *that can be detected by this method* are usually apparent on visual inspection of the electrophoretic strip: quantitation of the fractions adds little useful information.

Estimation of specific proteins is becoming increasingly available and information about its diagnostic value is growing. The most important of these proteins will be considered in relation to the routine electrophoretic strip (Fig. 25). The significance of the latter should thus be clarified: more importantly, it should become evident that significant protein alterations can occur without obviously changing the routine electrophoretic pattern. However, simple electrophoresis, critically approached, can sometimes be a valuable investigation. For instance, it is the only method for detecting a paraprotein.

In the following sections the approximate concentrations of specific proteins are given as a rough guide. Since "normal" ranges differ with age and sex as well as with methodology, reference ranges must be provided by the laboratory performing the estimations. The student should also remember that genetic variations occur, and that levels of

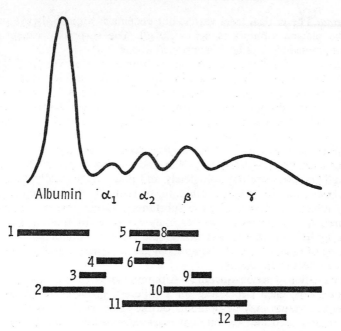

FIG. 25.—The plasma protein electrophoretic pattern and the twelve main plasma proteins (excluding fibrinogen).

1. Albumin;          2. α Lipoprotein;        3. $α_1$ Acid glycoprotein;
4. $α_1$ Antitrypsin;    5. $α_2$ Macroglobulin;   6. Haptoglobin;
7. β Lipoprotein;    8. Transferrin;          9. $C_3$ component of complement;
10. IgG;             11. IgA;                 12. IgM

some proteins are markedly affected by pregnancy and oral contraceptives.

CHANGES IN INDIVIDUAL PLASMA PROTEIN FRACTIONS

**Albumin**

There are two rare congenital anomalies of albumin synthesis. In one, *bisalbuminaemia*, two albumin types are present. This is a curiosity only, as there are no clinical consequences. In the other, *analbuminaemia*, there is deficient synthesis of the protein. Clinical consequences are slight, and oedema, though present, is surprisingly mild.

An abnormally high albumin level is found only with dehydration and all clinical interest centres on low albumin levels.

**Consequences of a low serum albumin.**—The role of plasma proteins in fluid distribution has been discussed. As albumin is the most important fraction in this respect, one of the consequences of a reduced level is *oedema*. The critical level varies, but oedema is almost always present if the plasma albumin is below 20 g/l. The surprising exception of analbuminaemia has been mentioned above.

About half the plasma calcium is bound to albumin (p. 235) and hypoalbuminaemia is accompanied by *hypocalcaemia*. As this involves only the protein bound (physiologically inactive) half, symptoms of tetany do not develop and calcium or vitamin D supplements are contra-indicated. As a rough guide there is about 0·25 mmol/l (1 mg/dl) of calcium per 8 g/l albumin.

Albumin also binds *bilirubin, free fatty acids,* and a number of *drugs* such as salicylates, penicillin and sulphonamides. The albumin bound fractions are physiologically and pharmacologically inactive. A marked reduction in serum albumin, by reducing the binding capacity, may affect free levels of these substances. Drugs which are albumin bound, if administered together, may compete for binding sites, also altering free concentrations: an example of this is simultaneous adminis-tration of salicylates and the anticoagulant warfarin, with potentiation of the effect of the latter.

**Estimation of serum albumin.**—Albumin can be estimated by a variety of methods, each depending on a different property of the protein. Different values may be obtained on the same specimen depending on estimation used and the student *must refer to the "normal" ranges of his own laboratory.*

The most usual routine method utilises the ability of albumin to bind certain dyes: since the substances mentioned above may compete for binding sites, falsely low values may be obtained in *jaundiced patients,* or in those taking *certain drugs.*

**Causes of hypoalbuminaemia.**—In normal subjects albumin accounts for almost half the total protein concentration. Significant changes in albumin levels therefore usually affect the latter and the causes of hypo-albuminaemia are very similar to those listed on p. 315 for low total protein levels. In addition, mild changes, such as are found non-specifically, may not significantly affect total protein levels.

(1) *Decreased synthesis.*—Albumin is catabolised at a steady rate of about 4 per cent of the body pool daily, and any impairment of synthesis soon causes hypoalbuminaemia. Albumin is synthesised in the liver, and *liver disease,* especially if chronic, causes low levels.

(2) *Loss.*—Albumin is a relatively small molecule with a molecular weight of about 70 000. It is the predominant fraction lost in protein-losing states such as the *nephrotic syndrome* (p. 338), *protein-losing enteropathy* (p. 280) or exudation from the skin after *extensive burns.* The lowest levels are usually reached in this category of disease.

(3) *Defective intake.*—*Malabsorption* or *malnutrition* due to severe

dietary deficiency cause hypoalbuminaemia because of the poor supply of amino acids.

(4) *Non-specific.*—One of the commonest reasons for a low serum albumin is less easily explained. In many acute conditions, including apparently *minor illnesses* such as colds and boils, the serum albumin level falls. The answer cannot lie entirely in decreased synthesis or increased breakdown as there is a rapid return to normal on recovery. It could be due partly to a switch to synthesis of other fractions, because the total protein level often remains the same. This phenomenon is seen commonly in hospitalised patients, and changes of concentration of up to 5–10 g/l may be a non-specific finding. *Posture* influences albumin levels which may be 5–10 g/l higher in the upright than in the recumbent position.

Low albumin levels are found in the *newborn*.

(5) *Haemodilution.*—In the late stages of pregnancy, or in overhydrated subjects, low concentrations of albumin and all protein fractions may be found due to dilution effects. This is *not* an abnormality of albumin metabolism.

(6) *Artefactual.*—Dilution may result artefactually if blood is taken from a "drip" arm (p. 462). The effect of bilirubin and drugs on the results of dye binding methods has already been mentioned.

### $\alpha_1$ and $\alpha_2$ Globulins

The proteins making up this group of globulins all tend to increase when there is active tissue damage ("acute phase reaction"). *Raised levels of both $\alpha_1$ and $\alpha_2$ globulins are therefore a non-specific finding in the serum of most ill patients.* They occur:

(i) in inflammatory conditions
(ii) associated with malignancy
(iii) after trauma
(iv) post-operatively
(v) in autoimmune disease.

$\alpha_1$ **Globulins.**—The main components of this fraction are:

$\alpha_1$ antitrypsin (approximately 2·9 g/l)
$\alpha_1$ glycoprotein (orosomucoid: seromucoid) (approximately 0·9 g/l)
$\alpha$ lipoprotein (p. 222)

Of these proteins only $\alpha_1$ *antitrypsin* stains well on routine electrophoresis: the $\alpha_1$ fraction virtually reflects the level of this protein.

*Low concentrations* are found in the rare inherited condition $\alpha_1$ *antitrypsin deficiency.* The homozygous state is easily diagnosed by the almost complete absence of $\alpha_1$ globulins on the electrophoretic strip.

Clinically it is associated with emphysema (with a family history), cirrhosis of the liver in young subjects, or neonatal cholestatic jaundice.

$\alpha_2$ **Globulins.**—The components of this fraction are:

$\alpha_2$ macroglobulin (approximately 2·5 g/l)
Haptoglobin (approximately 1·8 g/l)
Caeruloplasmin (approximately 0·4 g/l).

$\alpha_2$ *Macroglobulin* makes up most of the $\alpha_2$ globulin. It is a very large molecule (MW approximately 900 000) and is retained in the blood stream in the *nephrotic syndrome*, when there may be loss of all other protein fractions: levels may even *increase* in this condition, possibly due to feed-back stimulation of synthesis. *A relatively or absolutely raised $\alpha_2$ globulin, with reduction of other fractions is characteristic of the nephrotic syndrome and other protein-losing states.*

*Haptoglobin* binds free haemoglobin. The haptoglobin/haemoglobin complex is catabolised faster than haptoglobin alone. Low levels are therefore found in haemolytic conditions. This is rarely detectable on the routine electrophoretic strip.

*Caeruloplasmin* is normally present in such low concentration that changes are rarely reflected on the routine electrophoretic strip and special methods are used to measure this protein. Levels *increase* in *pregnancy* and during *oestrogen therapy* (p. 435) and in *cirrhosis of the liver*. Low levels occur in "Wilson's disease" (hepatolenticular degeneration) (p. 372).

$\alpha$ *fetoprotein* is a protein not found in significant concentration in the serum of normal adults. Its presence is usually associated with hepatocellular carcinoma (primary hepatoma) (p. 301). When present it may very occasionally be visible as a band between albumin and $\alpha_1$ globulin.

$\beta$ **globulins.**—The main components of the $\beta$ globulin fraction are:

$\beta$ lipoprotein (approximately 5 g/l). (On cellulose acetate this fraction overlaps both $\alpha_2$ and $\beta$ fractions.)
Transferrin (p. 392) (approximately 3 g/l.)
Several components of the complement system. (These are not detectable on routine electrophoresis.)

Changes in the $\beta$ globulin fraction visible on routine electrophoresis are uncommon. The rise in $\beta$ lipoprotein which occurs in the nephrotic syndrome contributes to the raised $\alpha_2$ globulin found in this condition, in which the $\beta$ fraction is very rarely raised. High levels of $\beta$ globulin are very occasionally found in *pregnancy* and in *biliary obstruction*.

**Serum complement.**—The complement system consists of 9 components

$(C_1-C_9)$, many of which are synthesised in the liver. These react sequentially after binding to antigen-antibody complex. This sequential reaction is important in the enhancement of phagocytosis and in many other features of the acute inflammatory response. For further details the student should consult textbooks of immunology.

*Low levels of* $C_3$ (the most commonly measured component) occur in many types of *nephritis* in the active phase: they may be persistently low in membranoproliferative glomerulonephritis and in systemic lupus erythematosus (SLE.). The low levels found in cases of chronic liver disease are not related to activity, and reflect reduced synthesis.

*High levels of* $C_3$ occur in any acute inflammatory condition.

### γ Globulins

This group of proteins is considered more fully in the section on immunoglobulins. Although all γ globulins are immunoglobulins, not all immunoglobulins are found in the γ region of the electrophoretic strip: some may be found in the $\beta$ and $\alpha_2$ regions.

Immunoglobulins are antibodies, and a general increase in levels occurs in chronic inflammatory states and in some autoimmune diseases. This general increase should be distinguished from the increase in a single immunoglobulin type (a paraprotein). Paraproteinaemia is discussed on p. 331.

A diffusely raised γ globulin occurs in:

1. Chronic infections, including subacute bacterial endocarditis. In the tropical diseases kala-azar and lymphogranuloma venereum very high values may be found.

2. Cirrhosis of the liver.

3. Sarcoidosis.

4. Autoimmune disease.
   Systemic lupus erythematosus (SLE).
   Rheumatoid arthritis.

Decreased γ globulin levels may be due to:

1. Increased loss.
   Nephrotic syndrome and other protein-losing states.

2. Decreased synthesis.
   Severe malabsorption and malnutrition.
   Primary immune deficiency (p. 331).

3. Secondary suppression of synthesis.
   Secondary immune deficiency (p. 331).

## ABNORMAL ELECTROPHORETIC PATTERNS

The commonest abnormal protein patterns seen on routine electrophoresis are illustrated in Fig. 26. This figure represents "scanned" strips (p. 317).

1. **Parallel changes in all fractions** (not shown in Fig. 26). This is a *normal pattern with an abnormal total protein concentration.* An *increase* in all protein fractions (including immunoglobulins) may be found in *dehydration* and a reduction in *overhydration* (p. 315). A reduction also occurs in severe protein *malnutrition* and *malabsorption,* unless accompanied by infection.

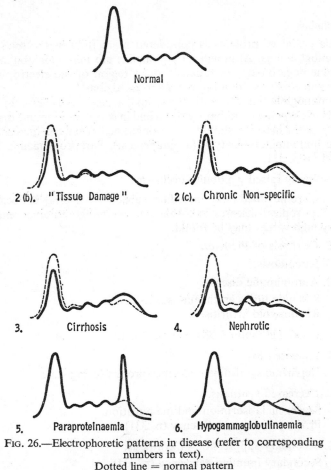

FIG. 26.—Electrophoretic patterns in disease (refer to corresponding numbers in text).
Dotted line = normal pattern

2. **Non-specific patterns.**—These occur in a variety of illnesses and are of no more importance in diagnosis than a raised ESR.

(*a*) *A low albumin level*, without changes in other fractions, occurs in many acute illnesses (not shown in Fig. 26).

(*b*) *A low albumin and raised $\alpha_1$ and $\alpha_2$ globulin* concentration occurs in more severe acute infections, inflammation and neoplasia ("tissue damage" pattern).

(*c*) *A low albumin with diffusely raised $\gamma$ and often raised $\alpha_2$ globulin levels* occur with chronic "tissue damage", as in chronic infections and some autoimmune diseases (for example, rheumatoid arthritis). This pattern may also be found in chronic liver disease.

3. **Cirrhosis of the liver.**—The changes in plasma proteins in liver disease are considered more fully on p. 299. They are usually "non-specific", but in cirrhosis a characteristic pattern is sometimes seen. Albumin and often $\alpha_1$ globulin levels are reduced and the $\gamma$ globulin concentration is markedly raised, with apparent fusion of the $\beta$ and $\gamma$ bands.

4. **Nephrotic syndrome.**—Plasma protein changes depend on the severity of the renal lesion (p. 338). In early cases a low albumin level may be the only abnormality, but the typical pattern in established cases is a reduced albumin, $\alpha_1$ and $\gamma$ globulin with an increase in $\alpha_2$ globulin. $\beta$ globulin levels are usually normal. If the syndrome is due to systemic lupus erythematosus the $\gamma$ globulin may be normal or raised.

Paraproteinaemia (5) and hypogammaglobulinaemia (b) are discussed in the section on immunoglobulins (pp. 331 and 330).

## IMMUNOGLOBULINS

It has long been known that the $\gamma$ globulin fraction of plasma proteins has antibody activity. The terms antibody and $\gamma$ globulin are often used synonymously. However, antibodies also occur in the $\beta$ and $\alpha_2$ fractions (Fig. 25), and the term immunoglobulin (Ig) is preferable. Bence Jones protein (light chain) and heavy chain are naturally occurring subunits of the Ig molecule.

### STRUCTURE OF THE IMMUNOGLOBULINS

The immunoglobulin unit is a Y-shaped molecule depicted schematically in Fig. 27.

The following points should be noted:

1. Four polypeptide chains are linked by disulphide bonds. There are two heavy (H) and two light (L) chains in each unit. The *H chains* in a

single unit are similar and determine the immunoglobulin *class* of the protein. H chains $\gamma$, $\alpha$, $\mu$, $\delta$ and $\epsilon$ occur in IgG, IgA, IgM, IgD and IgE respectively. *L chains* are of two *types* $\kappa$ or $\lambda$. In a single molecule the L chains are of the same type, although the Ig class as a whole contains both types.

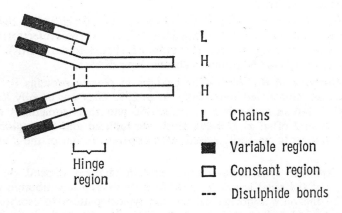

L
H
H
L    Chains

■   Variable region

□   Constant region

---   Disulphide bonds

Hinge region

FIG. 27.—Schematic representation of Ig subunit.

2. There are two antibody combining sites per unit. These lie at the ends of the arms of the Y: both H and L chains are necessary for full antibody activity. The amino acid composition of this part of the chain varies in different units (variable region). When this part of the molecule combines with antibody, the arms of the Y swing open at the hinge region (Fig. 27).

3. The rest of the H and L chains are less variable (constant region). The constant region of the H chains is responsible for such properties of the Ig unit as the ability to bind complement or actively to cross the placental barrier. The H chains are associated with a variable amount of carbohydrate; IgM has the highest content.

### NOMENCLATURE OF THE IMMUNOGLOBULINS

Two forms of nomenclature are in use.

### 1. Based on Immunochemical Studies

The division of immunoglobulins into classes based on the structure of the H chains and into types based on the structure of the L chains has already been discussed. The chains are identified by use of specific antisera. Several subclasses of the major Ig classes exist, but in routine work IgG, IgA and IgM are the most commonly measured.

## 2. Based on Ultracentrifuge Studies

The simplified account of the immunoglobulin unit given above ignores the fact that each molecule may contain more than one unit. The IgM molecule, for example, consists of 5 basic units. This variation in size, and therefore in density, enables the classes to be separated in the ultracentrifuge. Classification is by Svedberg coefficient (S), the S value of a protein increasing with increasing size. IgG, which circulates as a single unit, sediments as 7S, IgM as 19S and IgA, which can contain a variable number of units, as between 7S and 11S. By comparison, albumin sediments as 4.5S

### FUNCTION OF THE IMMUNOGLOBULINS

The immune response mechanism of the body consists of a cellular and a humoral component. Although we are concerned here only with the humoral component—the immunoglobulins—the student should remember that both are necessary: he should consult a textbook of immunology for details of the cellular mechanisms.

The specific functions of each class of immunoglobulin will be discussed briefly, and followed by an outline of changes in disease.

### IgG (MW 160000)

IgG accounts for about 75 per cent of circulating immunoglobulins and contains most of the normal plasma antibodies. It is a relatively small molecule, and is present in very low concentration throughout the extracellular compartment: its most important function seems to be protection of the tissue spaces. IgG synthesis is stimulated by soluble antigens such as bacterial toxins. *IgG deficiency* is characterised by *recurrent pyogenic infections* of tissue spaces by toxin-producing organisms such as staphylococci and streptococci: pulmonary and subcutaneous infections are common.

IgG can bind complement and cross the placental barrier. In the first few months of life endogenous IgG levels are very low. Maternal IgG which has crossed the placenta provides antibody cover during this period. Adult concentrations of this class of protein are not reached before the age of 3 to 5 years.

### IgA (MW 160000. Polymers also occur)

About 20 per cent of circulating immunoglobulin consists of this class. However, most IgA is synthesised beneath the mucosa of the gastro-intestinal and respiratory tracts. A "secretory piece", synthesised by epithelial cells, binds two IgA units to form secretory IgA. The latter appears in intestinal and bronchial secretions, sweat, tears and colostrum; it protects body surfaces, particularly against viral infection. The secretory piece prevents destruction by digestive enzymes and enables IgA to perform an important protective function in the gut. *Deficiency of IgA may be associated with mild recurrent respiratory tract infections* and *intestinal disease*, but is sometimes symptomless.

IgA does not bind complement or cross the placenta. Detectable levels are found a few months after birth, but adult levels may not be reached before the age of 15 years.

### IgM (MW 1 000 000)

Because of its very large size, IgM (macroglobulin) is almost entirely intravascular. It accounts for about 7 per cent of circulating immunoglobulins. IgM antibodies are usually the first to be synthesised in response to infection: particulate antigens, such as circulating organisms, are particularly effective in stimulating synthesis. In *IgM deficiency septicaemia* is common.

The major blood group isohaemagglutinins belong to this group of proteins.

Although IgM can bind complement it does not cross the placenta. Synthesis occurs even in the foetus, and elevated IgM values at birth indicate intra-uterine infection. Adult levels are reached by about 9 months of age.

### IgE (MW 200 000)

IgE is synthesised by plasma cells beneath the mucosae of the gastro-intestinal and respiratory tracts and by those in the lymphoid tissue of the nasopharynx. It is present in nasal and bronchial secretions. Circulating IgE is rapidly bound to cell surfaces, particularly to those of mast cells and circulating basophils, and plasma levels are therefore very low. Combination of antigen with this cell-bound antibody results in the cells releasing mediators and accounts for immediate hypersensitivity reactions such as occur in hay-fever. Desensitisation therapy of allergic disorders aims at stimulating production of circulating IgG against the offending antigen, to prevent it reaching cell-bound IgE. *Raised concentrations* are found in several diseases with an *allergic* component such as some cases of eczema, asthma and parasitic infestations.

### IgD (MW 190 000)

The function and physiological importance of IgD is still unknown.

TABLE XXVI

PROPERTIES AND FUNCTIONS OF IMMUNOGLOBULINS

| | IgG | IgA | IgM | IgE | IgD |
|---|---|---|---|---|---|
| Molecular weight | 160000 | 160000 and polymers | 1000000 | 200000 | 190000 |
| Sedimentation coefficient | 7 S | 7 S, 9 S, 11 S | 19 S | 8 S | 7 S |
| % total serum immunoglobulin | 73 | 19 | 7 | 0·001 | 1 |
| Able to fix complement | Yes | No | Yes | No | No |
| Able to cross placenta | Yes | No | No | No | No |
| Approximate mean normal adult concentration: g/l IU/ml | 9 to 12 140 | 2·5 150 | 1·0 140 | 0·0003 | 0·03 |
| Adult levels reached by | 3–5 years | 15 years | 9 months | ? 15 years | ? 15 years |
| Major function | Protects extravascular tissue spaces. Secondary response to antigen. Neutralises toxins. | Protects body surfaces as secretory IgA (11 S) | Protects blood stream. Primary response to antigen. Lyses bacteria. | Tissue bound antibodies of immediate hypersensitivity reactions | Not known |

## IMMUNOGLOBULIN RESPONSE IN DISEASE

Most infections produce a general immunoglobulin response and within a few weeks raised levels of all three major Igs are detectable. In such circumstances immunoglobulin estimation adds little to the observation of an increase in $\gamma$ globulin on routine electrophoresis. In certain conditions, some of which are listed in Table XXVII, one or more immunoglobulin classes predominate. Although there is considerable overlap, individual Ig estimation may help in diagnosis of such cases.

TABLE XXVII

| Predominant Ig | Examples of Clinical Conditions |
|---|---|
| IgG | Autoimmune diseases, such as SLE or chronic aggressive hepatitis |
| IgA | Disease of intestinal tract, e.g. Crohn's disease<br>Diseases of respiratory tract, e.g. tuberculosis, bronchiectasis<br>Early cirrhosis of the liver |
| IgM | Primary biliary cirrhosis<br>Viral hepatitis<br>Parasitic infestations, especially when there is parasitaemia<br>In the newborn, indicating intra-uterine infection |

### Immunoglobulin Deficiency

Immunological deficiency states may involve cellular, as well as humoral mechanisms. Only immunoglobulin (humoral) deficiency will be discussed here, but it is important to recognise that *normal immunoglobulin concentrations do not exclude an immunological deficiency state.*

Immunoglobulin deficiency (often referred to as hypogammaglobulinaemia), may be primary, or may be secondary to other disease. Any or all of the immunoglobulin classes may be affected. A history of recurrent infections in infancy or early childhood is suggestive of primary Ig deficiency, although the condition may first present in adult life.

As mentioned on p. 327, deficiency of IgG is associated with pyogenic infections of tissue spaces, of IgA with respiratory tract infections and gastro-intestinal disease, and of IgM with septicaemia.

Several classifications have been proposed for these deficiencies. We will, very briefly, discuss three categories presenting with Ig deficiency.

1. **Transient immunoglobulin deficiency.**—In the newborn infant circulating IgG is derived from the mother by placental transfer. Levels decrease over the first 3 to 6 months of life, and then gradually rise as

endogenous IgG synthesis increases. In some subjects onset of synthesis is delayed, and "physiological hypogammaglobulinaemia" may persist for several more months.

Most of the IgG transfer across the placenta takes place in the last 3 months of pregnancy. Severe deficiency may therefore develop in very premature babies as maternal IgG falls before endogenous levels rise.

2. **Primary immunoglobulin deficiency.**—Several rare syndromes, usually familial, have been described. In one, *infantile sex-linked agammaglobulinaemia* (Bruton's disease), which occurs only in males, there is almost complete absence of circulating immunoglobulins, while cellular immunity seems normal. Other syndromes have varying degrees of immunoglobulin deficiency and impaired cellular immunity, and can occur in either sex.

3. **Secondary (acquired) immunoglobulin deficiency.**—Ig deficiency coexisting with other disease is much commoner than primary forms. It may be found at any age in either sex, and may accompany *severe illness* such as diabetes mellitus or renal failure. It is seen relatively commonly in association with neoplasia of the lymphoid system, and in conditions such as myeloma and macroglobulinaemia.

### Investigation of Immunoglobulin Deficiency

Although severe immunoglobulin deficiency may be obvious by routine cellulose acetate or immunoelectrophoresis, determination of individual immunoglobulin classes should be performed on all suspected cases. Diagnosis during the first year of life may be difficult because of the "physiologically" low IgG levels: the condition should be suspected if IgA and IgM concentrations fail to rise normally.

PARAPROTEINAEMIA

Immunoglobulin-producing cells (plasma cells and lymphocytes) are very specialised in function. Each cell produces only immunoglobulins of a single class and type. Cells producing the same class and type form a *clone*. Immunocyte (immunoglobulin-producing cell) proliferation may involve many cell types synthesising a range of Ig classes and types (*polyclonal response*), or a single cell type producing a single class and type (*monoclonal response*).

The normal response of immunocytes to infection is polyclonal, resulting in a diffusely raised $\gamma$ globulin on routine electrophoresis. More rarely a narrow, dense band is seen, usually in the $\gamma$ region, but which may be anywhere from the $\alpha$ to $\gamma$ regions inclusive. If this is shown to consist of a single class and type of immunoglobulin it is termed a *paraprotein*, and is probably due to proliferation of a single clone of immunocytes. The protein is usually structurally normal: the monoclonal proliferation however, is pathological, producing abnormally high levels of the protein. Although paraproteins can occur in benign conditions, paraproteinaemia usually indicates malignancy of immunocytes.

Confusing terms are "M band", "M globulin" or "M protein". "M" originally denoted "myeloma" but now means "monoclonal". It does *not* mean IgM.

### Bence Jones Protein

Bence Jones protein (BJP) is found in the urine of many patients with malignant immunocytomata. It consists of free light chains, or fragments of L chains, which have been synthesised in excess of H chains. Its presence implies a degree of dedifferentiation of immunocytes. Because of its low molecular weight (20000 to 40000) the protein is filtered at the glomerulus and only accumulates in the plasma if there is glomerular failure. BJP may damage renal tubular cells and may form large casts, producing the "myeloma kidney". BJP may also pass into tissue spaces and is associated with amyloid formation.

BJP was originally detected by its peculiar heat properties. At a pH of about 4·9 it precipitates at a temperature below 60°C and redissolves on further heating, reappearing on subsequent cooling. The test is only positive if there is a considerable amount of BJP present. A more sensitive test is to perform electrophoresis on urine which has been concentrated 100 to 200 fold: BJP will appear as a narrow band in the globulin region. Immunological studies will show this protein to be free L chains of *either* κ or λ type. It should be noted that the commonly used tests for urinary protein, "Albustix" and precipitation with salicylsulphonic acid, may not detect BJP.

### Causes of Paraproteinaemia

Although the presence of a paraprotein is strongly suggestive of a malignant process, this is not always so. Paraproteins may be found in the following conditions:

#### Malignant paraproteinaemia

1. Associated with malignancy of immunocytes.

Myelomatosis (this accounts for the majority of paraproteinaemias).
Macroglobulinaemia.
Heavy chain diseases.
Malignancy of the lymphoreticular system.

2. Associated with malignancy of other tissues.

### Benign paraproteinaemia

1. Associated with conditions normally producing a polyclonal response, for example,

Autoimmune diseases, including rheumatoid arthritis.
Severe chronic infections.
Cirrhosis of the liver.

2. "Essential" paraproteinaemia (including "essential" cryoglobulinaemia).

Benign paraproteinaemias may be transient.

## Myelomatosis (*Synonyms*: multiple myeloma; plasma cell myeloma)

Myelomatosis is a condition occurring with increasing frequency after the age of 50; it is very rare before the age of 30. It occurs equally in both sexes. In the commonest form there is malignant proliferation of plasma cells throughout the bone marrow. In such cases the clinical and laboratory features are due to:

1. Malignant proliferation of plasma cells.
2. Disordered immunoglobulin production.

### 1. Malignant proliferation of plasma cells

(*a*) *Bone pain*, which may be severe, is due to pressure from the proliferating cells. *X-rays* may show discrete *punched-out areas* of radiotranslucency, most frequently in the skull, vertebrae, ribs and pelvis. There may be generalised osteoporosis. Histologically, there is little osteoblastic activity around the lesions.

(*b*) *Pathological fractures* may occur.

(*c*) *Anaemia*, the next most common presenting feature, is due to many factors, including marrow replacement. It may sometimes be aggravated by haemolysis, or by haemorrhage.

### 2. Disordered immunoglobulin synthesis

(*a*) *Infections* are common due to reduced synthesis of immunoglobulins other than those produced by the malignant cells.

(*b*) *The* "*hyperviscosity syndrome*", although more common in macroglobulinaemia than in myeloma, can occur: production of the paraprotein may be so excessive that the circulating protein levels are very high. Consequent sluggish blood flow leads to sludging and thrombotic episodes in small vessels. This may result in:

Retinal vein thrombosis with impairment of vision
Cerebral thrombosis
Peripheral gangrene.

(c) *Haemorrhages* may be due to complexing of coagulation factors by the myeloma protein.

(d) *Cold sensitivity and Raynaud's phenomenon* occur if the paraprotein is a cryoglobulin (p. 336).

(e) *Renal failure* may be due to damage by BJP.

(f) *Amyloidosis* may occur especially in cases with BJP production.

**Laboratory findings and diagnosis.**—The first clue to the diagnosis may be noticed during venepuncture. The blood may be very viscous and may clot in the syringe. Preparation of blood films can be extremely difficult.

1. *Serum protein changes.*—The total protein concentration is often raised although in early cases it may be within normal limits. On electrophoresis there is usually a narrow band, most commonly in the $\gamma$ region (paraprotein), with reduced levels of normal $\gamma$ globulins. The raised protein is usually IgG, less commonly IgA (about 3:1) and rarely Bence Jones protein (if renal failure is present). Occasionally IgD, IgM or IgE are found, the last two being very rare.

Rarely no abnormal serum protein band is seen. This may occur if the malignant cells are very undifferentiated and fail to synthesise immunoglobulins, or if they are only producing Bence Jones protein. In either case synthesis of other immunoglobulins is usually depressed and there is often no other $\gamma$ globulin visible than the paraprotein. In IgD myeloma $\gamma$ globulin increase may not be apparent on routine electrophoresis. A completely normal electrophoretic pattern may occur in myelomatosis, but is very uncommon.

2. *Bence Jones proteinuria.*—In about 70 per cent of cases of myeloma L chains are produced in excess of H chains and can be detected by electrophoresis of concentrated urine. Immunological methods will confirm that these L chains are of a single type. The heat test, even when performed with great care, is positive in less than a third of proven cases, and should not be relied upon to exclude the presence of Bence Jones proteinuria.

3. *Hypercalcaemia* may be present (p. 247). High levels usually suppress with cortisone (see cortisone suppression test, p. 260) and this fact can be used in treatment.

As there is little osteoblastic activity, the *alkaline phosphatase level is normal* unless there is liver involvement; in this case the raised level is accompanied by a raised level of 5'-nucleotidase. A normal alkaline phosphatase level in cases with bone lesions helps to differentiate the picture from bony metastases.

4. If there is significant renal damage, the biochemical *features of renal failure* (p. 16) such as a raised plasma urea, urate and phosphate concentration may be present. Renal failure is an indication of a poor prognosis.

5. The *erythrocyte sedimentation rate* (ESR) is usually raised and may be very high. This may be the first indication of the disease and is probably due to the abnormal protein pattern.

6. *Haematological abnormalities.*—There is usually anaemia. The leucocyte and platelet counts are usually normal, but may be low. Rarely, large numbers of plasma cells appear in the peripheral blood (plasma cell leukaemia).

7. *Bone marrow.*—In myelomatosis malignant plasma cells ("myeloma cells") are seen in a bone marrow aspirate or biopsy. In the early stages the distribution of these cells may be patchy, and puncture at several sites may be necessary to confirm the diagnosis. Inspection of the bone marrow is an essential step in the diagnosis of myelomatosis.

**Soft tissue plasmacytoma.**—Rarely myeloma involves soft tissues, without marrow changes (extramedullary plasmacytoma). Although the protein abnormalities of myeloma are often found in these cases, their behaviour and prognosis are different. Spread is slow and tends to be local. Local excision of a solitary tumour is often effective.

### Waldenström's Macroglobulinaemia

Like myeloma, macroglobulinaemia occurs in the older age group (between 60 and 80 years), but is more common in males than in females. Also like myeloma, it is due to malignancy of immunocytes, but the malignant cells resemble lymphocytes rather than plasma cells. Symptoms of the "hyperviscosity syndrome" are commoner than in myeloma, probably because of the large size of the IgM molecule, but skeletal manifestations are rare. There is anaemia and lymphadenopathy.

### Laboratory Findings and Diagnosis

1. *Serum protein changes.*—As in myelomatosis there is usually a raised total protein concentration and electrophoresis shows a paraprotein in the $\gamma$ region. This can be identified as monoclonal IgM. Serum IgA concentration is usually reduced, but that of IgG may be raised.

As in myeloma there may be reduced levels of the normal immunoglobulins.

2. *Bence Jones* protein can be identified in the urine by electrophoresis in most cases.

3. *Haematological findings.*—Anaemia is common and is due to several factors including bleeding, marrow replacement and haemolysis.

As in myeloma, the ESR is usually markedly raised.

The bone marrow aspirate or lymph node biopsy contains atypical lymphocytoid cells.

## Heavy Chain Diseases

This is a rare group of disorders characterised by the presence of an abnormal protein identifiable as part of the H chain ($\alpha$, $\gamma$ or $\mu$). The clinical picture is that of generalised lymphoma ($\gamma$ chain disease), intestinal lymphoma with malabsorption ($\alpha$ chain disease) or chronic lymphatic leukaemia ($\mu$ chain disease). In many, but not all, cases, a paraprotein is detectable in the serum.

## Cryoglobulinaemia

Proteins which precipitate when cooled below body temperature are called cryoglobulins. Most commonly they are IgG or IgM, or a mixture of the two. Their presence may be associated with a number of diseases in which there is a disturbance of immunoglobulin production; like other paraproteinaemias, they may be an isolated finding ("essential cryoglobulinaemia"). They may only be detected by laboratory testing. Occasionally they give rise to symptoms of hyperviscosity syndrome after exposure to cold, the temperature at which the particular protein precipitates and its concentration being factors determining the severity of the symptoms.

Cryoglobulins have been described as occurring in all the conditions which may be associated with paraproteinaemia (p. 332), heavy chain disease being the exception.

Diagnosis.—A paraprotein is often detectable, but its absence, even in a properly collected specimen, does not exclude cryoglobulinaemia. If the diagnosis is suspected blood should be collected in a syringe warmed to 37°C, and maintained at this temperature until it has been tested. Failure to observe this precaution may result in false negative findings, as the cryoprecipitate is incorporated into the blood clot on cooling.

## "Benign" and "Essential" Paraproteinaemia

The only difference between these two syndromes is that in "benign" paraproteinaemia the paraprotein is found in association with a disease normally producing a polyclonal response, while in "essential" paraproteinaemia the subject is apparently healthy. Such cases have been reported to account for between 10 per cent and 30 per cent of all paraproteinaemias.

The diagnosis of either "benign" or "essential" paraproteinaemia should be made provisionally; the patients should always be followed up because they may later be found to have had early myeloma or macroglobulinaemia. The following points strongly suggest that the condition is malignant.

1. Bence Jones proteinuria.
2. Reduced levels of other immunoglobulins ("immune paresis").
3. An initial level of paraprotein above 20 g/l of IgG, or 10 g/l of IgA or IgM.
4. Increasing levels of paraprotein. In malignant paraproteinaemia the concentration usually doubles in less than 2 years.

If after 5 years there is no evidence of malignancy the diagnosis of benign paraproteinaemia can probably be made.

## PROTEINURIA

Normal subjects excrete up to 0·08 g of protein a day in the urine, amounts undetectable by usual screening tests. Proteinuria of more than 0·15 g a day almost always indicates disease.

Proteinuria may be due to renal disease or more rarely may occur because large amounts of low molecular weight proteins are circulating.

### Renal Proteinuria

1. **Glomerular proteinuria** is due to increased glomerular permeability (*nephrotic syndrome*): this is discussed more fully below. *Albumin* is usually the predominant protein.

"*Orthostatic* (*postural*) *Proteinuria*".—Most proteinuria is more severe in the upright than the prone position. The term "orthostatic" or "postural" has been applied to proteinuria, often severe, which disappears at night. It appears to be glomerular in origin and is commonest in adolescents and young adults. Although often harmless, evidence of renal disease may occur after some years.

2. **Tubular proteinuria** may be due to renal tubular damage from any cause, especially pyelonephritis (p. 16). If glomerular permeability is normal, proteinuria is usually less than 1 g a day, and consists mainly of $\alpha_2$ *and* $\beta$ *globulins*.

### Proteinuria with Normal Renal Function

Proteinuria can be due to production of *Bence Jones protein* (p. 332), to severe haemolysis with *haemoglobinuria*, or to severe muscle damage with *myoglobinuria*.

Bence Jones proteinuria can be inferred by inspection of the zone electrophoretic pattern of urinary proteins. BJP is the only protein of a molecular weight lower than albumin likely to be found in significant amounts in the urine (in the absence of haemoglobinuria or myoglobinuria). The presence in the urine of a band denser than that of albumin, especially if it is not present in the serum, suggests such a protein.

## NEPHROTIC SYNDROME

In the nephrotic syndrome *increased glomerular permeability* causes protein loss, often massive, with consequent hypoalbuminaemia and oedema, and with hyperlipoproteinaemia. The renal disease may be primary, or secondary to other pathology. It has been reported:

1. In most types of glomerulonephritis (about 80 per cent of nephrotics). In children "minimal change" glomerulonephritis is the commonest cause.

2. Secondary to:

Diabetes mellitus
Systemic lupus erythematosus (SLE)
Inferior vena caval or renal vein thrombosis
Amyloidosis
Malaria due to *P. malariae* (in malarial districts).

### Laboratory Findings

1. **Protein abnormalities.**—*Proteinuria* in the nephrotic syndrome ranges from 5 to 50 g per day. The proportion of different proteins lost is an index of the severity of the glomerular lesion. In mild cases albumin (MW about 70 000) and transferrin (MW about 80 000) are the predominant urinary proteins, and $\alpha_1$ antitrypsin (MW 45 000) is also present. With increasing glomerular permeability IgG (MW 160 000) and larger proteins appear.

The *serum protein* electrophoretic picture reflects the pattern of loss and has been described on p. 325.

Electrophoresis of urine on cellulose acetate gives some idea of the severity of the lesion. The *differential protein clearance* is a more precise measure of the selectivity of the lesion. The clearance of a low MW protein, such as transferrin or albumin, is compared with that of a larger one, such as IgG. The result is usually expressed as a ratio, obviating the need for timed collections. A ratio of IgG to transferrin clearance of less than 0·2 indicates high selectivity (predominant loss of small molecules), with a more favourable prognosis than those cases in which the ratio is higher: such cases usually respond well to steroid or cyclophosphamide therapy.

The consequences of the protein abnormalities are:

1. Oedema due to hypoalbuminaemia (p. 52).

2. Reduction in protein-bound substances due to loss of carrier protein. It is important not to misinterpret low total levels of calcium, thyroxine, cortisol and iron.

2. **Lipoprotein abnormalities.**—In mild cases $\beta$ lipoprotein increases, with consequent *hypercholesterolaemia.* In more severe cases a rise in pre-$\beta$ lipoprotein and its associated *triglycerides* may cause turbidity of the plasma. The cause of these changes is not fully understood.

Fatty casts may appear in the urine.

3. **Renal function tests.**—In the early stages glomerular permeability is high, and the plasma urea concentration normal. Later glomerular failure may develop, with the features of uraemia. At this stage, protein loss is reduced and plasma levels of protein and lipid may revert to normal. In the presence of uraemia this does *not* indicate recovery.

## SUMMARY

**Plasma Proteins**

1. Total protein estimations are of limited use in the diagnosis of protein abnormalities, although they may be used to assess hydration.

2. Serum proteins are usually fractionated by zone electrophoresis on cellulose acetate. Five fractions are recognised by this method—albumin, $\alpha_1$, $\alpha_2$, $\beta$ and $\gamma$ globulins. These fractions are affected in different ways by disease.

3. A non-specific pattern is seen in many inflammatory diseases, including liver disease and in conditions characterised by tissue damage. Diagnostic patterns may be seen in the nephrotic syndrome, the paraproteinaemias, hypogammaglobulinaemias and in some cases of cirrhosis of the liver.

4. Estimation of one or more specific proteins is of value in selected conditions.

**Immunoglobulins**

1. The immunoglobulins (Ig) are a group of proteins that are structurally related. Five classes are described. The main ones are IgG, IgA and IgM. Paraproteins are immunoglobulins and Bence Jones protein is related to them.

2. The functions and properties of the Ig classes and their response to antigenic stimuli differ. Estimation of IgG, IgA and IgM may be helpful in diagnosis of a few conditions.

3. Immunoglobulin deficiency is only one aspect of immunological deficiency. *Normal Ig levels do not exclude immunological deficiency.* Ig deficiencies may be congenital or acquired and may involve one or all Ig classes.

4. A paraprotein is a narrow band, usually found in the $\gamma$ globulin region of the electrophoretic strip. It usually, but not always, indicates malignant proliferation of immunocytes.

5. Paraproteins are most commonly associated with myeloma. Myeloma presents clinically in a variety of ways, reflecting bone marrow replacement and abnormal plasma protein levels. The diagnosis is made by bone marrow examination, and by finding protein abnormalities in serum and urine.

6. Bence Jones protein (usually found only in the urine) consists of free light chains. Its presence indicates malignancy of immunocytes.

7. Cryoglobulins are proteins which precipitate when cooled below body temperature. They cause symptoms on exposure to cold. They may occur in any of the diseases associated with paraproteinaemia.

## Proteinuria

1. Proteinuria may be due to glomerular or tubular disease. Glomerular proteinuria is the commoner form: massive proteinuria is always of glomerular origin.

2. Proteinuria may occur with normal renal function if abnormally large amounts of low molecular weight proteins are being produced.

3. The nephrotic syndrome is characterised by proteinuria, with a low serum albumin, oedema and hyperlipoproteinaemia. The proteinuria is glomerular in type. The severity of the lesion may be assessed by differential protein clearance.

## FURTHER READING

BELLANTI, J. A., and SCHLEGEL, R. J. (1971). The diagnosis of immune deficiency diseases. *Pediat. Clin. N. Amer.*, **18**, 49.

Report of a Symposium on Plasma Proteins (1971). *Ann. gen. Pract.*

Review Articles on immunoglobulins and immune response in *Brit. J. hosp. Med.*, 1970, **3**, 669 *et seq.* and 1972, **8**, 648 *et seq.*

ROITT, I. M. (1974). *Essential Immunology*, 2nd edit. Oxford: Blackwell Scientific Publications. (This is a concise, readable, comprehensive account of immune mechanisms.)

### REFERENCE

Report of a WHO Committee (1971). Primary immunodeficiencies. *Pediatrics*, **47**, 927.

# APPENDIX

## BLOOD SAMPLING FOR PROTEIN ESTIMATION

Blood for protein estimation, including that for immunoglobulins, should be taken with a *minimum of stasis*, or falsely high results may be obtained.

*Clotted blood* is essential for routine electrophoresis, because the presence of fibrinogen may mask, or be interpreted as, an abnormal protein.

## TESTING FOR URINARY PROTEIN

Several rapid screening tests are in routine use. There are two common *limitations:*

1. Most of the tests were developed to detect albumin and may be negative in the presence of even large amounts of other proteins such as BJP.

2. As these tests depend on *protein concentration*, negative results may be found with very dilute urines despite significant proteinuria.

It is *essential* that the sample should be *fresh.*

**Albustix (Ames).**—The test area of the reagent strip is impregnated with an indicator, tetrabromphenol blue, buffered to pH 3. At this pH it is yellow in the absence of protein; because protein forms a complex with the dye, stabilising it in the blue form, it is green or bluish-green if protein is present. The colour after testing is compared with the colour chart provided, which indicates the approximate protein concentration. The strips should be kept in the screwtop bottles, in a cool place. The instructions on the container should be carefully followed.

*False positives* occur:

1. In strongly alkaline (infected or stale) urine, when the buffering capacity is exceeded. A green colour in this case is a reflection of the alkaline pH.

2. If the urine container is contaminated with disinfectants such as chlorhexidine.

*False negatives* occur if acid has been added to the urine as a preservative (for example, for estimation of urinary calcium).

**Salicylsulphonic acid test.**—The urine is filtered if turbid. 2 to 3 drops of 25 per cent salicylsulphonic acid are added to about 1 ml of clear urine. The appearance of turbidity or flocculation indicates the presence of protein.

*False positives* may be due to:

Radio-opaque media such as are used in intravenous pyelograms
The presence of metabolites of tolbutamide
Administration of large doses of penicillin
The presence of urate in high concentration.

*False negative* results may occur in very alkaline (stale) specimens.

**Boiling test.**—The urine is filtered if turbid. About 5 ml of clear urine is boiled in a test tube. If turbidity appears 33 per cent acetic acid is added drop by drop, while mixing thoroughly. Turbidity due to phosphates will redissolve on acidification, while that due to protein remains.

These three tests detect protein present in concentration above about 0·2 g/l. Rough quantitation of the salicylsulphonic acid and heat tests can be made by noting the degree of turbidity.

| *Appearance* | *Approximate protein concentration* |
|---|---|
| | *g/l* |
| + Turbidity | 0·1–0·3 |
| + + Turbidity with | 0·4–1·0 |
| granular precipitation | |
| + + + Heavy turbidity with | 2·0–5·0 |
| flocculation | |
| + + + + Marked flocculation. | More than 5·0 |
| Coagulation of | More than 20 |
| specimen | |

**Bradshaw's test for globulins, including Bence Jones protein.**—This test is a better side-room test for BJP than the heat test. *It must be stressed that the presence of BJP can only be confirmed or excluded in the laboratory.*

Urine is layered gently on a few ml of concentrated hydrochloric acid in a test tube. A thick, white precipitate at the interface indicates the presence of *globulin* in the urine (most commonly BJP). Although albumin does not give a positive reaction, in severe nephrotic syndrome the presence of globulins may cause a positive result.

# Chapter XIV

# PLASMA ENZYMES IN DIAGNOSIS

MOST enzymes are present in cells at much higher concentrations than in plasma: some enzymes occur predominantly in certain types of cell. "Normal" plasma levels reflect the balance between the *release of enzymes* during ordinary cell turnover and their *catabolism and excretion*. *Proliferation of cells,* an increase in the rate of cell turnover, or *cell damage* usually results in *raised plasma levels*; these are readily demonstrable, because very low concentrations produce easily measurable activity *in vitro*. Hence enzymes can be used as "markers" to detect and localise cell damage or proliferation.

Very occasionally plasma activities may be *lower than normal,* due either to *reduced synthesis* or to *congenital deficiency or variants;* an example is plasma cholinesterase.

## Assessment of Cell Damage and Proliferation

Plasma enzyme levels depend on:

1. The rate of release from cells, in turn dependent on the rate of cell damage;

2. The extent of cell damage, or the degree of proliferation.

These two factors are balanced against

3. The rate of enzyme excretion and catabolism.

The rapid damage of relatively few cells as, for example, in many cases of viral hepatitis, may produce very high enzyme levels, which fall as excretion and catabolism take place. By contrast, the liver may be much more extensively involved in advanced cirrhosis, but the rate of cell breakdown is often low and enzyme levels may be only slightly raised, or even be "normal". In very severe liver disease, if many cells have been destroyed, fewer are left to undergo further damage and plasma enzyme levels may even *fall* terminally.

If the fall of levels during recovery depends largely on renal excretion as in the case of the relatively small α amylase molecule, renal disease may delay this fall; it may even be the cause of high levels in the absence of pancreatic disease.

For these reasons it is unwise to predict the extent of cell damage from single or serial enzyme estimations without taking the clinical picture and other findings into account.

## Localisation of Damage

An increase in cell numbers (for instance in osteoblasts with calcium deficiency or with metastatic bony deposits) increases plasma levels of associated enzymes (in this example of alkaline phosphatase).

Many enzymes occur in nearly all cells, although some are associated predominantly with certain tissues. Even if the enzyme is relatively specific for a particular tissue, elevated levels indicate only damage to the cells, and not the type of disease process. For example, following *major surgical procedures* or after extensive *trauma*, levels of lactate dehydrogenase, aspartate transaminase and creatine kinase may be raised, as the enzymes leak out of traumatised cells, mainly of skeletal muscle. If *circulatory failure* is present, due to cardiac failure or shock, and especially after cardiac arrest, very high levels of several enzymes may occur: estimations cannot be used to detect the cause of the arrest. When unexplained raised levels (especially of lactate dehydrogenase) are found, the possibility of *malignancy* should be considered: some malignant cells contain very high concentrations of such enzymes. *Haemolysis* is often associated with increases in transaminases and other enzymes liberated from damaged erythrocytes.

Specificity may be improved in two ways:

(*a*) **By isoenzyme determination.**—Measured enzyme activity may be due to closely related but slightly different molecular forms of the enzyme. These different forms (isoenzymes) can be identified by physical or chemical means and assume clinical significance when isoenzymes have different tissues of origin: for instance, different lactate dehydrogenase isoenzymes predominate in heart and liver.

(*b*) **By estimation of more than one enzyme.**—Although many enzymes are widely distributed, their *relative concentrations in different tissues* may vary. For instance both alanine and aspartate transaminases are abundant in liver, while in heart muscle there is much more aspartate than alanine transaminase. This difference is reflected in the markedly higher plasma aspartate than alanine transaminase levels after myocardial infarction, whereas both are usually elevated if there is liver involvement.

The *distribution of enzymes within the cell* may differ. Some, such as alanine transaminase and lactate dehydrogenase, occur only in the *cytoplasm*. Others such as glutamate dehydrogenase are found only in *mitochondria*, while aspartate transaminase, for example, occurs in both of these cellular compartments. Different relative levels may indicate different types of disease (p. 297).

## Non-specific Causes of Raised Enzyme Levels

Before attributing enzyme changes to a specific disease process, it is advisable to consider more generalised causes. Some of these have already been mentioned.

Small rises in aspartate transaminase are common as a non-specific finding in a variety of illnesses, some of which may be minor.

Other causes which should be considered are:

### (i) Physiological

(a) **Newborn.**—The levels of some enzymes such as aspartate transaminase are moderately raised in the neonatal period.

(b) **Childhood.**—Alkaline phosphatase levels are high until after puberty.

(c) **Pregnancy.**—In the last trimester levels of alkaline phosphatase are raised. During and immediately after labour moderate rises are noted in several enzymes, such as transaminases.

### (ii) Enzyme induction by drugs

Some drugs, particularly diphenylhydantoin and barbiturates may stimulate the production of enzymes (induction). For instance, raised levels of alkaline phosphatase in patients taking these drugs do not necessarily imply a pathological process, but should be investigated further. γ Glutamyltransferase levels in plasma correlate well with alcohol intake, a change that may be a result of enzyme induction.

### (iii) Artefactual elevations of enzyme levels

**Haemolysed samples** in general are *unsuitable* for enzyme estimation. Even if the red cell does not contain the enzyme being measured, other released enzymes may interfere with the estimation to produce a falsely elevated result.

## Enzyme Units

The total amount of enzymes (as protein) in the blood is less than 1 g/l. Results are not expressed as concentrations, but as *activities* as measured in the laboratory. With the development of clinical enzymology many methods of estimation were introduced, each with its own units (for example, Reitman-Frankel units, King-Armstrong units, spectrophotometric units). In an attempt to achieve standardisation international units were introduced, but even now results differ with variations in methodology, particularly as there is no standardised temperature at which the tests are performed. *Enzyme activities must be interpreted in relation to the normal range of the issuing laboratory,* and "normal ranges" will not be given in this chapter.

## ALTERED PLASMA ENZYME LEVELS

The enzymes we will discuss, and the abbreviations recommended by the panel quoted on p. 357 are:

| | |
|---|---|
| Aspartate transaminase | AST |
| Alanine transaminase | ALT |
| Lactate dehydrogenase | LD |
| Creatine kinase | CK |
| α Amylase | AMS |
| Alkaline phosphatase | ALP |
| Acid phosphatase | ACP |
| Aldolase | ALS |
| 5'-nucleotidase | NTP |
| Isocitrate dehydrogenase | ICD |
| γ Glutamyltransferase | GGT |
| Glutamate dehydrogenase | GMD |

### TRANSAMINASES

The transaminases are enzymes that are involved in the transfer of an amino group from an α-amino to an α-oxo acid. They are widely distributed in the body.

### Aspartate Transaminase (AST)

AST (serum glutamate oxaloacetate transaminase, SGOT) is widely distributed, with high concentrations in the heart, liver, skeletal muscle, kidney and erythrocytes, and damage to any of these tissues may cause raised levels.

### Causes of Increased AST

(a) Physiological—newborn (approximately $1\frac{1}{2}$ times adult "normal").

(b) Markedly raised levels (10–100 times normal).
   1. Myocardial infarction (p. 353).
   2. Viral hepatitis (p. 301).
   3. Toxic liver necrosis (p. 304).
   4. Circulatory failure, with "shock" and hypoxia.

(c) Moderately raised levels.
   1. Cirrhosis (up to twice normal).
   2. Cholestatic jaundice (up to 10 times normal).

3. Malignant infiltrations of the liver.
4. Skeletal muscle disease (p. 354).
5. After trauma or surgery (especially cardiac surgery).
6. Severe haemolytic anaemia.
7. Infectious mononucleosis (liver involvement).

(*d*) Artefactually raised levels occur in a haemolysed specimen.

## Alanine Transaminase (ALT)

ALT (serum glutamate pyruvate transaminase, SGPT) is present in high concentration in the liver and to a lesser extent in skeletal muscle, kidney and heart.

## Causes of Raised ALT

(*a*) Markedly raised levels:
　　1. Viral hepatitis (p. 301).
　　2. Toxic liver necrosis (p. 304).
　　3. Circulatory failure, with "shock" and hypoxia.

(*b*) Moderately raised levels:
　　1. Cirrhosis (p. 304).
　　2. Cholestatic jaundice (p. 303).
　　3. Liver congestion secondary to cardiac failure.
　　4. Infectious mononucleosis (liver involvement).
　　5. Extensive trauma and muscle disease (much less than AST).

## LACTATE DEHYDROGENASE (LD)

LD catalyses the reversible interconversion of lactate and pyruvate. It is widely distributed with high concentrations in the heart, skeletal muscle, liver, kidney, brain and erythrocytes.

## Causes of Raised LD Levels

(*a*) Marked increase (more than five times normal):
　　1. Myocardial infarction (p. 353).
　　2. Haematological disorders.

In blood diseases such as pernicious anaemia and leukaemias, very high values (up to 20 times normal) may be found. Lesser increases occur in other states of abnormal erythropoiesis, such as thalassaemia and myelofibrosis.
　　3. Circulatory failure, with "shock" and hypoxia.

(b) Moderate increase:
1. Viral hepatitis.
2. Malignancy anywhere in the body.
3. Skeletal muscle disease.
4. Pulmonary embolism.
5. Infectious mononucleosis.
6. Acute haemolysis.
7. Cerebral infarction ⎱ occasionally.
8. Renal disease ⎰

(c) Artefactually raised levels are found in haemolysed specimens.

### Isoenzymes of LD

Five isoenzymes are normally detectable by electrophoresis and are referred to as $LD_{1-5}$. $LD_1$ is the fastest moving fraction on electrophoresis and is the form found in heart muscle and erythrocytes, blast cells and kidney. $LD_5$ occurs predominantly in the liver. The heart isoenzyme may be measured by two further characteristics. It is relatively heat-stable and it is active against the substrate hydroxybutyrate. As *hydroxybutyrate dehydrogenase* (HBD) activity measures mainly heart isoenzyme, it is used in some laboratories in place of LD. Levels are raised in myocardial infarction, megaloblastic anaemias, acute leukaemias, and severe active renal damage (such as rejection of transplants) and much less so in liver disease.

## CREATINE KINASE (CK)

CK (CPK) is found in heart muscle, brain and skeletal muscle.

### Causes of Raised CK Levels

(a) Physiological:
1. Newborn—slightly raised.
2. Parturition—for a few days.

(b) Marked increase:
1. Myocardial infarction (p. 353).
2. Muscular dystrophies (p. 354).

(c) Moderate increase:
1. Muscle injury.
2. After surgery (for about a week).
3. Severe physical exertion.
4. Hypothyroidism (thyroxine apparently influences the catabolism of the enzyme).
5. Alcoholism (possibly due partly to alcoholic myositis).

6. Acute psychotic episodes.
7. Some cases of cerebrovascular accident and head injury.
8. Some patients predisposed to malignant hyperpyrexia (p. 374).

(*d*) Artefactually high levels occur in haemolysed samples (certain methods only).

The possible contribution of moderate exercise (particularly in young subjects) and of intramuscular injections is difficult to assess, but should be borne in mind as possible causes of mildly increased values.

## α AMYLASE (AMS)

α Amylase is the enzyme concerned with the breakdown of dietary starch and glycogen to maltose. It is present in pancreatic juice and saliva as well as in liver, Fallopian tubes and muscle. The enzyme is excreted in the urine.

The main use of amylase estimations is in the diagnosis of acute pancreatitis (p. 276), when levels may be very high. Moderately raised levels are found in a number of conditions and are not diagnostic (p. 277).

### Causes of Raised Plasma Amylase Levels

(*a*) Marked increase (5–10 times normal):
1. Acute pancreatitis.
2. Severe uraemia.
3. Severe diabetic ketoacidosis

(*b*) Moderate increase:
1. Other acute abdominal disorders:
  (i) Perforated peptic ulcer.
  (ii) Acute cholecystitis.
  (iii) Intestinal obstruction.
  (iv) Abdominal trauma.
  (v) Ruptured ectopic pregnancy.

2. Salivary gland disorders:
  (i) Mumps—not usually required for diagnosis except in mumps encephalitis when it may be raised without obvious salivary gland enlargement.
  (ii) Salivary calculi.
  (iii) After sialography.

3. Morphine administration (spasm of sphincter of Oddi).
4. Severe uraemia (may be markedly raised).

5. Myocardial infarction (occasionally).
6. Acute alcoholic intoxication (transient).
7. Diabetic ketoacidosis (may be markedly raised).
8. Macroamylasaemia—a rare condition, apparently symptomless, characterised by an abnormally large amylase molecule (or an amylase-plasma protein complex) that cannot be excreted by the kidneys.

Low levels of plasma amylase may be found in infants up to one year and in some cases of hepatitis. They are of no diagnostic importance.

α Amylase is a relatively small molecule, rapidly cleared by the kidneys. Raised *urinary* levels occur whenever plasma levels are raised, unless these are due to renal failure or macroamylasaemia.

## ALKALINE PHOSPHATASE (ALP)

Alkaline phosphatases are a group of enzymes which hydrolyse phosphates at an alkaline pH. The activity measured by routine methods includes that of several isoenzymes. They are found in bone, liver, kidney, intestinal wall, lactating mammary gland and placenta. In bone the enzyme is found in *osteoblasts* (less so in osteoclasts) and is probably important for normal bone formation.

In adults, the normal levels of alkaline phosphatase are derived largely from the liver. In children with high osteoblastic activity there is an additional contribution from bone and this accounts for the higher levels of total activity found at this age. In both there is a variable contribution from the intestine. Pregnancy raises the "normal" values because of the production of a heat-stable alkaline phosphatase by the placenta.

The source of pathologically elevated plasma ALP levels is almost always *liver* disease with a cholestatic element, or *bone* disease with increased osteoblastic activity. The raised levels in liver disease are not due to failure of excretion of the plasma enzyme but to increased production by cells lining bile canaliculi and regurgitation into the blood (p. 299).

Isoenzyme separation by electrophoresis is not routinely available, but the liver isoenzyme can be readily assessed by estimation of the *5'-nucleotidase activity*. 5'-nucleotidase (NTP) is an enzyme, the activity of which closely parallels that of liver ALP and which appears to have a similar origin. Bone disease does not produce elevation of 5'-NT.

Estimation is particularly valuable in the detection of liver disease in children, in whom the normal levels are lower than in adults. Moderate elevations of ALP may be masked by the wide normal range, while elevations of NTP are readily detectable.

## Causes of Raised Plasma ALP

(a) Physiological:

1. Children—until about the age of puberty (up to 2–2½ times adult normal).
2. Pregnancy—in the last trimester, values may be in the upper normal range or moderately increased.

(b) Bone disease (raised ALP; normal NTP):

1. Osteomalacia and rickets (p. 244).
2. Primary hyperparathyroidism with bone disease (p. 243).
3. Paget's disease of bone (may be very high).
4. Secondary carcinoma in bone.
5. Some cases of osteogenic sarcoma.

(c) Liver disease (raised ALP *and* NTP):

1. Cholestasis (p. 303).
2. Hepatitis (p. 301).
3. Cirrhosis of the liver (sometimes).
4. Space-occupying lesions—tumours, granulomata, infiltrations.

(d) Induction by drugs, such as barbiturates.

## Low Levels of Plasma ALP

1. Arrested bone growth:
   Achondroplasia.
   Cretinism.
   Vitamin C deficiency.
2. Hypophosphatasia.

## ACID PHOSPHATASE (ACP)

Acid phosphatase is found in the prostate, liver, red cells, platelets and bone. The main use of the estimation is in the diagnosis of prostatic carcinoma, and the level is usually little, if at all, affected in liver and bone disease. Many methods have been devised to measure the prostatic fraction only in plasma. There is no entirely satisfactory routine technique, but one of the best is based on the fact that the prostatic fraction is inhibited by l-tartrate ("tartrate labile").

Normally acid phosphatase in prostatic secretion drains via the prostatic ducts and very little appears in the blood. In prostatic carcinoma, particularly if it has metastasised, plasma acid phosphatase levels rise, probably because of increased numbers of prostatic cells. False negative results may occur when the tumour is too undifferentiated to secrete acid phosphatase at all.

It is important to remember, in assessing possible prostatic malignancy, to take blood before doing a rectal examination. Following rectal examination, passage of a catheter, or even in a constipated patient, the tartrate labile acid phosphatase may rise and values above normal may persist for up to a week. In any case in which a marginal rise is found estimations should be repeated on more than one occasion.

## Causes of Raised Plasma Acid Phosphatase

(a) Tartrate labile:
1. Prostatic carcinoma with extension.
2. Following rectal examination.
3. Acute retention of urine.
4. Passage of a catheter.

(b) Total:
1. Paget's disease and some cases of metastatic malignant disease (only if the ALP levels are very high, and therefore of no diagnostic significance).
2. Gaucher's disease (probably from Gaucher cells).
3. Occasionally in thrombocythaemias.

Acid phosphatase is unstable, and specimens must be sent to the laboratory without delay.

### γ GLUTAMYLTRANSFERASE (GGT)

γ Glutamyltransferase (γ glutamyltranspeptidase), occurs mainly in liver, kidney and pancreas. Plasma levels are *raised* in the same conditions that affect those of liver ALP. It is, however, a more sensitive indicator of liver disease, particularly for the detection of cirrhosis, metastatic carcinoma and hepatic infiltrations. Plasma GGT is elevated in chronic alcoholism and correlates with alcohol intake. Whether this represents early liver disease or enzyme induction is uncertain. Values in patients on anticonvulsant therapy are often raised.

### ALDOLASE (ALS)

Estimation of aldolase, another widely distributed enzyme, is used mainly in the diagnosis of muscle disorders (p. 354) and liver disease, in which it follows the same pattern as the transaminases. Raised levels are also found after myocardial infarction, extensive muscle trauma, haemolysis and generalised malignancy.

## ISOCITRATE DEHYDROGENASE (ICD)

Isocitrate dehydrogenase is yet another widely distributed enzyme. Its main diagnostic use is as an index of liver cell damage in which it follows the changes in the transaminases. Despite high concentrations In heart muscle, the level in myocardial infarction is normal. Other causes of moderately raised levels include placental infarction, megaloblastic anaemia and infectious mononucleosis.

## GLUTAMATE DEHYDROGENASE (GMD)

Glutamate dehydrogenase is found in the *mitochondria* of liver cells. It is not normally detectable in blood. Its appearance in the plasma in liver disease indicates cellular necrosis and the level is an index of the severity of the pathological process.

# ENZYME PATTERNS IN DISEASE
## MYOCARDIAL INFARCTION

The enzyme estimations of greatest value in the diagnosis of myocardial infarction are CK, AST and LD (or HBD). The choice of estimation depends on the time interval after the suspected infarction. A guide to the sequence of changes is given below:

TABLE XXVIII

| Enzyme | Time after infarction | | |
|---|---|---|---|
| | Starts to rise (hours) | Peak elevation (hours) | Duration of rise (days) |
| CK | 4–8 | 24–48 | 3–5 |
| AST | 6–8 | 24–48 | 4–6 |
| LD (HBD) | 12–24 | 48–72 | 7–12 |

Enzyme elevations are present in about 95 per cent of cases of myocardial infarction, and may reach very high levels. The height of the rise is a very rough index of the extent of damage and as such is of some value in prognosis. It is, however, only one of many factors (p. 343). A second rise in enzyme levels after return to normal indicates extension of the infarction, or the development of congestive cardiac failure, in which case CK does not increase. Levels in angina are usually normal, any rise probably being indicative of some myocardial necrosis. If a patient is first seen after the total LD has returned to normal, diagnosis may still be possible on the basis of a raised heart isoenzyme as detected by HBD estimation or isoenzyme electrophoresis.

Some confusion may arise if the patient presents with cardiac failure and secondary liver involvement, with raised transaminase and LD levels. Evaluation of the two processes is assisted by estimation of HBD (minimal rise if only the liver is involved) and the rise in ALT accompanying that in AST. Pulmonary embolism with right-sided cardiac failure usually produces this "liver pattern".

## LIVER DISEASE

Enzyme changes in liver disease are discussed in context on p. 297. Only a summary is appended here.

1. Rises of aspartate transaminase, alanine transaminase, isocitrate dehydrogenase, glutamyltransferase and, to a lesser extent, lactate dehydrogenase are indices of liver cell damage.

(a) Levels are raised in the prodromal phase of viral hepatitis and reach their maximum soon after the onset of jaundice, thereafter to return to normal. The magnitude of the increase reflects the extent of the process, but is not necessarily of prognostic significance.

(b) Very high levels occur associated with toxic hepatic necrosis.

(c) In early extrahepatic cholestasis enzyme increases are minimal, but if obstruction persists cell damage with consequent rise in enzyme levels occurs.

(d) Levels in cirrhosis may be moderately raised, but only if the process is active.

2. Alkaline phosphatase, $\gamma$ glutamyltransferase and 5'-nucleotidase are useful mainly as indicators of cholestasis.

(a) Intrahepatic cholestasis, for example, in hepatitis, is usually associated with moderately (2–3 times normal) raised levels. Biliary cirrhosis produces very high levels.

(b) Extrahepatic cholestasis may cause very high enzyme levels.

(c) Space-occupying lesions in the liver may produce raised ALP, NTP and GGT without a corresponding rise in bilirubin (p. 304).

3. Cholinesterase levels are decreased with liver cell dysfunction. This estimation parallels plasma albumin as a guide to prognosis.

## MUSCLE DISEASE

In the muscular dystrophies, probably because of increased leakage of enzymes from the damaged cells, plasma levels of muscle enzymes are increased. The relevant enzymes are CK, aldolase and the transaminases. Of these, CK is the most valuable, followed by aldolase. Points to consider in interpretation are:

1. Levels are highest (up to 10 times normal or more) in the early stages of the disease. Later, when much of the muscle has wasted, they are lower and may even be normal.

2. Levels are higher on activity immediately after rest (build up of muscle CK) than after prolonged activity.

3. In detecting affected newborns, it must be remembered that levels at this age are higher than in adults.

Similar, but less marked, changes are found in many subjects with myositis.

Carriers of the Duchenne type muscular dystrophy can often be detected by raised CK levels. As elevations are moderate, non-specific causes of raised enzyme activity (p. 345) must be excluded. Although enzyme levels are usually normal in *neurogenic muscular atrophy*, the number of exceptions make such tests an unreliable index in the differential diagnosis from primary muscle disease.

## ENZYMES IN MALIGNANCY

1. Tartrate labile acid phosphatase in prostatic carcinoma (p. 351) is the only truly specific enzyme for malignancy.

2. Malignancy anywhere in the body may be associated with a non-specific increase in LD, HBD and occasionally transaminases.

3. For follow-up of treated cases of malignancy, alkaline phosphatase estimations are of value. Raised levels may indicate secondaries in bone (normal NTP) or liver (raised NTP) (see p. 350). Liver secondaries may, in addition, produce increases in transaminases or LD.

GGT is also a sensitive indicator of liver involvement.

## HAEMATOLOGICAL DISORDERS

The extreme elevation of LD and HBD in megaloblastic anaemia and leukaemia has been mentioned (p. 348). Similar elevations may be found in other conditions with abnormal erythropoiesis. Typically there is much less change in the level of AST than in that of LD and HBD. Severe haemolysis produces changes in transaminases and LD.

## PLASMA CHOLINESTERASE AND SUXAMETHONIUM SENSITIVITY

There are two cholinesterases, one found predominantly in erythrocytes and nervous tissue (acetyl cholinesterase) and the other in plasma.

This latter enzyme, cholinesterase ("pseudocholinesterase") is synthesised chiefly in the liver and is the one routinely measured.

Cholinesterase has been used as a test of liver function. Low levels also result from poisoning by anticholinesterases such as the organophosphates.

### Causes of Decreased Plasma Cholinesterase

1. Hepatic parenchymal disease:
   (a) Hepatitis.
   (b) Cirrhosis.
2. Anticholinesterases (organophosphates).
3. Inherited abnormal cholinesterase.
4. Myocardial infarction.

### Causes of Increased Plasma Cholinesterase

1. Recovery from liver damage.
2. Nephrotic syndrome.

### Suxamethonium Sensitivity

The muscle relaxant suxamethonium is broken down by cholinesterase and this limits its period of action. In certain subjects administration is followed by a prolonged period of apnoea. Rarely no enzyme is detectable, but most patients have been found to have low levels of plasma cholinesterase, and the enzyme present is qualitatively different from the normal. The abnormal cholinesterase may be classified by the degree of enzyme inhibition produced by dibucaine (a spinal anaesthetic) or by fluoride. Results are expressed as *dibucaine* and *fluoride numbers*. For example, normal cholinesterase is 80 per cent inhibited by dibucaine (dibucaine number 80) while in subjects homozygous for the defective gene the dibucaine number is about 20. Heterozygotes have intermediate numbers.

Ten possible combinations have been described. In general, marked sensitivity is found only in homozygotes. Discovery of a patient with this abnormality requires investigation of the whole family and issuing of appropriate warning cards to all affected individuals.

### SUMMARY

1. Enzyme activities in cells are high. Natural decay of cells releases enzymes into the plasma, where activities are measurable, but usually low.

2. The main use of plasma enzyme estimations is the detection of raised levels due to cell damage.

3. Few enzymes are specific for any one tissue, but isoenzyme studies may increase specificity. In general, patterns of enzyme changes together with clinical findings are used for interpretation.

4. Non-specific causes of raised plasma enzyme activity include peripheral circulatory insufficiency, trauma, malignancy and surgery.

5. Enzyme estimations are of value in:

(a) Myocardial infarction—AST, CK, LD and isoenzymes.
(b) Liver disease—transaminases, ALP, ICD, GGT and NTP.
(c) Bone disease—ALP.
(d) Prostatic carcinoma—tartrate labile ACP.
(e) Acute pancreatitis—amylase.
(f) Muscle disorders—CK, aldolase.

6. Artefactual increases may occur in haemolysed samples.

## FURTHER READING

BARON, D. N., MOSS, D. W., WALKER, P. G., WILKINSON, J. H. (1975). Revised list of abbreviations for names of enzymes of diagnostic importance. *J. clin. Path.*, **28**, 592.

ROSALKI, S. B. (1969). *Diagnostic Enzymology*, 2nd edit. Miami: Dade Division, American Hospital Supply Corp.

## Chapter XV

# INBORN ERRORS OF METABOLISM

THE chemical make-up of an individual is determined by the 20 000 to 40000 gene pairs transmitted from generation to generation on the chromosomes. Random selection and recombination during meiosis, as well as occasional mutation, introduces individual variations. Such variations may at one extreme be incompatible with life or, at the other, produce biochemical differences detectable only by special techniques, if at all. In the latter category are the genetic variations of plasma proteins such as the transferrins or haptoglobins: such differences are useful in population and inheritance studies but do not necessarily impair function. Between the two extremes there are many variations that do produce functional abnormalities. It is to these variations that the term "inborn errors of metabolism" applies.

## GENERAL PRINCIPLES

Because the sequence of bases making up the DNA strands in the genes codes, via RNA, for protein structure, it is not surprising that most, if not all, inherited biochemical abnormalities can be explained by defective synthesis of a single peptide. This abnormality may occur in structural proteins, such as in the abnormal globins of the haemoglobinopathies, or in an enzyme, in which case chemically detectable metabolic consequences may result. The abnormality in the protein may be quantitative or qualitative: the haemoglobinopathies and cholinesterase variants fall, at least partly, into the latter group.

### Possible Metabolic Consequences

Deficiency of a single enzyme in a metabolic chain may produce its effects in several ways. Let us suppose that substance A is acted on by enzyme X to produce substance B, and that substance C is on an alternative pathway.

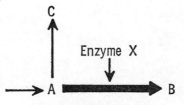

1. The effect may be due to deficiency of the products of the enzyme reaction, B. Examples of this are cortisol deficiency in congenital adrenal hyperplasia (p. 153), and the hypoglycaemia of some forms of glycogen storage disease (p. 210).

2. The effect may be due to the accumulation of the substance acted on by the enzyme (A) (for instance, phenylalanine in phenylketonuria, p. 367).

3. If a substance cannot be metabolised by the normal route because of an enzyme deficiency it may pass through an alternative pathway, and some product of the latter may produce effects (C). The virilisation due to androgens in congenital adrenal hyperplasia falls into this group.

The effects of the last two types of abnormality will be aggravated if the whole metabolic pathway is controlled by feed-back from the final product. For instance, in congenital adrenal hyperplasia, cortisol deficiency stimulates steroid synthesis and the accumulation of androgens, with consequent virilisation, is accentuated.

The clinical effects of some inborn errors are evident only in artificial situations. Patients with cholinesterase variants develop prolonged paralysis only if the muscle relaxant suxamethonium is administered (p. 356), and in some subjects with glucose-6-phosphate dehydrogenase deficiency haemolysis occurs only if they ingest drugs such as primaquine. Such cases appear "normal" without the intervention of modern therapeutics.

### When to Suspect an Inborn Error of Metabolism

The possibility of an inherited metabolic defect should be considered in the presence of any bizarre clinical or biochemical picture in infancy or childhood, especially if more than one baby in the family has been affected. The following situations, without obvious cause, are particularly suggestive:

(*a*) Failure to thrive; vomiting.
(*b*) Hypoglycaemia.
(*c*) A peculiar smell or staining of napkins.
(*d*) Hepatosplenomegaly.
(*e*) Retarded mental development; fits; spasticity.
(*f*) Renal calculi.
(*g*) Rickets which is refractory to treatment.

### Clinical Importance of Inborn Errors of Metabolism

The recognition of many inborn errors of metabolism is of academic interest only, either because the abnormality produces no clinical effect, or because no effective treatment is available. However, there is a group of such diseases in which recognition in early infancy is vital, since *treatment may prevent irreversible clinical consequences or death.* Some of the most important of these are:

Phenylketonuria (p. 367).
Galactosaemia (p. 371).
Maple syrup urine disease (p. 370).

Examples of conditions which should be sought *in relatives of affected patients,* either because *further ill effects may be prevented,* or because a *precipitating factor should be avoided,* are:

Cholinesterase abnormalities (p. 356).
Glucose-6-phosphate dehydrogenase deficiency (p. 182).
Acute intermittent porphyria (p. 408) and porphyria variegata (p. 409).
Haemochromatosis (p. 399).
Cystinuria (p. 365).
Wilson's disease (p. 372).

Other conditions can be *treated symptomatically.* Examples are:

Hereditary nephrogenic diabetes insipidus (p. 46).
Congenital disaccharidase deficiency (p. 279).

Some inborn errors of metabolism are completely, or almost completely, harmless. Their importance lies in the fact that they produce *effects which may lead to misdiagnosis* or which may alarm the patient. Examples of these are:

Renal glycosuria (p. 211).
Alkaptonuria (p. 369).
Gilbert's disease (p. 306).

Finally, the clinical effects of some inborn errors of metabolism may not appear until after reaching child-bearing age, and in these cases *genetic counselling* of relatives is desirable. An example of this type of disease may be:

Wilson's disease (p. 372).

## Laboratory Diagnosis of Inborn Errors of Metabolism

The enzyme deficiency is usually demonstrated *indirectly* by detecting a high concentration of the substance normally metabolised by the deficient enzyme, such as phenylalanine in phenylketonuria. *Direct* enzyme assay is available only in special centres but, if possible, all cases should be confirmed by this method. *Prenatal diagnosis* of some metabolic defects is becoming possible by studying cells cultured from the amniotic fluid in early pregnancy.

## Screening for Inborn Errors

Because of the importance of early treatment many countries have instituted programmes of screening of all newborn infants for inherited metabolic disorders, particularly for phenylketonuria. Tests may be performed on blood (obtained from a heel prick) or on urine. The sample is often collected on filter paper to facilitate transport to the laboratory.

The timing of the sample collection is important to avoid false negative results. Substances on the metabolic pathway before the enzyme block (for example, phenylalanine in phenylketonuria or galactose in galactosaemia) only accumulate once the infant begins to ingest the precursor (such as protein or milk products). Blood samples for screening are usually collected on the 6th to 9th day of life. Abnormal metabolites may not be found in the urine before 4 to 6 weeks after birth if their "renal threshold" is relatively high.

A positive result on screening should be confirmed by quantitative analysis or by repeat testing. Many abnormalities detected are transient.

## Treatment of Inborn Errors

Some inborn errors of metabolism are amenable to treatment by supplying the missing metabolite or limiting the dietary intake of precursors in the affected metabolic pathway. Accumulated products may occasionally be removed (for example, iron in haemochromatosis).

### PATTERNS OF INHERITANCE

The following account is intended for quick revision only. Details are available in books on genetics, a selection of which is listed at the end of the chapter.

Every inherited characteristic is governed by a pair of genes on homologous chromosomes (one received from each of the parents). Different genes governing the same characteristic are called alleles. If an individual has two identical alleles he is *homozygous* for that gene or characteristic; if he has two different alleles (AN) he is *heterozygous*. Genes may be carried on the sex chromosomes (X and Y) or on the autosomes (similar in both sexes) and the patterns of inheritance differ.

## Autosomal Inheritance

1. Let us suppose that one parent (parent 1 in the example below) carries an "abnormal" gene (A). If N is a normal gene, the possible gene combinations in the offspring are shown in the square.

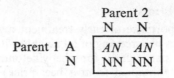

It will be seen that, on a *statistical basis*, half the offspring will carry one abnormal gene (AN): they are *heterozygous* for this gene, like Parent 1. None will be homozygous for the abnormal gene (AA).

2. Similarly, if both parents are heterozygous, a quarter of the offspring (in a large series) will be homozygous (AA) and half heterozygous.

$$
\begin{array}{c}
\text{Parent 2} \\
\text{A} \quad \text{N}
\end{array}
$$

Parent 1 A $\quad$
N

| | A | N |
|---|---|---|
| A | AA | AN |
| N | AN | NN |

3. If one parent is homozygous and the other "normal" all offspring would be heterozygous.

Since most genes producing clinical abnormalities are rare, example 1 above has the highest statistical likelihood. With consanguineous marriages example 2 becomes more probable, since blood relatives are more likely to carry the same abnormal genes than two unrelated people.

The *consequences* of the carriage of the abnormal gene depend on its potency compared to that of the normal one.

A *dominant* gene produces the abnormality in heterozygotes and homozygotes alike. Thus, in example 1, Parent 1 and half the offspring would be affected, and in example 2, both parents and 75 per cent of the offspring would be affected. Characteristically, cases appear in *successive generations*.

A *recessive* gene produces the abnormality only in homozygous individuals. Thus, in example 1, neither parents nor offspring would be affected, and in example 2 the parents would appear normal but 25 per cent of the offspring would be affected. Characteristically, one or more cases appear in a *single generation* with apparently normal parents.

The terms "dominant" and "recessive" are relative. A dominant gene may fail to manifest itself (*incomplete penetrance*) and so appear to skip a generation. A gene may vary in *expressivity*, that is, in the degree of abnormality it produces. Some genes manifest both in the homozygote and the heterozygote (in a less severe form)—*intermediate* inheritance. Finally, a recessive gene, while it produces disease only in the homozygote, may nevertheless be detectable by biochemical tests in the heterozygote.

## Sex-linked Inheritance

Some abnormal genes are carried only on the sex chromosomes, almost always on the X chromosome.

**X linked recessive inheritance.**—Females carry two X chromosomes and males one X and one Y. In X linked recessive inheritance an abnormal X chromosome (Xa) is latent when combined with a normal X chromosome, but active when combined with a Y. If the mother carries Xa she will appear to be normal, but, statistically, half the sons would be affected (YXa). Half the daughters would be carriers (XXa), but all daughters would appear clinically normal.

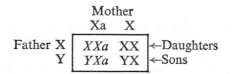

If the father is affected and the mother carries two normal genes, none of the sons will be affected, but all daughters will be carriers.

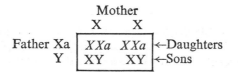

Inherited disease manifesting only in male offspring, but carried by females, is typical of X linked recessive inheritance. The female is only clinically affected in the extremely rare circumstance when she is homozygous for the abnormal gene. This would only occur if there were an affected father and carrier mother.

Haemophilia is the classical example of X linked recessive inheritance.

**X linked dominant inheritance.**—In this type of inheritance both XXa and YXa (males and females) are affected. An example is familial hypophosphataemia (p. 249).

## Multiple Alleles

Occasionally there may be several possible alleles governing the same characteristic. Different pair combinations may then produce different disease patterns (e.g. some haemoglobinopathies) or the variation may only be detectable by biochemical testing (e.g. plasma protein variants).

# DISEASES DUE TO INBORN ERRORS OF METABOLISM

The diseases discussed below are merely a fraction of the known inborn errors of metabolism. Selection must be biased by what the authors feel to be important, and others might disagree. On p. 376 some of the more clinically important abnormalities have been listed, and in the Appendix a fuller (but by no means complete) list is included under systematic headings; this includes the mode of inheritance, where known. Many of these conditions have been mentioned briefly in the relevant chapters. A few remain, which the authors feel to be of relative importance.

## Incidence

All the inborn errors of metabolism are very rare. The approximate incidence of disorders has been established by screening programmes in several countries. Of the disorders discussed below phenylketonuria, Hartnup disease, cystinuria, familial iminoglycinuria and histidinaemia are the commonest (1 in 10 000–20 000); galactosaemia is less common (about 1 in 100 000) and maple syrup urine disease much rarer (about 1 in 350000).

## AMINOACIDURIA

As disturbances of amino acid metabolism or excretion occur in many inborn errors of metabolism this manifestation is one of the first to be sought in suspected cases. It will be discussed briefly before considering specific diseases.

Amino acids are normally filtered at the glomerulus, and reach the proximal renal tubule at concentrations equal to those in plasma: almost all are actively reabsorbed at this site. Aminoaciduria may therefore be of two types:

1. *Overflow aminoaciduria* in which, because of raised blood levels, amino acids reach the proximal tubule at concentrations higher than the reabsorptive capacity of the cells.

2. *Renal aminoaciduria* in which plasma levels are low because of urinary loss due to defective tubular reabsorption.

A further subdivision may be made based on the pattern of the excreted amino acids.

(*a*) *Specific aminoaciduria* is the excessive excretion of either a single amino acid or a group of related amino acids. It may be overflow or renal in type and is almost always due to a genetic defect.

(*b*) *Non-specific aminoaciduria* is the excessive excretion of a number of unrelated amino acids. It is almost always due to an acquired lesion. It may be overflow in type, such as occurs in severe hepatic disease, when failure of deamination of amino acids causes raised plasma levels; more commonly renal aminoaciduria results from nonspecific proximal tubular damage from any cause (p. 14). In the latter, known as the *Fanconi syndrome*, other substances reabsorbed in the proximal tubule are lost in excessive amounts (phospho-glucoaminoaciduria): its occurrence in inborn errors of metabolism is much more commonly due to secondary tubular damage by the substance not metabolised normally (for instance, copper in Wilson's disease—hepatolenticular degeneration) than to a direct primary genetic defect. Acquired lesions will not be discussed further in this chapter.

### Inherited Abnormalities of Transport Mechanisms

Groups of chemically similar amino acids are often reabsorbed in the renal tubule by a single mechanism. In several cases similar group-specific mechanisms are involved in intestinal absorption and defects involve both the renal tubule and intestinal mucosa. Inborn errors involving the following group pathways have been identified:

(*a*) The dibasic amino acids (with two amino and one carboxyl group), cystine, ornithine, arginine and lysine (COAL is a useful mnemonic) (cystinuria).

(*b*) Many neutral amino acids (with one amino and one carboxyl group) (Hartnup disease).

(*c*) The imino acids, proline and hydroxyproline, probably shared with glycine (familial iminoglycinuria).

### Cystinuria

Cystinuria is due to an inherited abnormality of tubular reabsorption of the dibasic amino acids, cystine, ornithine, arginine and lysine, resulting in excessive urinary excretion of these four amino acids. A similar transport defect is present in the intestinal mucosa, but, although cystine absorption is diminished, failure of renal tubular reabsorption results in a high urinary excretion of the endogenously produced amino acid.

Many cases of cystinuria are asymptomatic. Cystine is non-toxic, but is relatively insoluble and the danger is due to its precipitation in the renal tract with calculus formation and its attendant complications. The solubility of cystine is such that only in homozygotes do urinary concentrations reach levels at which precipitation may occur and result in crystalluria and stone formation, although increased excretion can be demonstrated in heterozygotes.

The *diagnosis* of cystinuria is made by the demonstration of excessive urinary excretion of cystine and the other characteristic amino acids. The demonstration of the latter is necessary to distinguish the stone-forming homozygote from heterozygous cystine-lysinurias and from cystinuria occurring as part of a generalised aminoaciduria.

The *management* of cystinuria is aimed at the prevention of calculi by a high fluid intake, thus reducing the urinary cystine concentration. Alkalinising the urine also increases the solubility of cystine. If these measures are inadequate, administration of D-penicillamine may be tried.

Several genetic forms of cystinuria exist and the condition follows an autosomal recessive pattern of inheritance.

The relatively harmless condition described above must not be confused with *cystinosis*. This is a very rare inherited disorder of cystine metabolism characterised by accumulation of intracellular cystine in many tissues. In the kidney this produces tubular damage and consequently the Fanconi syndrome. The aminoaciduria is non-specific and of renal origin. Death occurs at an early age.

### Hartnup Disease

Hartnup disease, named after the first described patient, is a rare but interesting disorder in which there is a renal and intestinal transport defect involving neutral amino acids.

As in cystinuria the defect is present in both the proximal renal tubule and the intestinal mucosa. Most, if not all, the clinical manifestations can be ascribed to the reduced intestinal absorption and increased urinary loss of tryptophan. The amino acid is normally partly converted to nicotinamide, this source being especially important if the dietary intake of nicotinamide is marginal (p. 423). The clinical features of Hartnup disease are intermittent and resemble those of pellagra, namely:

1. A red, scaly rash on exposed areas of skin.
2. Reversible cerebellar ataxia.
3. Mental confusion of variable degree.

The thesis that nicotinamide deficiency is the cause of the clinical picture is supported by the response to administration of the vitamin and the fact that the features of the disease are frequently preceded by a period of dietary inadequacy.

In spite of the generalised defect of amino acid absorption protein malnutrition is not seen: this may possibly be due to absorption of intact peptides by a different pathway.

An associated chemical feature is the excretion of excessive amounts of *indole* compounds in the urine. These originate in the gut from the action of bacteria on the unabsorbed tryptophan.

Hartnup disease has a recessive mode of inheritance.

*Diagnosis* is made by demonstrating the characteristic amino acid pattern in the urine. Heterozygotes are not readily detectable by present techniques.

### Familial Iminoglycinuria

An abnormality of the transport mechanism of the imino acids leads to increased urinary excretion of proline, hydroxyproline and glycine, but with normal plasma levels. The condition is apparently harmless but must be distinguished from other, more serious causes, of iminoglycinuria. It is inherited as an autosomal recessive.

## DISORDERS OF AMINO ACID METABOLISM

A number of inborn errors of amino acid metabolism have been described. Most are characterised by raised levels of the relevant amino acid(s) in the blood, with overflow aminoaciduria. Only a few of the better known conditions are described here.

### Disorders of Aromatic Amino Acid Metabolism

The main chemical reactions of this pathway are outlined in Fig. 28, together with the site of known enzyme defects. It will be seen that *tyrosine*, normally produced in the body from phenylalanine, is the precursor of several important substances.

The inherited defects in thyroid synthesis are considered on p. 171.

**Phenylketonuria.**—This condition, if untreated, leads to mental retardation. Extensive screening of newborns is carried out in many countries to detect early cases. The problems arising from such surveys are discussed later.

The clinical features are:

1. Irritability, feeding problems, vomiting and fits in the first few weeks of life.

2. Mental retardation developing at between 4 and 6 months with unusual psychomotor irritability.

3. Generalised eczema in many cases.

4. A tendency to reduced melanin formation. Many patients have a pale skin, fair hair and blue eyes.

The abnormality in phenylketonuria is a deficiency of the enzyme *phenylalanine hydroxylase*. Phenylalanine accumulates in the blood and

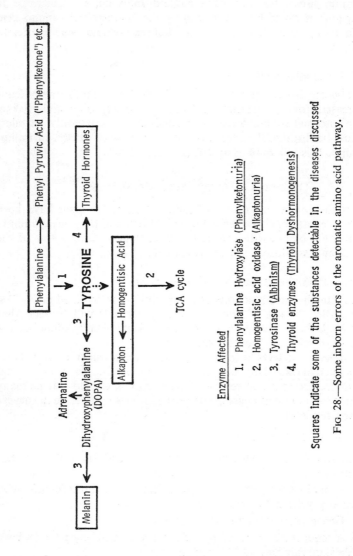

Enzyme Affected

1. Phenylalanine Hydroxylase (Phenylketonuria)
2. Homogentisic acid oxidase · (Alkaptonuria)
3. Tyrosinase (Albinism)
4. Thyroid enzymes (Thyroid Dyshormonogenesis)

Squares indicate some of the substances detectable in the diseases discussed

FIG. 28.—Some inborn errors of the aromatic amino acid pathway.

is excreted in the urine together with its derivatives, such as phenyl-pyruvic acid: the disease acquires its name from the recognition of this (phenylketone) in the urine. The cerebral damage is thought to be due to the high circulating levels of phenylalanine or one of its metabolites.

*Diagnosis.*—1. The *phenylalanine concentration* may be measured in *blood* taken from a heel-prick. The technique is suitable for mass screen-ing and tests are best performed in special centres. The timing of the test is critical (p. 361). It is recommended that tests be performed between 6 and 10 days after birth (just before the infant leaves hospital).

2. Detection of phenylpyruvic acid in the urine with *ferric chloride or Phenistix* (Ames) is the classical method of detecting phenylketonuria. This test may only be positive after about six weeks, when blood phenylalanine levels are very high. Since at this stage the infant has been discharged from hospital and possibly lost to the screening pro-gramme, measurement of blood phenylalanine levels is now the recom-mended procedure.

3. The absence of the enzyme phenylalanine hydroxylase may be demonstrated in liver biopsy material. This is not usually required for diagnosis.

Since the introduction of screening tests it has become apparent that raised blood levels of phenylalanine occur in other conditions than phenylketonuria, particularly in premature babies, when it is possibly due to delayed maturation of enzyme systems. These cases may be distinguished by repeated blood testing and by estimation of blood tyrosine levels (not raised in phenylketonuria, but raised in many of the other cases).

A variant, persistent hyperphenylalaninaemia, without mental re-tardation has been described.

A newly recognised problem is *maternal phenylketonuria*. Babies exposed *in utero* to high phenylalanine levels by unrecognised or untreated phenylketonuric mothers are mentally retarded and show other congenital abnormalities, even though they themselves are not phenylketonuric.

*Management.*—The aim of management is the reduction of blood phenylalanine levels and to this end a low phenylalanine diet is pre-scribed. Such treatment is difficult, expensive and tedious for the patient and parents, and requires careful biochemical monitoring of blood phenylalanine levels. Phenylalanine deficiency itself has deleterious effects. Tyrosine must also be included in the diet as it is the precursor of many important metabolites (Fig. 28).

Phenylketonuria is inherited as an autosomal recessive. *Heterozy-gotes* are clinically normal, but may be detected by biochemical tests.

**Alkaptonuria.**—An inherited deficiency of *homogentisic acid oxidase* results in alkaptonuria. Homogentisic acid accumulates in blood, tissues and urine. Oxidation and polymerisation of this substance produces the pigment "alkapton", in much the same way as polymerisation of DOPA (see Fig. 28) produces melanin. Deposition of alkapton in cartilages, with consequent darkening, is called *ochronosis*: this may cause *arthritis* in later life and may be clinically visible as darkening of the ears. Conversion of homogentisic acid to "alkapton" is speeded up in alkaline conditions and the most obvious abnormality in alkaptonuria is passage of *urine which is black*, or which darkens as it becomes more alkaline on standing. This finding may, however, be absent in a significant number of cases. The condition is often first noticed by the mother who is worried by the black nappies, which only become blacker when washed in alkaline soaps or detergents. The condition is compatible with a normal life span and treatment is unnecessary, but arthritis in middle and later life is common. Homogentisic acid is a *reducing substance* and reacts with Benedict's solution or Clinitest tablets.

Alkaptonuria is inherited as an autosomal recessive. Heterozygotes are not detectable by clinical or by biochemical findings.

**Albinism.**—A deficiency of *tyrosinase* in melanocytes causes one form of albinism and is inherited as a recessive character. The patient lacks pigment in skin, hair and iris (the eyes appear pink) and the condition is especially striking in negroes. Acute photosensitivity occurs because of pigment lack in the skin and iris. The tyrosinase involved in catecholamine synthesis is a different enzyme, controlled by a different gene. Adrenaline metabolism is normal in albinos.

### DISORDERS OF OTHER AMINO ACIDS

#### Maple Syrup Urine Disease

In maple syrup urine disease there is deficient decarboxylation of the three *branched-chain amino acids*, leucine, isoleucine and valine. These accumulate in the blood and are excreted in the urine together with their corresponding oxoacids. The odour of the urine, which resembles that of maple syrup, gives the disease its name.

The disease presents in the first week of life and, if untreated, severe neurological lesions develop with death in a few weeks or months. If, on the other hand, the condition is recognised and a diet low in branched-chain amino acids is given, normal development seems possible.

*Diagnosis* is made by demonstrating the raised levels of branched-chain amino acids in blood and urine. It may be confirmed by demonstrating the enzyme defect in the leucocytes.

The condition has a recessive mode of inheritance.

## Histidinaemia

Histidinaemia is associated with deficiency of the enzyme *histidase* which is required for the normal metabolism of histidine. Blood levels of histidine are raised and histidine and a metabolite, imidazole pyruvic acid, appear in increased amounts in the urine. Like phenylpyruvic acid (excreted in phenylketonuria) imidazole pyruvic acid reacts with ferric chloride to give a blue-green colour with ferric chloride or Phenistix (Ames). About half the cases described have shown mental retardation and speech defects but the rest appear normal. The results of dietary therapy are as yet inconclusive.

The condition is probably inherited as an autosomal recessive trait.

### DISORDERS OF CARBOHYDRATE METABOLISM

Many of the inborn errors of carbohydrate metabolism have been discussed in Chapter VIII. Only galactosaemia will be considered here.

## Galactosaemia

Galactose is necessary for the formation of cerebrosides, some glycoproteins and, during lactation, of milk. Any excess is rapidly converted to glucose (via glucose-1-phosphate).

The normal pathway of galactose metabolism is shown in a simplified form in Fig. 29.

Deficiency of either of the two enzymes shown results in an inability to convert galactose to glucose-1-phosphate with *galactosaemia* and *galactosuria*.

The commonest form of galactosaemia is due to deficiency of *hexose-1-phosphate uridylyltransferase* (previously known as *galactose-1-phosphate uridyltransferase*).

The condition becomes apparent only after milk has been added to the infant's diet. The main features are:

1. Vomiting and diarrhoea with failure to thrive.
2. Hepatomegaly leading to jaundice and cirrhosis.
3. Cataract formation.
4. Mental retardation.
5. Renal tubular damage (Fanconi syndrome).
6. Hypoglycaemia.

Galactose in the urine is a cause of a positive reaction with Benedict's solution or Clinitest tablets. This feature is dependent on ingestion of galactose and may be absent if the subject is not receiving milk. Tubular damage may result in generalised aminoaciduria.

The *diagnosis* is made by identifying the urinary sugar as galactose by chromatography and by demonstrating a deficiency of the relevant

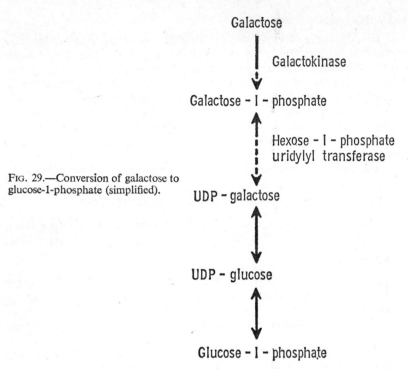

FIG. 29.—Conversion of galactose to glucose-1-phosphate (simplified).

enzyme in the erythrocytes. The latter test should be done on cord blood in all newborn infants with affected siblings.

*Treatment.*—Galactose (in milk and milk products) should be eliminated from the diet. Sufficient galactose (as UDP-galactose) is synthesised endogenously for the body's needs.

### INHERITED DEFICIENCIES OF CARRIER PROTEINS

Many hormones and trace metals are carried in the blood stream bound to albumin or to a specific carrier protein. Deficiencies of transferrin (p. 403) and of thyroxine-binding globulin (p. 166) have been mentioned. Caeruloplasmin deficiency occurs in Wilson's disease.

### Wilson's Disease (Hepatolenticular Degeneration)

Wilson's disease is an inherited condition characterised by excessive deposition of copper in the basal ganglia of the brain, liver, renal tubules and the eye. These accumulations produce:

1. Neurological symptoms due to basal ganglia degeneration.

2. Liver damage leading to cirrhosis.

3. Renal tubular damage with any or all of the biochemical features of this condition, including aminoaciduria (Fanconi syndrome).

4. Kayser-Fleischer ring at the edge of the cornea due to copper deposition in Descemet's membrane.

Like haemochromatosis this disease does not usually present clinically in childhood, but the condition should always be sought in "idiopathic" cirrhosis in children. Cirrhosis may be present without the neurological abnormalities. Some cases only develop clinical symptoms in early adult life, often after childbearing, and for this reason genetic counselling of relatives is necessary.

The basic defect in Wilson's disease is unknown. Most cases show a reduced concentration of the copper-binding protein *caeruloplasmin* in the plasma. This may possibly allow increased passage of free copper into the tissues. However, cases can occur with normal levels of apparently normal caeruloplasmin, and in most patients there is poor correlation between the levels of the protein and the severity of the disease. A low concentration of caeruloplasmin remains, however, one of the diagnostic features of the disease. Other biochemical abnormalities noted in Wilson's disease include low plasma copper levels and increased urinary copper excretion, as well as those produced by the cirrhosis and renal tubular damage. Copper concentrations in liver biopsy specimens are high.

The clinical condition has a recessive mode of inheritance but heterozygotes may also have reduced caeruloplasmin levels. Distinction between presymptomatic homozygotes and heterozygotes is important as the former require treatment.

In some cases a liver biopsy may be necessary for the diagnosis. In assessing caeruloplasmin concentration it is important to remember that *low levels* may also occur during the first few months of life, with malnutrition and in the nephrotic syndrome due to loss in the urine. *Raised levels* are found in active liver disease (this may account for some "normal" levels in patients with Wilson's disease), in the last trimester of pregnancy, in women taking oral contraceptives and non-specifically in states of tissue damage (for example, inflammation and neoplasia).

*Treatment* with agents chelating copper, such as D-penicillamine, aim to reduce tissue copper concentration.


### Drugs and Inherited Metabolic Abnormalities

The variation in individual response to identical drugs may be partly due to genetic variation. There are, however, a number of well-defined inherited disorders that are aggravated by, or which only become apparent on, administration of certain drugs. These may be classified into two groups.

### 1. Disorders Resulting from Deficient Metabolism of a Drug

The muscle relaxant suxamethonium (succinylcholine) normally has a very brief action as it is rapidly broken down by plasma cholinesterase. In *suxamethonium sensitivity* (p. 356) an abnormal cholinesterase is present, breakdown of the drug is impaired and prolonged respiratory paralysis may occur.

Two other inherited disorders are characterised by defective metabolism of the drugs *isoniazid* and *diphenylhydantoin* respectively. In both, toxic effects appear more frequently and at lower dosages than in normal individuals.

### 2. Disorders Resulting from Abnormal Response to a Drug

A form of haemolytic anaemia, common in many parts of the world, is due to a deficiency of *glucose-6-phosphate dehydrogenase* in the erythrocyte. This is the first enzyme in the hexose monophosphate shunt (p. 182) and is required for the formation of NADPH. This, in turn, is probably essential for the maintenance of an intact red cell membrane. Numerous variants of G-6-PD deficiency have been described. In many cases haemolysis is precipitated by drugs, notably certain antimalarial drugs, such as primaquine, sulphonamides and vitamin K analogues.

In the inherited *hepatic porphyrias* (p. 408) acute attacks may be precipitated by any of a number of drugs, particularly barbiturates.

Some people react to general anaesthesia (most commonly halothane with suxamethonium) with a rapidly rising temperature, muscular rigidity and acidosis; the majority die as a result (*malignant hyperpyrexia*). Many, but not all, susceptible subjects in affected families have an elevated creatine kinase (CK) level.

This short section should serve to remind readers to consider the possibility of an inborn error when an abnormal reaction to a drug is encountered. A reference to this growing field of *pharmacogenetics* is given at the end of the chapter.

### SUMMARY

1. Inborn errors of metabolism are diseases due to inherited defects of protein synthesis. Most of those presenting with clinical symptoms are due to abnormalities of enzyme synthesis.

2. Inborn errors of metabolism may produce no clinical effects, may only produce them under certain circumstances (for example, cholines-

terase variants) or, at the other extreme, may produce severe disease. Some are incompatible with life.

3. Recognition of some inherited abnormalities is of academic interest only. Diagnosis is important if the condition is serious but treatable, if precipitating factors can be avoided, or if confusion with other diseases is possible.

4. Inheritance may be autosomal or sex-linked, dominant or recessive. In diseases producing severe clinical effects inheritance is most commonly autosomal recessive, and they are most common in the offspring of consanguineous marriages.

5. In many cases in which the clinical disease is inherited in a recessive manner, lesser degrees of the abnormality can be detected by chemical testing.

6. Some inborn errors of metabolism not mentioned elsewhere in the book are discussed in this chapter.

## FURTHER READING

EMERY, A. E. H. (1971). *Elements of Medical Genetics*, 2nd edit. Edinburgh: Livingstone.

FRASER ROBERTS, J. A. (1973). *An Introduction to Medical Genetics*, 6th edit. London: Oxford University Press.

Many similar books on the principles of inherited disease exist. The above two are comprehensive and readable.

RAINE, D. N. (1972). Management of inherited metabolic disease. *Brit. med. J.*, **2**, 329.

This includes a useful summary of the clinical features of many inborn errors of metabolism.

PRICE EVANS, D. A. (1968). Clinical Pharmacogenetics. In *Recent Advances in Medicine*, 15th edit., p. 203. Eds. Baron, D. N., Compston, N., and Dawson, A. M. London: Churchill.

VESELL, E. S. (1973). Advances in Pharmacogenetics. In *Progress in Medical Genetics*, *IX*, p. 291. Eds. Steinberg, A. G., and Bearn, A. G. New York: Grune and Stratton.

The last two references deal with drugs and inherited abnormalities.

REFERENCE

STANBURY, J. B., WYNGAARDEN, J. B., and FREDRICKSON, D. S., Eds. (1972). *The Metabolic Basis of Inherited Disease*, 3rd edit. New York: McGraw-Hill.

# APPENDIX

The following list of inborn errors of metabolism is far from complete. It is meant for reference only, and the student should not attempt to learn it. Most of the abnormalities have been discussed in this book, and a page reference is given. Where it is known the mode of inheritance is given, unless the heading applies to a group of diseases of different modes in inheritance.

D = Autosomal Dominant
R = Autosomal Recessive
X linked D = X linked Dominant
X linked R = X linked Recessive

| | Inheritance | Page |
|---|---|---|
| I. DISORDERS OF CELLULAR TRANSPORT | | |
| Most of these are recognised as renal tubular transport defects, and in some defective intestinal transport can also be demonstrated. | | |
| *Generalised Proximal Tubular Transport* | | |
| Phosphoglucoaminoaciduria | ? | 365 |
| *Amino Acids* | | |
| Dibasic amino acids—Cystinuria | R | 365 |
| Neutral amino acids—Hartnup disease | R | 366 |
| Familial iminoglycinuria | R | 367 |
| *Glucose* | | |
| Renal glycosuria | D | 211 |
| *Water (Failure to Respond to ADH)* | | |
| Hereditary nephrogenic diabetes insipidus | X linked R | 46 |
| *Sodium (Failure to Respond to Aldosterone)* | | |
| Pseudo-Addison's disease | ? | 48 |
| *Potassium (all cells)* | | |
| Familial periodic paralysis | D | 67 |
| *Calcium (Failure to Respond to PTH)* | | |
| Pseudohypoparathyroidism | X linked D | 245 |
| *Phosphate* | | |
| Familial hypophosphataemia | X linked D | 249 |
| *Hydrogen Ion* | | |
| Renal tubular acidosis | D | 95 |
| *Bilirubin (Liver Cells)* | | |
| Congenital hyperbilirubinaemias | — | 306 |

| | Inheritance | Page |
|---|---|---|
| II. DISORDERS OF AMINO ACID METABOLISM | | |
| *Aromatic Amino Acids* | | |
| Phenylketonuria | R | 367 |
| Alkaptonuria | R | 369 |
| Albinism | R | 370 |
| Thyroid dyshormonogenesis | All R | 171 |
| *Sulphur Amino Acids* | | |
| Cystinosis | R | 366 |
| Homocystinuria | R | |
| *Branched-Chain Amino Acids* | | |
| Maple syrup urine disease | R | 370 |
| | | |
| III. DISORDERS OF CARBOHYDRATE METABOLISM | | |
| Glycogen storage diseases | R | 210 |
| Galactosaemia | R | 371 |
| Hereditary fructose intolerance | R | 206 |
| Essential pentosuria | R | 212 |
| Essential fructosuria | R | 212 |
| Diabetes mellitus | ? | 191 |
| | | |
| IV. DISORDERS OF LIPID METABOLISM | | |
| Hyperlipoproteinaemias | — | 227 |
| Hypolipoproteinaemias | R | 222 |
| Plasma LCAT deficiency | ?R | 222 |
| | | |
| V. ABNORMALITIES OF PLASMA PROTEINS | | |
| Immunoglobulin deficiencies | — | 330 |
| Carrier protein abnormalities | | |
| Transferrin | ?R | 403 |
| Thyroxine binding globulin deficiency | X linked | 166 |
| Wilson's disease (Hepatolenticular degeneration) | R | 372 |
| Cholinesterase variants | R | 356 |
| $\alpha_1$ Antitrypsin deficiency | | 321 |
| | | |
| VI. ERYTHROCYTE ABNORMALITIES | | |
| Haemoglobinopathies (see haematology textbooks) | | |
| Glucose-6-phosphate dehydrogenase deficiency | X linked | 374 |
| NADP methaemoglobin reductase deficiency (see Haematology textbooks) | R | |
| | | |
| VII. DISORDERS OF PORPHYRIN AND IRON METABOLISM | | |
| Porphyrias | — | 408 |
| Haemochromatosis | ? | 399 |
| | | |
| VIII. DISORDERS OF STEROID METABOLISM | | |
| Congenital adrenal hyperplasias | All R | 153 |

| | Inheritance | Page |
|---|---|---|
| **IX. DISORDERS OF PURINE METABOLISM** | | |
| Primary gout | ? | 383 |
| Xanthinuria | ?R | 386 |
| Lesch-Nyhan syndrome | ? X linked R | 385 |
| | | |
| **X. DISORDERS OF DIGESTION** | | |
| Disaccharidase deficiencies | R | 279 |
| Cystic fibrosis of the pancreas | R | 272 |
| | | |
| **XI. DISORDERS OF OXALATE METABOLISM** | | |
| Primary hyperoxaluria | ?R | 25 |
| | | |
| **XII. DISORDERS INDUCED BY DRUGS** | | |
| Suxamethonium sensitivity | R | 356 |
| Slow inactivation of isoniazid | R | 374 |
| Diphenylhydantoin toxicity | D | 374 |
| Glucose-6-phosphate dehydrogenase deficiency | X linked | 374 |
| Malignant hyperpyrexia | D | 374 |

# Chapter XVI

# PURINE AND URIC ACID METABOLISM

## HYPERURICAEMIA AND GOUT

HYPERURICAEMIA may be asymptomatic, or may give rise to the clinical syndrome of gout. In either case it should be treated; the relative insolubility of urate means that there is the danger of precipitation in all tissues. If this takes place in the kidney renal damage can result (compare the danger of hypercalcaemia, p. 239). Hyperuricaemia may be due to a primary lesion of purine metabolism or be secondary to a variety of other conditions. The primary syndrome has a familial incidence.

Plasma urate is in the form of the monosodium salt. Uric acid is less soluble than its sodium salts.

### NORMAL URIC ACID METABOLISM

Uric acid is the end product of purine metabolism in man. In most other mammals it is further broken down to the soluble compound, allantoin, and it is the poor solubility of urates which makes man prone to clinical gout and renal damage by urate. The purines, adenine and guanine, are constituents of both types of *nucleic acid* (DNA and RNA). The purines used by the body for nucleic acid synthesis may be derived from the breakdown of ingested nucleic acid (mainly in meat which is rich in cells), or may be synthesised in the body from small molecules *de novo*.

### Synthesis of Purines

The synthetic pathway of purines is complex, and involves the incorporation of many small molecules into the relatively complex purine ring. The upper part of Fig. 30 summarises some of the more important steps in this synthesis. Cytotoxic drugs such as 6-mercaptopurine and folic acid antagonists inhibit various stages in this pathway, so preventing DNA formation and cell growth.

The following stages in Fig. 30 should be especially noted.

Step (a) is the first one in purine synthesis. It involves condensation of phosphate with phosphoribose to form phosphoribosyl pyrophosphate (PRPP).

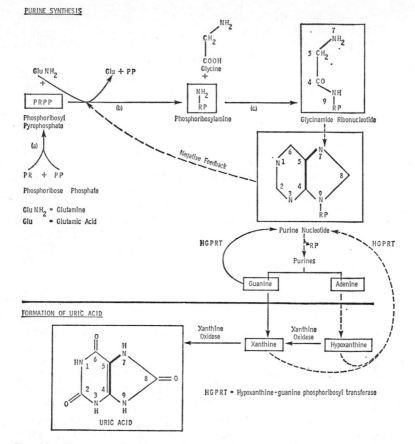

PURINE SYNTHESIS

FORMATION OF URIC ACID

HGPRT = Hypoxanthine-guanine phosphoribosyl transferase

FIG. 30.—Summary of purine synthesis and breakdown to show steps of clinical importance. (See text for explanation of small letters.)

In Step (b) the amino group of glutamine is incorporated into the ribose phosphate molecule and pyrophosphate is released. This is the *rate limiting* or controlling step in purine synthesis. It is subject to feedback inhibition from increased levels of purine nucleotides: thus the rate of synthesis is slowed when its products increase. This step may be at fault in primary gout.

Step (c) shows how the *glycine* molecule is added to phosphoribosylamine. Labelled glycine can be used to study the rate of purine synthesis. The atoms in the glycine molecule have been numbered in the diagram to correspond with those of the purine and uric acid molecules, and the heavy lines further indicate the final position of the amino acid in these molecules. By use of labelled glycine it has been shown that purine synthesis is increased in primary gout.

After many complex steps purine ribonucleotides (purine ribose phosphates) are formed and, as has already been stated, the level of these controls Step (b). Ribose phosphate is split off, thereby releasing the purines.

## Fate of Purines

Purines arising from *de novo* synthesis, those derived from the diet, and those liberated by endogenous breakdown of nucleic acids may follow one of two pathways:

1. They may be synthesised into new nucleic acid.
2. They may be oxidised to uric acid.

**Formation of uric acid from purines.**—As shown in the lower part of Fig. 30, some of the adenine is oxidised to hypoxanthine and hypoxanthine is further oxidised by the liver enzyme, *xanthine oxidase*, to xanthine. Guanine can also form xanthine. Xanthine, in turn, is oxidised by xanthine oxidase to form uric acid. Thus the formation of uric acid from purines depends on xanthine oxidase activity, a fact of importance in the treatment of gout.

**Reutilisation of purines.**—Some xanthine, hypoxanthine and guanine can be resynthesised to purine nucleotides by pathways involving, amongst other enzymes, hypoxanthine–guanine phosphoribosyl transferase (HGPRT).

**Excretion of urate.**—75 per cent of the urate leaving the body is excreted in the urine and 25 per cent passes into the intestine, where it is broken down by intestinal bacteria (*uricolysis*). The urate filtered at the renal glomerulus is probably completely reabsorbed in the tubules and the urinary urate is derived from active tubular secretion: urinary excretion may be enhanced by various drugs used in the treatment of gout.

Renal excretion of urate is inhibited by such organic acids as lactic and oxoacids.

### Causes of Hyperuricaemia

Figure 31 summarises the factors which may contribute to hyperuricaemia. These are:

1. Increased rate of formation of uric acid.
   Increased synthesis of purines (a).
   Increased intake of purines (b).
   Increased turnover of nucleic acids (c).
2. Reduced rate of excretion of urate (e).

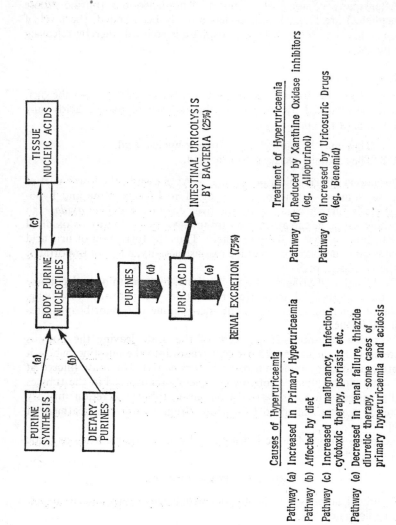

FIG. 31.—Origin and fate of uric acid in normal subjects.

Causes of Hyperuricaemia

Pathway (a) Increased in Primary Hyperuricaemia
Pathway (b) Affected by diet
Pathway (c) Increased in malignancy, infection, cytotoxic therapy, psoriasis etc.
Pathway (e) Decreased in renal failure, thiazide diuretic therapy, some cases of primary hyperuricaemia and acidosis

Treatment of Hyperuricaemia

Pathway (d) Reduced by Xanthine Oxidase inhibitors (eg. Allopurinol)
Pathway (e) Increased by Uricosuric Drugs (eg. Benemid)

Steps (b), (c) and (e) are causes of secondary hyperuricaemia. Increased synthesis is probably the basic fault in primary hyperuricaemia.

## DANGERS OF HYPERURICAEMIA

The solubility of urate in plasma is limited. Precipitation in tissues may be favoured by a variety of local factors of which the most important are probably tissue pH and trauma. Crystallisation in *joints*, especially those of the foot, produces the classical picture of gout, first described by Hippocrates in 460 B.C. It is thought that local inflammation due to urate precipitation produces an increase in leucocytes in the area, and that lactic acid production by these lowers the pH locally: this reduces the solubility of urate and sets up a vicious circle in which further precipitation occurs. It should be noted that in attacks of acute gouty arthritis local factors are of more importance than the plasma urate levels; the latter may even be normal during the attack.

Precipitation can also occur in other tissues, and the subcutaneous collections of urate, which are especially common in the ear and in the olecranon and patellar bursae and tendons, are called *gouty tophi*.

Gouty attacks are extremely painful and unpleasant and may lead to permanent joint deformity. Tophi are disfiguring but harmless. The more serious effect of hyperuricaemia is due to precipitation of urate in the kidney, leading to *renal failure*. For this reason it has been recommended that all cases with plasma urate levels consistently above 0.55 mmol/l (9 mg/dl) should be treated.

## PRIMARY HYPERURICAEMIA AND GOUT

### Familial Incidence

In A.D. 150 Galen said that gout was due to "debauchery, intemperance and an hereditary trait". "Intemperance", as we shall see, may aggravate the condition. The striking familial incidence of hyperuricaemia confirms that there is probably "an hereditary trait", but in this respect we know little more than Galen did, since the mode of inheritance (like that of diabetes mellitus) is still obscure.

### Sex and Age Incidence

Gout and hyperuricaemia are very rare in children, and rare in women of child-bearing age. The difference in incidence in males and females is not due to a sex-linked inheritance, because it can be transmitted by males. Plasma urate levels are low in children and rise in both sexes at puberty, more so in males than females. Women become more prone to hyperuricaemia and gout in the post-menopausal period (compare plasma cholesterol and iron levels).

## Precipitating Factors

The classical image of the gouty subject is the red-faced, good living, hard drinking Squire depicted in novels and paintings of the 18th century. Galen mentioned "debauchery and intemperance" as causes of gout. Two factors probably account for the high incidence of clinical gout in this type of subject.

1. Alcohol has been shown to decrease renal excretion of urate. This may be because it is partly metabolised to lactate, which inhibits urate excretion.

2. A high meat diet contains a high proportion of purines.

Neither of these factors are likely to precipitate gout in a normal person, but may do so in a subject with a gouty trait.

For reasons which are not clear, hyperuricaemia, and even clinical gout, are relatively common in patients with hypercalcaemia from any cause, and in patients with recurrent calcium-containing renal calculi, even when not accompanied by hypercalcaemia.

## Biochemical Lesion of Primary Hyperuricaemia

Use of labelled glycine has shown that *purine synthesis is increased* in about 25 per cent of cases of primary hyperuricaemia. There may be overactivity of the enzyme controlling the formation of phosphoribosylamine (Fig. 30). This could be due to failure of normal feedback suppression by nucleotides.

*Reduced renal secretion of urate* has also been demonstrated in other cases of primary hyperuricaemia. Both increased synthesis and decreased excretion may be present in many subjects.

## Principles of Treatment of Hyperuricaemia

Treatment may be based on:

**Reducing purine intake** (Step (b) Fig. 31).—This is not very effective by itself.

**Increasing renal excretion of urate** with *uricosuric drugs*, such as probenecid and salicylates in large doses (Step (e) Fig. 31). These are very effective if renal function is normal, but are useless in the presence of renal failure. Fluid intake must be kept high. *Low doses* of salicylate *inhibit* urate secretion.

**Reducing uric acid production** by drugs which inhibit xanthine oxidase (Step (d) Fig. 31), such as *allopurinol* (hydroxypyrazolopyrimidine). This compound is structurally similar to hypoxanthine and acts as a competitive inhibitor of the enzyme. *De novo* synthesis may also be decreased by this drug.

**Colchicine,** which has an anti-inflammatory effect in acute gouty arthritis, does not affect uric acid metabolism.

## Juvenile Hyperuricaemia (Lesch-Nyhan Syndrome)

This is a very rare inborn error, probably carried on an X-linked recessive gene, in which severe hyperuricaemia has been reported to occur in young male children. A deficiency of the enzyme *hypoxanthine-guanine phosphoribosyl transferase (HGPRT)* has been demonstrated in affected subjects. Hypoxanthine and other purines cannot be recycled to form purine nucleotides, and uric acid production from them is probably increased. The syndrome is associated with mental deficiency, a tendency to self-mutilation, aggressive behaviour, athetosis and spastic paraplegia.

### SECONDARY HYPERURICAEMIA

High plasma urate levels may be the result of a variety of conditions.

1. Increased turnover of nucleic acids ((c) in Fig. 31).
    (*a*) Rapidly growing malignancy, especially leukaemias and polycythaemia rubra vera.
    (*b*) Treatment of malignant tumours.
    (*c*) Psoriasis.
    (*d*) Increased tissue breakdown in
        Acute starvation
        Tissue damage.
2. Reduced excretion of urate (Step (e) in Fig. 31).
    (*a*) Glomerular failure.
    (*b*) Thiazide diuretics.
    (*c*) Acidosis.

**Increased turnover of nucleic acids in malignancy** can cause hyperuricaemia. *Treatment* of large tumours by radiotherapy or cytotoxic drugs can cause massive release of urates and has been known to cause acute renal failure due to tubular blockage by crystalline urate. During such treatment allopurinol should be used, and, if renal function is good, fluid intake should be kept high.

**Starvation and tissue damage.**—In acute starvation and with tissue damage endogenous tissue breakdown is increased. Increased amounts of uric acid are produced. In both these conditions acidosis is probably present (due to ketosis in starvation, and due to tissue catabolism in both) and these acids probably inhibit renal excretion of urate, aggravating the hyperuricaemia. Levels may reach 0.9 mmol/l (15 mg/dl) or more in complete starvation. Note that in chronic starvation urate levels, like those of urea (p. 20), tend to be low.

**Glomerular failure** causes retention of urate as well as retention of other waste products of metabolism. When estimating plasma urate, urea should always be estimated on the same specimen to exclude renal failure as a cause of hyperuricaemia. It has already been mentioned that hyperuricaemia may *cause* renal failure and, in the presence of uraemia, it may be difficult to know which is cause and which is effect. As a rough guide the plasma urate concentration would be expected to be about 0.6–0.7 mmol/l (10–12 mg/dl) at a urea level of about 50 mmol/l (300 mg/dl); if it is much higher than this hyperuricaemia should be suspected as the primary cause of the renal failure. Clinical gout is rare in secondary hyperuricaemia due to renal failure.

Increased intestinal secretion and bacterial uricolysis have been claimed to occur in renal failure and may account for the fact that, although plasma urea and urate levels rise in parallel, the rise in urtae is less on a molar basis than that of urea.

Clinical gout, is a rare complication of therapy with *thiazide diuretics*, although hyperuricaemia is relatively common. These drugs inhibit renal excretion of urate.

## PSEUDOGOUT

Pseudogout, while not a disorder of purine metabolism, produces a similar clinical picture to gout. Calcium pyrophosphate precipitates in joint cavities and calcification of the cartilages is seen radiologically. The crystals may be identified under a polarising microscope. The plasma urate is normal.

## HYPOURICAEMIA

Hypouricaemia is rare, except as a result of treatment of hyperuricaemia. It is an unimportant finding in the *Fanconi syndrome* (p. 14) when there is decreased tubular reabsorption of urate. It may occur in severe chronic starvation.

**Xanthinuria** is a very rare inborn error in which there is a deficiency of liver xanthine oxidase. Purine breakdown stops at the xanthine-hypoxanthine stage. Plasma and urinary urate levels are very low. The increased urinary excretion of xanthine may lead to the formation of xanthine stones (the reason why this does not happen during therapy with xanthine oxidase inhibitors is not clear, but may be due to reduction of *de novo* synthesis). The mode of inheritance is probably autosomal recessive.

## SUMMARY

1. Urate is the end product of purine metabolism.

2. Hyperuricaemia may be the result of:
   Increased nucleic acid turnover (malignancy, tissue damage, starvation).
   Increased synthesis of purines (primary gout).
   Reduced rate of renal excretion of uric acid (glomerular failure, thiazide diuretics, acidosis).

3. Hyperuricaemia may be aggravated by:
   High purine diets.
   Acidosis and a high alcohol intake.

4. Primary hyperuricaemia and gout have a familial incidence and are rare in women of child-bearing age.

5. Because severe hyperuricaemia may cause renal damage it should be treated, even if asymptomatic.

6. Hypouricaemia is rare and usually unimportant. It occurs in the very rare inborn error, xanthinuria.

## FURTHER READING

RASTEGAR, A., and THIER, S. O. (1972). The physiologic approach to hyper-uricemia. *New Engl. J. Med.*, **286**, 470.

# Chapter XVII

# IRON METABOLISM

IN man the circulating iron-containing pigment, haemoglobin, carries oxygen from the lungs to metabolising tissues, and in muscle myoglobin increases the local supply of oxygen. The ability to carry oxygen depends, among other factors, on the presence of ferrous iron in the haem molecule; iron deficiency is associated with deficient haem synthesis, and the symptoms of anaemia are due to tissue anoxia. Certain enzymes necessary for electron transfer reactions (and therefore, among other things, for oxidative phosphorylation) and the cytochromes also contain iron: it is doubtful if clinical iron deficiency, unless very severe, affects these.

## NORMAL IRON METABOLISM

### DISTRIBUTION OF IRON IN THE BODY

Figure 32 represents diagrammatically the distribution of iron in the body. The total body iron is about 50 to 70 mmol (3 to 4 g).

1. About 70 per cent of the total iron is circulating in erythrocyte *haemoglobin.*

2. Up to 25 per cent of the body iron is stored in the reticulo-endothelial system, in the liver, spleen and bone marrow. This storage iron is complexed with protein to form *ferritin* and *haemosiderin*. Unlike ferritin, haemosiderin is visible under the light microscope. Haemosiderin synthesis is stimulated by the presence of iron.

An appreciable amount of this storage iron is found in the reticuloendothelial cells of the *bone marrow*, where it acts as a reserve supply for erythropoiesis.

Storage iron, unlike haem iron, can be stained with potassium ferrocyanide (Prussian Blue reaction).

3. *Only about* 50 to 70 $\mu$mol (3 to 4 mg: 0·1 per cent) *of the total body iron is circulating in the plasma,* bound to protein. This is the fraction measured in *plasma iron* estimations.

4. The remainder of the body iron is incorporated in myoglobin, cytochromes and iron-containing enzymes.

Iron can only cross cell membranes by active transport in the ferrous form: it is in this reduced state in both oxyhaemoglobin and "reduced"

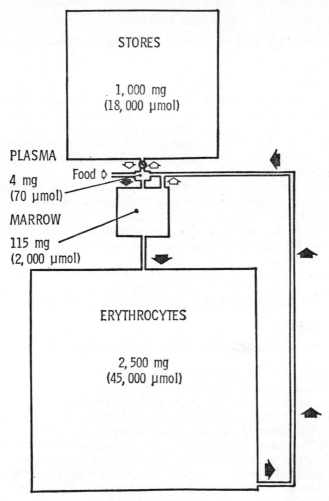

FIG. 32.—Body iron compartments.

haemoglobin. In ferritin and haemosiderin, and when bound to trans-ferrin, it is in the ferric state.

The *control* of iron distribution in the body is poorly understood. Plasma iron concentrations can vary by 100 per cent or more for purely physiological reasons, and are also affected by a variety of pathological factors other than the amount of iron in the body. These variations in plasma iron are probably due to redistribution between stores and plasma.

## IRON BALANCE

As shown in Fig. 32 iron, once in the body, is virtually in a closed system.

### Iron Excretion

There appears to be no control of iron excretion and loss from the body probably depends on the iron content of desquamated cells. Negligible amounts appear in the urine, reflecting the fact that it is entirely protein bound in the circulation. Most of the loss is probably into the intestinal tract and from the skin. The total daily loss by these routes is about 18 $\mu$mol (1 mg).

In women the mean monthly menstrual loss of iron is about 290 $\mu$mol (16 mg) and averages 10–18 $\mu$mol (0·5–1 mg) a day over the month above the basal 18 $\mu$mol (1 mg) daily, although it may be much higher in those with menorrhagia, who may become iron deficient. During pregnancy the mean extra daily loss to the foetus and placenta is about 27 $\mu$mol (1·5 mg).

It should be noted for comparison that a male blood donor losing a pint of blood every 4 months averages an extra loss of 36 $\mu$mol (2 mg) daily above the basal loss of 18 $\mu$mol (1 mg) (Table XXIX).

### Iron Absorption

The control of body content of iron depends upon control of absorption.

Iron is absorbed by an active process in the upper small intestine and passes rapidly into the plasma. It can cross cell membranes, including those of intestinal cells, only in the ferrous form. Within the intestinal cell some of the iron is combined with the protein, apoferritin, to form ferritin: this, like ferritin elsewhere, is a storage compound. Its function, if any, in the process of iron absorption is not well understood: formerly it was thought to control this by "blocking" further absorption when it was saturated with iron ("mucosal block"), but this theory has been abandoned. Any remaining iron stays in the cell as ferritin and is lost into the intestinal tract when the cell desquamates.

Absorption normally amounts to about 18 $\mu$mol (1 mg) of iron a day and just replaces loss. The percentage absorption of dietary iron depends to some extent on the food with which it is taken, but is usually about 10 per cent.

Iron absorption appears to be influenced by oxygen tension, by marrow erythropoietic activity, or by the size of iron stores: all these factors may affect it. From the clinical point of view it is important to note that *iron absorption is increased in many anaemias not due to iron deficiency.*

TABLE XXIX

COMPARISON OF IRON LOSSES

| | Source of loss | Extra loss | Daily extra loss | Daily total loss |
|---|---|---|---|---|
| Men and non-menstruating women | Desquamation | — | — | 18 $\mu$mol (1 mg) |
| Menstruating women (Mean value) | Desquamation + menstruation | 290 $\mu$mol (16 mg)/ month | 9 $\mu$mol (0·5 mg) | 27 $\mu$mol (1·5 mg) |
| Pregnancy | Desquamation + loss to foetus and in placenta | 7,000 $\mu$mol (380 mg)/ 9 months | 27 $\mu$mol (1·5 mg) | 45 $\mu$mol (2·5 mg) |
| Male blood donors | Desquamation + 1 pint blood | 4,500 $\mu$mol (250 mg)/ 4 months | 36 $\mu$mol (2·0 mg) | 54 $\mu$mol (3·0 mg) |

If an adequate diet is taken most normal women probably absorb slightly more iron than men to replace their greater losses.

## IRON TRANSPORT IN PLASMA

Iron is carried in the plasma in the ferric form, attached to the specific binding protein, *transferrin* (siderophilin), at a concentration of about 18 $\mu$mol/l (100$\mu$g/dl). The protein is normally capable of binding about 54 $\mu$mol/l (300 $\mu$g/dl) of iron, and is therefore about a third saturated. Transferrin bound iron is carried to stores and to bone marrow: in stores it is laid down as ferritin and haemosiderin, and in the marrow some may pass directly from transferrin into the developing erythrocyte to form haemoglobin.

Little, if any, of the iron in the body is free. In intestinal cells and stores it is bound to protein in ferritin and haemosiderin, in the plasma it is bound to transferrin, and in the erythrocyte it is incorporated in haemoglobin. Free iron is toxic.

## FACTORS AFFECTING PLASMA IRON LEVELS

Plasma iron estimation is frequently requested, but is rarely of clinical value, and results are often misinterpreted. As we have seen, plasma iron, like plasma potassium, represents a very small proportion of the total body content, and for this reason alone is likely to be a poor index of it: but, while plasma potassium levels normally are relatively constant, those of plasma iron are much less so and can vary greatly under physiological conditions.

### PHYSIOLOGICAL FACTORS AFFECTING PLASMA IRON LEVELS

The causes of physiological changes in plasma iron concentrations are not well understood, but they are very rapid, and almost certainly represent shifts between plasma and stores. The following factors are known to affect levels and some may cause changes of 100 per cent or more.

### 1. Sex and Age Differences

Plasma iron levels are higher in men than in women, like those of haemoglobin and the erythrocyte count. This difference is probably hormonal in origin. It is first evident at puberty, before significant menstrual iron loss has occurred, and disappears at the menopause. Androgens tend to increase plasma iron concentration and oestrogens to lower it.

## 2. Cyclical Variations

(a) **Circadian (diurnal) rhythm.**—Plasma iron is higher in the morning than in the evening. If subjects are kept awake at night this difference may be less marked or absent, and it is reversed in night workers.

(b) **Monthly variations in women.**—Plasma Iron may reach very low levels just before or during the menstrual period. This reduction is probably due to hormonal factors rather than to blood loss.

## 3. Random Variations

Very large day to day variations (occasionally as much as threefold) occur in plasma iron, and these usually overshadow cyclical changes. They may sometimes be associated with physical or mental stress, but more usually a cause cannot be found.

## 4. Effect of Pregnancy and Oral Contraceptives

In women taking oral contraceptives the plasma iron rises to levels similar to those found in men. A similar rise occurs in the first few weeks of pregnancy: if iron deficiency develops in late pregnancy this effect may be masked.

### PATHOLOGICAL FACTORS AFFECTING PLASMA IRON LEVELS

1. Iron deficiency and iron overload usually cause low and high plasma iron levels respectively.

*Iron deficiency* is associated with a hypochromic microcytic anaemia, and with reduced amounts of stainable marrow iron.

*Iron overload* is associated with increased amounts of stainable iron in liver biopsy.

2. Many illnesses including *infection (acute or chronic, mild or severe)*, *renal failure, malignancy and autoimmune diseases such as rheumatoid arthritis* cause hypoferraemia. Many of these chronic conditions are associated with normocytic, normochromic anaemia. Iron stores are normal or even increased, and the anaemia does not respond to iron therapy.

3. In conditions in which the *marrow cannot utilise iron*, either because it is hypoplastic, or because some other factor necessary for erythropoiesis (such as vitamin $B_{12}$ or folate) is absent, plasma iron levels are often high. Blood and marrow films may show a typical picture: in the case of, for instance, pyridoxine responsive anaemia and in thalassaemia, this may be similar to that of iron deficiency (sideroblastic anaemia), but iron stores are increased, and this can be shown on the marrow film.

4. In *haemolytic anaemia* the iron from the haemoglobin of broken down erythrocytes is released into the plasma and reticulo-endothelial system. Plasma iron may be high during the haemolytic episode and is usually normal in quiescent periods. Marrow iron stores are usually increased in chronic haemolytic conditions.

5. In *acute liver disease* disruption of cells may release ferritin iron into the blood stream and cause a transient rise of plasma iron. In *cirrhosis* plasma iron levels may sometimes be high. Although the reasons for this are not clear, it may sometimes be due to increased iron absorption associated with a high iron intake.

## TRANSFERRIN AND TOTAL IRON BINDING CAPACITY (TIBC)

It will be seen that plasma iron levels by themselves give no useful information about the state of iron stores. In the rare situations in which doubt remains about this after haematological investigations have been carried out, diagnostic precision may be improved by measuring the iron binding capacity of the plasma at the same time as the plasma iron. It is a waste of time and money to estimate only plasma iron.

Transferrin is usually measured indirectly by measuring the iron binding capacity of the plasma. An excess of inorganic iron is mixed with the plasma and any which is not bound to transferrin is removed, usually with a resin. The remaining iron is estimated on the plasma sample and the result expressed as a total iron binding capacity (TIBC).

### PHYSIOLOGICAL CHANGES IN TIBC

The TIBC is less labile than the plasma iron. However, it rises rapidly in subjects on *oral contraceptives*, and this point should be remembered when interpreting results in women. It also increases after about the 28th week of *pregnancy*, even in those women with normal iron stores.

### PATHOLOGICAL CHANGES IN TIBC

1. The *TIBC rises in iron deficiency* and *falls in iron overload*.

2. The *TIBC falls* in chronic infection, malignancy and the *other pathological conditions associated with a low plasma iron concentration* other than iron deficiency. This includes the nephrotic syndrome, in which the protein is lost in the urine.

3. The TIBC is unchanged in acute infection.

Thus the low plasma iron of iron deficiency is associated with a high TIBC. That of anaemia not due to iron deficiency is associated with a low TIBC.

TABLE XXX

CHANGES IN SERUM IRON (Fe) AND TOTAL IRON BINDING CAPACITY (TIBC)

| | Fe | TIBC | % Saturation | Marrow stores |
|---|---|---|---|---|
| *Low Iron Levels*<br>Iron deficiency | → | ↑ | ↓↓ | ↓ or absent |
| Chronic illnesses (e.g. infection and malignancy) | → | → | Variable | Normal or ↑ |
| Acute illnesses (e.g. infection) | → | Normal | → | Normal |
| Premenstrual | → | Normal | → | Normal |
| *High Iron Levels*<br>Iron overload | ← | → | ↑↑ | ← |
| Oral contraceptives and late pregnancy | ↑ (to male level) | ← | Normal | Normal |
| Early pregnancy | ↑ (to male level) | Normal | ← | Normal |
| Hepatic cirrhosis | Variable. May be ↑ | → | ↑ to ↑↑ | May be ↑ |
| Failure of marrow utilisation and haemolysis | ← | Normal or ↓ | ← | ← |

## PERCENTAGE SATURATION OF TIBC

The statement that the TIBC is normally a third saturated with iron is a very approximate one: physiological variations of plasma iron level are rarely associated with much change in TIBC and the saturation of the protein varies widely; the percentage saturation is, of course,

$$\frac{\text{Plasma Iron Concentration} \times 100}{\text{TIBC}}$$

It has been claimed that percentage saturation is a better index of iron stores than plasma iron concentration alone, and that if it is below 16 per cent iron deficiency is likely to be present. The first part of the statement is obviously true, because the low plasma iron of iron deficiency is accompanied by a high TIBC; this will result in a lower percentage saturation than with the same level of plasma iron in other conditions. However, saturation as low as 16 per cent can be found, for instance premenstrually and in acute infections, with no change in TIBC or in iron stores. It is probably more useful to take account of the results of both plasma iron and TIBC rather than to calculate the percentage saturation.

The findings in various conditions which may affect plasma iron levels and TIBC are summarised in Table XXX.

## PLASMA FERRITIN LEVELS

Normal plasma ferritin levels are about 100 $\mu$g/l. It has been suggested that circulating ferritin is in equilibrium with that in stores, and that levels less than 10 $\mu$g/l indicate iron deficiency, while high levels are found in iron overload or liver disease. This estimation is not yet in general use, and its value is still being assessed.

## INVESTIGATION OF ANAEMIA

Anaemia may be due to iron deficiency, or to a variety of other conditions. The subject of the diagnosis of anaemia is covered more fully in textbooks of haematology. However, so that we may see the value of plasma iron estimations in perspective, it is worth considering the order in which anaemia may be usefully investigated.

1. The clinical impression of anaemia should be confirmed by haemoglobin estimation. Iron deficiency can, however, exist with haemoglobin levels within the "normal" range.

2. A blood film should be examined, or absolute values estimated. Iron deficiency anaemia is hypochromic and microcytic in type, and hypochromia may be evident before the haemoglobin level has fallen

below the accepted normal range. Normocytic, normochromic anaemia is non-specific and usually associated with other disease: it is not due to iron deficiency unless there has been very recent blood loss. Typical appearances of other anaemias may be seen on the blood film.

In most cases of anaemia, consideration of these findings with the clinical picture will give the cause. Anaemias such as that of thalassaemia and of the pyridoxine responsive type, although rare, are most likely to confuse the picture, since they too are hypochromic, but are not due to iron deficiency.

3. A marrow film may be required to confirm the diagnosis (for example, of megaloblastic anaemia). If such a film is available staining with potassium ferrocyanide will give by far the best index of the state of iron stores, if this information is still required.

If marrow puncture is not felt to be justified, and *in the rare cases* in which diagnosis is not yet clear, biochemical investigations may occasionally help. If these are necessary, plasma iron *and* TIBC should be estimated. Plasma iron estimation alone is uninformative.

## "IRON DEFICIENCY WITHOUT ANAEMIA"

It has been claimed, but never proven, that the symptoms of iron deficiency can occur when haemoglobin levels are within the "normal" range. Iron deficiency can certainly exist under these conditions, and in most, if not all, such cases the diagnosis may be made on the appearances of the blood film: haemoglobin levels will rise after a short course of oral iron. A low plasma iron level is a poor index of such iron deficiency and subjective, symptomatic response to treatment is extremely difficult to assess in an individual subject. In clinical trials statistically significant symptomatic relief (associated with a rise of haemoglobin concentration) has not been demonstrated unless depleted marrow iron stores could be demonstrated before the trial started. In subjects with low plasma iron levels and vague symptoms, with normal marrow stores, symptomatic improvement occurred as frequently (and in some groups more frequently) when a placebo was given as when iron was administered. Some workers even claim that at haemoglobin levels above 10 g/dl iron therapy, although it may correct the blood picture, does not affect symptomatology. In other words, it is very unlikely that symptomatic iron deficiency exists without anaemia.

## IRON THERAPY

Because the body does not control iron excretion, and because body content is controlled by absorption, *parenteral iron therapy* may easily lead to iron overload. Repeated blood transfusions carry the same danger, as a pint of blood contains 4·5 mmol (250 mg) of iron. In anaemias other than that of true iron deficiency, stores are normal or even increased, and parenteral iron should not be given unless the diagnosis of iron deficiency is beyond doubt: even when this is so the oral is preferable to the parenteral route. Repeated blood transfusion may be necessary to correct severe anaemia in, for instance, chronic renal disease and hypoplastic anaemia, but, in such cases, the danger of overload should be remembered and blood should not be given indiscriminately.

Anaemia increases the rate of iron absorption even in the presence of increased iron stores. Treatment of, for instance, chronic haemolytic anaemia with *oral iron supplements* may occasionally cause iron overload; it does not improve the anaemia, which is due to the rate of breakdown of erythrocytes exceeding the rate of production, and not to deficiency of iron. The released iron stays in the body. In other noniron deficient anaemias the danger is similar.

Although iron absorption is controlled to some extent, this control is inefficient and is "swamped" by large loads, even in the absence of anaemia. Iron overload has been reported in a non-anaemic woman who continued to take oral iron (against medical advice) over a matter of years.

Iron therapy is potentially dangerous: it should be prescribed with care, and only when iron deficiency is proven.

## IRON OVERLOAD

As emphasised on p. 390, the excretion of iron from the body is limited. Iron absorbed from the gastro-intestinal tract, or administered parenterally, in excess of daily loss accumulates in body stores. If such "positive balance" is maintained over long periods, iron stores may exceed 350 mmol (20 g) (about five times the normal amount).

Iron overload may occur under the following circumstances:

1. Due to increased absorption from the intestinal tract

(*a*) Normal iron intake
    Idiopathic haemochromatosis
    Rarely in alcoholic cirrhosis,
      and in anaemias with        } Exaggerated by excess
      increased, but ineffective,     iron intake
      erythropoiesis.

(*b*) High iron intake
    Bantu siderosis
    Excessive oral iron therapy in the absence of iron deficiency
      over long periods of time.

2. Increased parenteral administration of iron.
   Transfusion siderosis.
   Prolonged parenteral administration in the absence of iron deficiency.

3. Deficiency of plasma iron-binding protein (transferrin). (Very rare.)

The effect of the accumulated iron depends on the distribution in the body. This in turn is influenced partly by the route of entry. Two main patterns are seen at post-mortem examination or in biopsy specimens.

(a) **Parenchymal iron overload** is typified by idiopathic haemochromatosis. Iron accumulates in the parenchymal cells of the liver, pancreas, heart and other organs. There is usually associated functional disturbance or tissue damage.

(b) **Reticulo-endothelial iron overload** is seen after excessive *parental administration of iron* or *multiple blood transfusions*. The iron accumulates in the reticulo-endothelial cells of the liver, spleen and bone marrow. There are few harmful effects other than a possible interference with haem synthesis, but under certain circumstances (p. 402), the pattern of distribution may change to that of the parenchymal type.

Two terms in common usage require definition. *Haemosiderosis* is defined as an increase in iron stores as haemosiderin. It does not necessarily mean that there is an increase in total body iron; for example, many anaemias have reduced haemoglobin iron (less haemoglobin) but increased storage iron. If the increase in storage iron in an organ is associated with tissue damage the condition is called *haemochromatosis*.

### IDIOPATHIC HAEMOCHROMATOSIS

Idiopathic haemochromatosis is a genetically determined disease in which increased intestinal absorption of iron over many years produces large iron stores of parenchymal distribution. It presents, usually in middle age, as cirrhosis with diabetes mellitus, hypogonadism and increased skin pigmentation. Because of the darkening of the skin, due to an increase in melanin rather than to iron deposition, the condition has been referred to as "bronzed diabetes", although the colour is more grey than bronze. Cardiac manifestations may be prominent, particularly in younger patients, many of whom die in cardiac failure. In about 10 to 20 per cent of cases hepatocellular carcinoma (primary hepatoma) develops.

The above concept of idiopathic haemochromatosis as a genetically determined disorder of iron absorption has been disputed by some workers. The main points of disagreement are the failure to demonstrate increased iron absorption in many cases, and the uncertain mode of inheritance. The difficulty in proving increased absorption is not surprising: the accumulation of 350 mmol (20 g) of iron over a period of 30–40 years requires an increased absorption of less than 36 $\mu$mol (2 mg) a day, not easily detectable by existing methods because of a wide normal range. In patients who present in the second or third decades, in whom the defect is presumably more severe, an increased rate of absorption is usually demonstrable. A further point is the depressant effect on iron absorption of the massive body stores in the overt cases (p. 390). Increased absorption is demonstrable after removal of the excess in many cases.

The mode of inheritance is uncertain. Evidence for the genetic nature of the disease is found in the study of families of patients, in whom the role of such environmental factors as a high iron intake can be properly assessed. In addition, evidence of iron overload (see below) has been found in a significant proportion of close relatives of patients with idiopathic haemochromatosis.

### Diagnosis of Idiopathic Haemochromatosis

1. **Plasma iron concentration and TIBC.**—The plasma iron concentration is almost invariably high, often above 36 $\mu$mol/l (200 $\mu$g/dl). This is associated with a reduced transferrin level (as shown by a lowered TIBC) and the percentage saturation is usually over 80 per cent, and often 100 per cent: in the presence of infection or malignancy, however, the plasma iron level and percentage saturation may be lower than expected; the TIBC remains low.

These findings may also be present in some cases of cirrhosis of the liver due to other causes, and diagnosis may be very difficult.

2. **Liver biopsy.**—Histological examination of a liver biopsy will show large amounts of stainable iron, predominantly within the parenchymal cells. Assessment of iron stores requires careful interpretation as variations in staining technique may render iron visible at very different concentrations. Within a single laboratory with a standardised method, excessive iron should be readily recognised.

3. **Demonstration of increased iron stores.**—The final diagnosis of iron overload can only be made after proof has been obtained of increased iron stores by a method other than liver histology.

(a) *Response to venesection.*—The lack of response of the patient to a therapeutic course of venesection offers the most convincing proof of increased iron stores, albeit retrospectively. Removal of a pint of blood (4·5 mmol [250 mg] iron) repeated at short intervals produces a rapid fall in plasma iron, soon followed by iron deficiency anaemia, in a subject with normal iron stores. In patients with idiopathic haemochromatosis, however, 350 mmol (20 g) or more of iron may be removed in this way before evidence of iron deficiency develops.

(b) *Use of chelating agents.*—An alternative method uses the chelating action of substances, such as desferrioxamine, which bind iron and are subsequently excreted in the urine. Following administration of desferrioxamine, subjects with increased iron stores excrete more iron in the urine than do normals.

4. **Other tests.**—(a) *Bone marrow aspiration* with staining for iron is not of great diagnostic value in this condition as this can only assess reticulo-endothelial stores.

(b) Examination of a *centrifuged urine specimen* for haemosiderin may indicate an excess of iron of renal tubular cell origin. If positive this test is of value, although positive results also occur in paroxysmal nocturnal haemoglobinuria, due to localised haemosiderosis. A negative result does not exclude the diagnosis of idiopathic haemochromatosis.

**Differential Diagnosis**

The distinction between idiopathic haemochromatosis and alcoholic cirrhosis may present difficulty. In both conditions diabetes mellitus and hypogonadism may occur and although the incidence is higher in idiopathic haemochromatosis this does not help in the individual case. Examination of liver biopsy specimens may further confuse the issue. The liver in cases of alcoholic cirrhosis not infrequently shows increased stainable iron. Not only do some alcoholic drinks, notably wines, contain significant amounts of iron, but there is evidence that in cirrhosis there may be increased iron absorption due possibly to an effect of alcohol.

The following points may aid in the separation of the two conditions:

1. The majority of patients with cirrhosis do not have increased iron stores despite the histological appearance of the liver biopsy. Such patients become rapidly anaemic during repeated venesection: after administration of desferrioxamine less iron is excreted in the urine than in patients with idiopathic haemochromatosis.

2. The amount of stainable iron in the liver biopsy may not be sufficient to lead to suspicion of idiopathic haemochromatosis. When it is, it is situated predominantly in portal tracts. It may be possible to assess the relative severity of the cirrhosis and of the iron overload. Massive iron accumulation with less marked cirrhosis is in favour of idiopathic haemochromatosis and vice versa.

3. Functional impairment of the liver is frequently much less evident in idiopathic haemochromatosis than in alcoholic cirrhosis with apparently equivalent iron overload.

4. There may be evidence, either clinical or from a previous liver biopsy, that the cirrhosis preceded other features of iron overload. Rare

cases of cirrhosis may have true iron overload and the distinction between the two conditions may be extremely difficult in the absence of such information. A family history or investigation of near relatives may help in diagnosis. The treatment of iron overload is the same in either case.

The diagnosis of idiopathic haemochromatosis *must* be followed by investigation of other members of the family, and treatment of those in whom increased iron stores are found.

## BANTU SIDEROSIS

A well-described form of alimentary iron overload is seen in the African population of Southern Africa. Unlike idiopathic haemochromatosis in which excessive iron is absorbed from a diet of normal iron content, the iron intake in cases of Bantu siderosis is grossly excessive. The main source of this iron is local beer, often brewed in iron containers: daily intakes of 1·4–1·8 mmol (80–100 mg) (about 4 or 5 times normal) are not uncommon. At high oral intakes of iron control of absorption (the only means of controlling body iron, p. 390) is imperfect and daily absorption of 35–55 $\mu$mol (2–3 mg) of iron leads to iron overload: many African males over the age of 40 have heavy deposition of iron in the liver.

In most cases the excess iron is confined to the reticulo-endothelial system and the liver (both portal tracts and parenchymal cells), and there is no tissue damage. There is, in many cases, associated ascorbic acid deficiency and osteoporosis. Symptomatic cutaneous hepatic porphyria has been associated with haemosiderosis (p. 410). In a small number of cases, usually those with the heaviest iron deposits and therefore presumably the highest intake of alcohol, cirrhosis develops. In such cases deposition in the parenchymal cells of other organs occurs and the clinical picture now closely resembles that of idiopathic haemochromatosis: it may be distinguished by the high concentrations of iron in the reticulo-endothelial system such as bone marrow and spleen (at autopsy).

The diagnostic approach is similar to that for idiopathic haemochromatosis. Liver biopsy shows the dual involvement and bone marrow aspiration demonstrates increased reticulo-endothelial stores.

Plasma iron levels may initially be low in patients with ascorbate deficiency, but rise after administration of the vitamin.

## OTHER CAUSES OF IRON OVERLOAD

Several types of anaemia may be associated with iron overload. In some, such as aplastic anaemia and the anaemia of chronic renal failure, the cause is multiple blood transfusions, and the iron accumulates in the reticulo-endothelial system. With massive overload (over 100 pints of blood), true haemochromatosis may develop with parenchymal overload.

In anaemias characterised by erythroid marrow hyperplasia, but with ineffective erythropoiesis, there is increased absorption of iron from the intestine (p. 390). This may be aggravated by oral iron medication and in a few cases true haemochromatosis develops. It must be stressed again that prolonged iron therapy for anaemia other than that due to iron deficiency carries the risk of overload.

The very rare condition of congenital transferrin deficiency presents as a refractory anaemia and massive iron overload.

## SUMMARY

1. No significant excretion of iron can occur from the body. Control of body stores is by control of absorption. For this reason parenteral iron therapy should be given with care.

2. Absorption of iron is increased by anaemia even in the absence of iron deficiency. For this reason oral iron therapy should not be given in anaemia other than that due to iron deficiency.

3. Plasma iron levels vary considerably under physiological circumstances.

4. Plasma iron levels fall in many cases of anaemia not due to iron deficiency.

5. For these two reasons (3 and 4) plasma iron levels alone are a very poor indication of body iron stores.

6. Iron is carried in the plasma bound to the protein transferrin. Transferrin is usually measured indirectly by measuring the total iron binding capacity (TIBC) of plasma.

7. The TIBC rises in iron deficiency and falls in iron overload.

8. The TIBC falls in many cases of anaemia associated with a low plasma iron, but not due to iron deficiency.

9. A low plasma iron concentration with a high TIBC is more suggestive of iron deficiency than a low plasma iron alone.

10. The quickest, cheapest and most informative tests for iron deficiency are simple haematological ones. These should be performed before requesting plasma iron estimation.

11. The factors governing the distribution of excessive iron are not fully understood. A feature common to all forms of parenchymal overload is a high percentage saturation of transferrin.

12. Iron overload may develop as a result of excessive absorption from a normal diet (idiopathic haemochromatosis) or from a high iron intake (Bantu siderosis); or it may develop as a result of excessive parenteral iron administration. The distribution of the iron in the body differs in the various forms.

13. Iron overload can be demonstrated by the response to repeated venesection or by chelating agents. The diagnosis of idiopathic haemochromatosis can only be made if massive iron overload is present.

14. Virtually all cases of parenchymal iron overload show a high plasma iron concentration with a high percentage saturation of transferrin.

## FURTHER READING

JACOBS, A., and WORWOOD, M. (Eds.) (1974). *Iron in Biochemistry and Medicine.* New York: Academic Press. (Especially Chapters 11, 13, 15, 16 and 17).

BOTHWELL, T. H., and CHARLTON, R. W. (1972). Haemosiderosis. *Brit. J. hosp. Med.*, **8**, 437.

# Chapter XVIII

# THE PORPHYRIAS

THE porphyrias are a group of diseases that have in common a disturbance of porphyrin metabolism. All types are uncommon in most parts of the world and several are extremely rare. Most are inherited diseases and the detection of a case must be followed by investigation of other members of the family. The peculiar importance of several of the commoner porphyrias is that, apart from presenting a diagnostic problem, the administration of certain drugs, notably barbiturates, can have serious or even fatal consequences.

In this chapter the porphyrias are dealt with in greater detail than is required for undergraduate examinations. We hope, however, that it may serve as a reference, particularly in areas of high incidence, when the diagnosis of porphyria is considered.

## PHYSIOLOGY

### Chemistry

Porphyrins are formed during the biosynthesis of haem and of related compounds such as cytochromes. They are tetrapyrroles and the general structure is shown diagrammatically in Fig. 33. Four pyrrole rings, linked by methene bridges, form the basic unit. Further chemical and biological differences are due to variation in the side chains. In addition the rings may vary in their relationship to one another, giving rise to four types of isomers. In nature only types I and III occur.

The most striking characteristic of the porphyrins is the fluorescence they exhibit on exposure to near ultraviolet light (maximal at a wavelength of about 400 nm). This property is used in their detection and measurement.

### Biosynthesis

The main steps are outlined below and in Fig. 34.

1. The first colourless compound formed is δ *aminolaevulinic acid* (ALA) by condensation of glycine and succinate, catalysed by the enzyme *ALA-synthetase*. Normally this is a rate-limiting step controlled partly by the end product haem. ALA-synthetase activity in porphyria is further considered on p. 413.

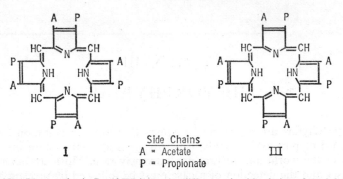

Side Chains
A = Acetate
P = Propionate

I      III

FIG. 33.—Uroporphyrin I and III (note the difference in side chains of the left-hand ring).

2. Two molecules of ALA condense to form the colourless mono-pyrrole, *porphobilinogen* (PBG).

3. Four molecules of PBG are combined to form uroporphyrinogen (tetrapyrrole). Both uroporphyrinogen I and III are formed.

4. Uroporphyrinogen III forms haem by the steps shown in Fig. 34. Uroporphyrinogen I forms coproporphyrinogen I and is not converted further.

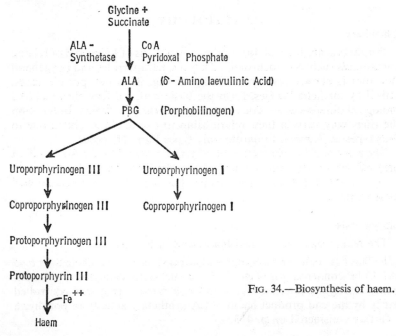

Glycine +
Succinate

ALA –
Synthetase     Co A
Pyridoxal Phosphate

ALA     (δ - Amino laevulinic Acid)

PBG     (Porphobilinogen)

Uroporphyrinogen III     Uroporphyrinogen I

Coproporphyrinogen III     Coproporphyrinogen I

Protoporphyrinogen III

Protoporphyrin III

$Fe^{++}$

Haem

FIG. 34.—Biosynthesis of haem.

The porphyrinogens are colourless compounds that do not fluoresce under ultraviolet light but which may be oxidised spontaneously to the corresponding porphyrins, which, with the exception of protoporphyrin, are not in the direct line of haem synthesis.

All the above steps are controlled by enzymes.

The breakdown product of haem is bilirubin and this is discussed in Chapter XII. The liberated iron recirculates.

### Excretion

Synthesis of haem in the bone marrow, and of related compounds in other tissues, is under fine control and normally very little porphyrin remains to be excreted.

1. *Urine* contains a small amount of porphyrin (mostly coproporphyrin I), insufficient to be detected by screening tests (see below).

One of the main factors influencing excretion of coproporphyrin is (as in the case of urobilinogen) urinary pH. Alkaline urine enhances excretion.

Small amounts of ALA and PBG are also present.

2. *Faeces* contains a mixture of protoporphyrin and coproporphyrin, mostly from biliary excretion. These may be present in sufficient quantity to impart a slight fluorescence to extracts of faeces (see below).


### SCREENING TESTS IN PORPHYRIA

Quantitative tests for porphyrins and porphyrin precursors are technically demanding and not generally available. There are, however, a number of screening tests used in most laboratories for the rapid diagnosis of porphyria (see Appendix for details). Such tests are not always adequate for the detection of latent cases.

1. Porphyrins in urine and faeces may be extracted into organic solvents where they are detected by the red fluorescence under ultraviolet light.

2. PBG may be detected by the reaction with Ehrlich's reagent.

There is no simple test for ALA.

The following points are important:

1. It is essential to test the *correct type of sample* (either urine or faeces) for the type of porphyria suspected.

2. During the *latent phase screening tests are not sensitive* enough to diagnose *acute intermittent porphyria*. Properly performed, *negative tests for faecal* porphyrin almost certainly *exclude the diagnosis of porphyria variegata*.

3. Tests are *negative before puberty*, even in children who will become porphyrics.

4. *Positive results must be confirmed* by quantitative analysis and typing of the porphyrins. There are other causes of increased porphyrin excretion (p. 413).

## DISORDERS OF PORPHYRIN METABOLISM

### CLASSIFICATION OF THE PORPHYRIAS

The porphyrias are classified according to whether the abnormality is in the liver or in the erythropoietic system. Erythropoietic porphyrias are very rare.

**Erythropoietic**

Congenital erythropoietic porphyria.

**Erythrohepatic**

Protoporphyria.

}Inherited

**Hepatic**

Acute intermittent porphyria.
Porphyria variegata.
Hereditary coproporphyria.
Symptomatic cutaneous hepatic porphyria.—Acquired

### HEPATIC PORPHYRIAS

**Acute Intermittent Porphyria** (Swedish genetic porphyria: Pyrrolloporphyria)

This is the commonest form of porphyria seen in the British Isles.

Acute intermittent porphyria is an inherited disease characterised, as the name implies, by intermittent acute attacks, with symptom-free intervening periods. The disease occurs more commonly in females than males and appears only after the onset of puberty. The main presenting features include:

1. *Abdominal* symptoms, especially colicky pain, vomiting and constipation. Differentiation from surgical causes of an "acute abdomen" requires recognition of the underlying porphyria.

2. *Neurological* manifestations may predominate and vary from peripheral neuritis to severe paralysis. Psychiatric disturbances are not uncommon.

Severe acute attacks are often accompanied by electrolyte disturbances, notably hyponatraemia.

The cause of the acute attack is not always obvious, although the greater incidence in women, and the absence of symptoms before puberty, indicate a hormonal influence. There are, however, a number of drugs that can precipitate an acute attack and which must be avoided by the patient. Important ones are *barbiturates, oestrogens, sulphonamides* and *griseofulvin*. A particular risk to the porphyric patient is the use of thiopentone during laparotomy for the supposed "acute abdomen". Severe neurological lesions may result.

The disease is transmitted as a Mendelian dominant.

**Biochemical abnormalities and diagnosis.**—The essential biochemical disturbance in acute intermittent porphyria is excessive production of the precursors ALA and PBG, and demonstration of this is necessary for diagnosis.

1. The *urine* is usually of normal colour when first passed, and contains greatly increased amounts of *ALA* (no screening test) and *PBG* which gives a strong positive test in the acute phase. Non-enzymatic conversion of PBG to porphyrin in the urine specimen leads to gradual darkening of the urine to a red-brown or deep red "port wine" colour. Tests for excess porphyrins will therefore also be positive at this stage.

2. Faecal porphyrin levels may be slightly increased during the acute phase.

Although screening tests for PBG in the urine may remain positive after the acute attack, quantitative estimation of ALA and PBG may be necessary to detect latent cases.

**Porphyria Variegata** (South African genetic porphyria: Protocoproporphyria)

This type is relatively common among the white population of South Africa. Most cases can be traced back to a single ancestor pair of the early settlers.

Porphyria variegata has many features in common with acute intermittent porphyria. It is also inherited as a dominant characteristic and only presents after puberty. It is distinguished clinically by the occurrence of *skin lesions*. The main features are:

1. *Acute attacks* which occur more commonly in females. The attacks are similar to those of acute intermittent porphyria, with abdominal or neurological symptoms, or both, and may be precipitated by the same drugs. *Severe electrolyte disturbances*, especially hyponatraemia, may occur. Hypokalaemia is often present.

2. *Skin lesions* are seen more commonly in males and range in

severity from annoying photosensitivity to blistering and scarring of the skin in exposed areas, notably the backs of the hands.

There is often increased skin pigmentation and, in women, facial hirsutism.

As in acute intermittent porphyria the main danger of the condition lies in administration of certain drugs. We have seen patients in these two groups rendered quadriplegic, with respiratory paralysis, by the use of barbiturates during anaesthesia for exploration of the "acute abdomen".

**Biochemical abnormalities and diagnosis.**—The major biochemical abnormality is the *excessive excretion of porphyrins in the faeces*. This is usually the only detectable laboratory abnormality in non-symptomatic cases. During an acute attack large amounts of *ALA*, *PBG* and *porphyrins* appear in the urine, but frequently disappear as the attack subsides. Raised *faecal porphyrin*, however, persists and diagnosis of latent cases is made on this finding.

### Hereditary Coproporphyria

Hereditary coproporphyria resembles acute intermittent porphyria. Photosensitivity, although uncommon, has been reported: it may be due to associated liver disease. Patients may develop acute attacks and are sensitive to similar drugs. The condition is apparently inherited as a Mendelian dominant.

The major biochemical abnormality is a marked increase in *faecal coproporphyrin* in both the acute and latent phases. During the acute attack *ALA*, *PBG* and *porphyrins* are present in the *urine*.

*Relatives of patients with inherited porphyria must be investigated*, and those with the disease warned about the risks of some drugs. They should always carry a warning card.

### Symptomatic Cutaneous Hepatic Porphyria (Porphyria Cutanea Tarda Symptomatica)

Unlike all other forms of porphyria this condition is acquired. It presents with cutaneous lesions, again in sun-exposed areas, which vary from photosensitivity to severe blistering and scarring. Hirsutism and hyperpigmentation are common. Acute attacks such as occur in the inherited hepatic porphyrias do not occur and the drugs precipitating these are harmless. Exacerbation of skin symptoms and mild abdominal pain may, however, occur after chloroquine administration. This is thought to be due to release of accumulated porphyrins from the liver.

The condition has several aetiologies, amongst which are:

1. *Severe liver disease*, notably alcoholic cirrhosis. There is often associated haemosiderosis. The photosensitivity can be relieved for

long periods by venesection. This form of porphyria is relatively common among the African and Cape Coloured population of Southern Africa.

2. *Ingestion of toxins.* An outbreak of porphyria of this type occurred in Turkey in 1955 and was traced to wheat treated with hexachlorobenzene. The pigmentation and hirsutism earned for the unfortunate victims the title of "monkey children".

**Biochemical abnormalities and diagnosis.**—The major abnormality is a marked increase in the excretion of porphyrin in the *urine*, mostly *uroporphyrin*. Faecal porphyrin levels are normal or only slightly raised. ALA and PBG are not found.

### Porphyrias Involving the Erythropoietic System

This group, although spectacular, is extremely rare and will only be described briefly.

### Congenital Erythropoietic Porphyria

This type of porphyria has two unique characteristics. It has a recessive mode of inheritance and the abnormality in porphyrin metabolism involves type I isomers (all others involve type III).

The condition presents at, or shortly after, birth with severe photosensitivity. Extensive blistering with secondary infection leads to mutilation of exposed areas. The teeth (and bones) may be red due to porphyrin and may fluoresce in ultraviolet light. Hirsutism may be present and there is haemolytic anaemia and splenomegaly. An interesting, if fanciful, theory is that these unfortunate individuals with hairy faces and shiny teeth who only ventured forth at night (to avoid the sun) gave rise to at least some of the werewolf legends.

The main biochemical abnormality is an excess of *porphyrins of type I in urine*, faeces and red cells. The carrier state is usually undetectable by chemical tests.

### Erythrohepatic Protoporphyria

This condition, of dominant inheritance, is characterised by increased photosensitivity from childhood onwards. Blistering, however, does not occur and there is no anaemia. Cirrhosis of the liver may develop.

The biochemical feature is an excess of protoporphyrin in faeces and red cells. The urine is usually normal.

Neither of these last two conditions is liable to the acute attacks of the hepatic porphyrias and there is no increased excretion of the porphyrin precursors, ALA and PBG.

## TABLE XXXI
### The Major Clinical and Biochemical Features of the Porphyrias

| | Hepatic porphyrias | | | | | | | Porphyrias involving the erythropoietic system | |
| | Acute intermittent porphyria | | Porphyria variegata | | Hereditary coproporphyria | | Symptomatic cutaneous hepatic porphyria | Congenital erythropoietic porphyria | Erythrohepatic protoporphyria |
| | acute | latent | acute | latent | acute | latent | | | |
|---|---|---|---|---|---|---|---|---|---|
| *Clinical Features* | | | | | | | | | |
| 1. Abdominal and neurological symptoms | + | − | ++ | − | + | − | | | |
| 1. Skin lesions | − | − | + | + | rarely | − | + | + | + |
| *Chemical Abnormalities* | | | | | | | | | |
| 1. Urine PBG and ALA | ++ | + | +++ | − | +++ | − | − | − | − |
| 2. Urine porphyrins | + | − | + | − | − | − | + | ++ | − |
| 3. Faecal porphyrins | − | − | + | + | + | + | − | + | + |
| | Acute attacks precipitated by drugs (for instance barbiturates, oestrogens, sulphonamides) | | | | | | Symptoms may be relieved by venesection | Erythrocyte porphyrins increased | |

### THE NATURE OF THE PORPHYRIAS

Increased ALA-synthetase activity has been demonstrated in all hepatic porphyrias. It is possible that defects later in the synthetic pathway reduce haem synthesis sufficiently to activate a negative feed-back mechanism. The varying symptomatology may be due to different defects. Porphyrias with an increase in porphyrin precursor (PBG and ALA) excretion are subject to acute attacks; those with increased porphyrin excretion are photosensitive. It is interesting that some of the drugs known to precipitate attacks of acute intermittent porphyria have been shown to further increase ALA-synthetase activity.

### OTHER CAUSES OF EXCESSIVE PORPHYRIN EXCRETION

Porphyria is not the only cause of disordered porphyrin metabolism and positive screening tests must be *confirmed* by quantitative analysis with *identification* of the porphyrin present. Three conditions must be considered.

1. *Lead poisoning* causes abnormalities at several stages of haem synthesis and eventually produces anaemia. The *urine* contains increased amounts of *ALA* (an early and sensitive test), sometimes of PBG and *coproporphyrin*. This last is used as a screening test for lead poisoning. Some of the symptoms of lead poisoning, such as abdominal pain, are similar to those of the acute porphyric attack, and this may cause difficulty in differential diagnosis.

2. *Liver disease* may produce an increase in *urinary coproporphyrin*, possibly due to decreased biliary excretion. This is probably the commonest cause of porphyrinuria. Occasionally there is mild photosensitivity (in symptomatic cutaneous hepatic porphyria (p. 410) there are more severe skin lesions due to uroporphyrin excess).

3. *Ulcerative lesions of the upper gastro-intestinal tract* may produce raised levels of *faecal porphyrin* by degradation of haemoglobin. If there is bleeding from the lower part of the tract the haemoglobin does not have time for conversion: this may help roughly to localise the site of bleeding.

### SUMMARY

1. Porphyrins are by-products of haem synthesis. ALA and PBG are precursors.

2. Screening tests for porphyrins are based on the fluorescence excited by ultraviolet light.

3. The porphyrias are diseases associated with disturbed porphyrin metabolism. Most are inherited. The main clinical and biochemical features are outlined in Table XXXI.

4. Acute attacks with abdominal or neurological symptoms are a feature of the inherited hepatic porphyrias. Such attacks are potentially fatal and may be provoked by a number of drugs. The diagnosis of porphyria in the acute phase depends on the demonstration of ALA, PBG and porphyrins in the urine.

5. The diagnosis of inherited porphyria must be followed by investigation of all relatives to detect asymptomatic cases. Screening tests may be negative in some types and quantitative estimations are necessary. Both urine and faeces should be examined.

6. Other causes of abnormalities in porphyrin excretion are lead poisoning, liver disease and upper gastro-intestinal bleeding.

7. The very rare erythropoietic porphyrias show excessive red cell porphyrin.

## FURTHER READING

ELDER, G. H., GRAY, C. H., and NICHOLSON, D. C. (1972). The porphyrias: a review. *J. clin. Path.*, **25**, 1013. (A full discussion including a list of the drugs known to precipitate acute attacks.)

EALES, L., LEVEY, M. J., and SWEENEY, M. B. (1966). The place of screening tests and quantitative investigations in the diagnosis of the porphyrias, with particular reference to variegate and symptomatic porphyria. *S. Afr. med. J.*, **40**, 63.

# APPENDIX

The screening tests for porphyria are usually carried out in the laboratory. In areas of high incidence, such as South Africa, however, simple side-room tests should be available.

The requirements are:

    1. A near ultraviolet (Wood's) lamp.

    2. An extracting solvent (equal parts of ether, glacial acetic acid and amyl alcohol).

    3. Ehrlich's reagent (2 per cent *p*-dimethylaminobenzaldehyde in 5 N HCl).

    4. *n*-Butanol.

## (*a*) Test for Porphobilinogen

(i) Equal parts of *fresh* urine and Ehrlich's reagent are mixed. If PBG is present a red colour develops.

(ii) If a red colour develops, 3 ml of *n*-butanol is added, the tube shaken and the phases allowed to separate. If the red colour does not enter the butanol (upper layer) it is PBG.

It is important to use *fresh* urine as PBG disappears on standing. A number of other substances, especially urobilinogen, may also give a red colour with Ehrlich's reagent. All such known substances are, however, extracted into the butanol layer.

## (*b*) Test for Porphyrins

(i) **Urine.**—About 10 ml of fresh urine is mixed with 2 ml of the porphyrin extracting solvent (item 2) and the phases allowed to separate. A red fluorescence in the upper layer under ultraviolet light indicates the presence of porphyrin.

(ii) **Faeces.**—A pea-sized piece of faeces is mixed into 2 ml of the solvent. A red fluorescence under ultraviolet light denotes either porphyrin or chlorophyll (with a similar structure and derived from the diet). These may be distinguished by adding 1·5 N HCl which extracts porphyrin but not chlorophyll.

# Chapter XIX

# VITAMINS

VITAMINS, like protein, carbohydrate and fat, are organic compounds which are essential dietary constituents (the name "vitamines" originally meant amines necessary for life): unlike most other constituents they are required only in very small quantities. They must be taken in the diet because the body either cannot synthesise them at all or, under normal circumstances, not in sufficient amounts for its requirements: "vitamin D", for instance, can be synthesised in the skin under the influence of ultraviolet light, but in temperate climates the amount of sunlight is insufficient to provide the required amount, while in hot countries the inhabitants tend to avoid the sun.

A normal mixed diet contains adequate quantities of vitamins and deficiencies are rarely seen in affluent populations except in those with intestinal malabsorption, in those on unsupplemented artificial diets, and in food faddists. Vitamin supplementation of a normal diet is unnecessary. Some vitamins (notably A and D) produce toxic effects if taken in excess.

The biochemical function of many vitamins is now understood, but it is not always easy to relate this knowledge to the clinical picture seen in deficiency states.

## CLASSIFICATION OF VITAMINS

In 1913 McCollum and Davis first showed that two growth factors are required for normal health, one being fat soluble and the other water soluble. Since then each of these two groups has been shown to consist of many compounds, but classification can still be made on the basis of this solubility. The distinction is important clinically, because steatorrhoea is associated with deficiency of fat soluble vitamins, but with relatively little clinical evidence of lack of most water soluble vitamins (the exception being vitamin $B_{12}$, p. 425).

### FAT SOLUBLE VITAMINS

The fat soluble vitamins are:

A
D
K
(E)

Each of these has more than one active chemical form, but variations in structure are very slight, and in the following discussion we will refer to each vitamin as a single substance.

## VITAMIN A

### Source of Vitamin A

Precursors of vitamin A (the carotenes) are found in the yellow and green parts of plants and are especially abundant in carrots: for this reason this vegetable has the reputation of improving night vision, but it is doubtful if it has any effect on a subject eating a normal diet. The vitamin is formed by hydrolysis of the $\beta$ carotene molecule, each of these producing a possible total of two molecules of vitamin A. In the body this hydrolysis occurs in the intestinal mucosa, but the yield of vitamin is much below the theoretical one, especially in children in whom a source of the vitamin itself is essential. It is stored in animal tissues, particularly in the liver, and these are the only source of preformed vitamin A.

### Stability of Vitamin A

Vitamin A is rapidly destroyed by ultraviolet light and should be kept in dark containers.

### Function of Vitamin A

1. Vitamin A is essential for normal *mucopolysaccharide synthesis* and deficiency causes drying up of mucus secreting epithelia.

2. Vitamin A is required for normal *mucus secretion*.

3. The *retinal pigment*, rhodopsin (visual purple), is necessary for vision in dim light (scotopic vision). Rhodopsin consists of a protein (opsin) combined with vitamin A. In bright light rhodopsin is broken down. It is partly regenerated in the dark, but, because this regeneration is not complete, vitamin A is needed to maintain the levels in the retina. The complete cycle is complex. Vitamin A deficiency is associated with poor vision in dim light, especially if the eye has recently been exposed to bright light.

### Clinical Effects of Vitamin A Deficiency

The clinical effects of vitamin A deficiency are:
"Night blindness".
Drying and metaplasia of ectodermal tissues.
Xerosis conjunctivae and xerophthalmia. Keratomalacia.
?Follicular hyperkeratosis.

*"Night blindness"*.—In vitamin A deficiency the rate of light adaptation can be shown to be reduced, although it is uncommon for the patient to complain of this.

*Xerosis conjunctivae and xerophthalmia*.—The conjunctivae and cornea become dry and wrinkled with squamous metaplasia of the epithelium and keratinization of the tissue, resulting from deficiency of mucus secretion. *Bitot's spots*, seen in more advanced cases, are elevated white patches found in the conjunctivae and composed of keratin debris. If deficiency continues *keratomalacia* occurs with ulceration and infection and consequent scarring of the cornea, causing blindness. Keratomalacia is an important cause of blindness in the world as a whole.

Squamous metaplasia occurs in other epithelial tissues. Skin secretion is diminished, and hyperkeratosis of hair follicles may be seen (*follicular hyperkeratosis*): these dry horny papules, varying in size from a pinhead to quarter-inch diameter, are found mainly on the extensor surfaces of the thighs and forearms. It is not certain if follicular hyperkeratosis is directly due to vitamin A deficiency.

Squamous metaplasia of the bronchial epithelium has also been reported and may be associated with a tendency to chest infection.

### Causes of Vitamin A Deficiency

Hepatic stores of vitamin A are so large that clinical signs only develop after a year or more of dietary deficiency.

Deficiency is very rare in affluent communities. In steatorrhoea overt clinical evidence is rarely seen, but blood levels have been shown to be low. By contrast, deficiency is relatively common in underdeveloped countries, especially in children, and is a common cause of blindness.

### Laboratory Diagnosis of Vitamin A Deficiency

This depends on the demonstration of low blood vitamin A levels. Specimens for this estimation should be kept in dark containers to prevent destruction of vitamin A by ultraviolet light.

### Treatment of Vitamin A Deficiency

In treatment of xerophthalmia doses of 50 000 to 75 000 international units of vitamin A in fish liver oil should be given. Response to treatment of "night blindness" and of early retinal and corneal changes is very rapid. Once corneal scarring has occurred blindness is irreversible.

### Hypervitaminosis A

Vitamin A in large doses is toxic. Acute intoxication has been reported in Arctic regions as a result of eating polar bear liver, which has a very high vitamin A content, but a more common cause is overdosage with vitamin preparations: the symptoms of acute poisoning are nausea and vomiting, abdominal pain, drowsiness and headache. In chronic hypervitaminosis A there is fatigue, insomnia and bone pains, and loss of hair with desquamation and pigmentation of the skin.

## VITAMIN D (CALCIFEROL)

### Source of Vitamin D

Vitamin D, like vitamin A, is stored in animal liver and fish liver oils are rich in this factor.

The metabolism and functions of "vitamin D", and the effects and treatment of its deficiency are discussed in Chapter X.

### Causes of Vitamin D Deficiency

Liver stores of vitamin D are usually adequate for several months' needs and clinical deficiency is slow to develop. The commonest cause in affluent communities is steatorrhoea, and rickets is now rare in Britain, except in recent immigrants.

### Hypervitaminosis D

Overdosage with vitamin D causes hypercalcaemia, with all its attendant dangers (p. 239). In chronic overdosage, because of increased liver stores of cholecalciferol, this may persist for several weeks after the therapy is stopped.

## VITAMIN K

Vitamin K cannot be synthesised by man but, like many of the B vitamins, it can be manufactured by the bacterial flora of the colon: unlike them it can probably be absorbed from this site and dietary deficiency is therefore not seen. In steatorrhoea the vitamin, whether taken in the diet or produced by bacteria, cannot be absorbed normally and deficiency may occur (p. 273).

Vitamin K is necessary for the synthesis of prothrombin and other coagulation factors in the liver and deficiency is accompanied by a bleeding tendency with a prolonged prothrombin time. If these findings are due to deficiency of the vitamin they are cured by parenteral administration (p. 299).

## VITAMIN E (TOCOPHEROLS)

Vitamin E is a fat soluble vitamin. In experimental animals deficiency has been reported to cause foetal death and sterility in both sexes, and also to be related to muscular dystrophy. Deficiency has not been shown to produce clinical effects in man, and therapeutic trials for various conditions have produced disappointing results.

## WATER SOLUBLE VITAMINS

The water soluble vitamins are:

The B Complex

Thiamine (Aneurine: $B_1$)
Riboflavin ($B_2$)
Nicotinamide (Pellagra Preventive (PP) factor: niacin)
Pyridoxine ($B_6$)
(Biotin and Pantothenic acid)
Folic Acid (Pteroylglutamic acid)
Vitamin $B_{12}$ (Cyanocobalamin)

Ascorbic Acid (Vitamin C)

### THE B COMPLEX

This group of food factors were originally lumped together in the B group (with the exception of vitamin $B_{12}$ and folate, which were later discoveries). The biochemical function of most of them is now well understood, and they mostly act as coenzymes. It is not easy to relate the clinical findings to the underlying biochemical lesion.

Many of these vitamins are synthesised by colonic bacteria. Opinions vary as to the importance of this source in man, but since the absorption of water soluble vitamins from the large intestine is poor, most of it is probably unavailable. However, clinical deficiency is rare in affluent communities. When deficiency does occur it is usually multiple, involving most of the B group and protein: for this reason it may be difficult to decide which signs and symptoms are specific for an individual vitamin and which are part of a general malnutrition syndrome.

### THIAMINE

#### Source of Thiamine and Cause of Deficiency

Thiamine cannot be synthesised by animals: it is found in most dietary components, and wheat germ, oatmeal and yeast are particularly rich in the vitamin. Adequate amounts are present in a normal diet, but the deficiency syndrome is still prevalent in rice eating areas: polished rice has the husk removed and this is the only source of thiamine in this food. In other areas thiamine deficiency occurs most commonly in alcoholics.

#### Function of Thiamine

Thiamine is a component of thiamine pyrophosphate, which is an essential coenzyme for *decarboxylation of α-oxoacids* (cocarboxylase):

one of these important reactions is the conversion of pyruvate to acetyl coenzyme A (acetyl CoA). In thiamine deficiency pyruvate cannot be metabolised and accumulates in the blood. Thiamine pyrophosphate is also an essential cofactor for transketolation reactions. One such reaction is that catalysed by *transketolase* in the pentose phosphate pathway; the keto (oxo) group is transferred from xylulose-5-phosphate to ribose-5-phosphate to produce glyceraldehyde-3-phosphate and sedoheptulose-7-phosphate.

### Clinical Effects of Thiamine Deficiency

Deficiency of thiamine causes the syndrome known as *beri-beri*, which includes anorexia and emaciation, neurological lesions (motor and sensory polyneuropathy, Wernicke's encephalopathy), and cardiac arrhythmias. In the so-called "wet" form of the disease there is oedema, sometimes with cardiac failure. Some of these findings may be due to associated protein deficiency rather than to that of thiamine.

Beri-beri can be aggravated by a high carbohydrate diet, possibly because this leads to an increased rate of glycolysis and therefore of pyruvate production.

### Laboratory Diagnosis of Thiamine Deficiency

The most reliable test is probably the estimation of *erythrocyte transketolase activity*, with and without added thiamine pyrophosphate. Blood pyruvate levels have been used, but are too variable and non-specific to be reliable.

### Treatment of Thiamine Deficiency

True beri-beri responds to 5–10 mg thiamine daily, although occasionally higher dosages may be required. In cases where multiple deficiency is suspected a mixture of the vitamins in the "B Complex" should be given.

## RIBOFLAVIN

### Source of Riboflavin

Riboflavin is found in large amounts in yeasts and germinating plants such as peas and beans.

### Function of Riboflavin

There are about 15 flavoproteins, mostly enzymes incorporating riboflavin in the form of flavin mononucleotide (FMN) and flavin adenine dinucleotide (FAD). FMN and FAD are reversible *electron carriers* in biological oxidation systems and are in turn oxidised by cytochromes.

## Clinical Effects of Riboflavin Deficiency

*Ariboflavinosis* causes a rough scaly skin, especially on the face, cheilosis (red, swollen, cracked lips), angular stomatitis and similar lesions at the mucocutaneous junctions of anus and vagina, and a swollen tender, red tongue, which is described as magenta coloured. Congestion of conjunctival blood vessels may be visible clinically on microscopic examination of the eye with a slit lamp.

## Laboratory Diagnosis of Riboflavin Deficiency

Recently, assay of erythrocyte glutathione reductase activity has been used. Microbiological assay of riboflavin is unsatisfactory.

NICOTINAMIDE

## Source of Nicotinamide

Nicotinamide can be formed in the body from nicotinic acid. Both substances are plentiful in animal and plant foods, although much of that in plants is bound in an unabsorbable form. Some nicotinic acid can also be synthesised in the mammalian body from tryptophan. Probably both dietary and endogenous sources are necessary to provide sufficient nicotinamide for normal metabolism.

## Function of Nicotinamide

Nicotinamide is the active constituent of the important coenzymes in *oxidation-reduction reactions*, nicotinamide adenine dinucleotide (NAD), and its phosphate (NADP). Reduced NAD and NADP are, in turn, re-oxidised by flavoproteins, and the functions of riboflavin and nicotinamide are closely linked. NAD and NADP are essential, for amongst other things, glycolysis and oxidative phosphorylation.

## Clinical Effects of Nicotinamide Deficiency

Nicotinamide deficiency produces a clinical syndrome often remembered by the mnemonic "three Ds"—diarrhoea, dermatitis and dementia. The dermatitis is a sunburn-like erythema, especially marked in areas exposed to the sun, and later leading to pigmentation and thickening of the dermis. The "dementia" takes the form of irritability, depression and anorexia, with loss of weight. As in riboflavin deficiency, there may be glossitis and stomatitis. Many of the symptoms of the multi-deficiency disease *pellagra* may be due to nicotinamide deficiency, and nicotinic acid has been called "pellagra preventive factor" (PP factor).

## Causes of Nicotinamide Deficiency

Dietary deficiency of nicotinamide, like that of the other B vitamins, is rare in affluent communities.

**Hartnup disease** is due to a rare inborn error of renal, intestinal and other cellular transport mechanisms for the monoamino monocarboxylic acids, including tryptophan (p. 366). Subjects with the disease may present with a pellagra-type rash, which can be cured by nicotinamide therapy of between 40 and 200 mg daily. Probably, if the supply of tryptophan for synthesis in the body is reduced, dietary nicotinic acid is insufficient to supply the body's needs over long periods of time: under these circumstances only a slight reduction of intake may precipitate pellagra. A similar clinical picture has been reported in the *carcinoid syndrome*, when tryptophan is diverted to the synthesis of large amounts of 5-hydroxytryptamine (p. 440).

*Laboratory assessment* of nicotinamide deficiency is unreliable.

## PYRIDOXINE

## Source of Pyridoxine and Cause of Deficiency

Pyridoxine is widely distributed in food and dietary deficiency is very rare. The antituberculous drug *isoniazid* (isonicotinic hydrazide) has been reported to produce the picture of pyridoxine deficiency, probably by competition with it in metabolic pathways.

## Function of Pyridoxine

Pyridoxal phosphate is a coenzyme for the *transaminases*, and for *decarboxylation of amino acids*.

## Clinical Effects of Pyridoxine Deficiency

Deficiency may produce roughening of the skin. A pyridoxine-responsive anaemia can occur (p. 393).

## Laboratory Diagnosis of Pyridoxine Deficiency

Pyridoxal phosphate is required for conversion of tryptophan to nicotinic acid. In pyridoxine deficiency this pathway is impaired. *Xanthurenic acid* is the excretion product of 3-hydroxykynurenic acid, the metabolite before the "block", and in pyridoxine deficiency is found in abnormal amounts in the urine after an oral *tryptophan load*.

## BIOTIN AND PANTOTHENIC ACID

Lack of these two vitamins of the B group probably almost never produces clinical deficiency syndromes.

**Biotin** is present in eggs, but large amounts of raw egg white in experimental diets have caused loss of hair and dermatitis thought to be due to biotin deficiency. Probably the protein avidin, present in the egg white, combines with biotin and prevents its absorption. Biotin is a coenzyme in carboxylation reactions.

**Pantothenic acid** is a component of coenzyme A (CoA), which is essential for fat and carbohydrate metabolism. It is very widely distributed in foodstuffs.

Figure 35 summarises some of the biochemical interrelationships of

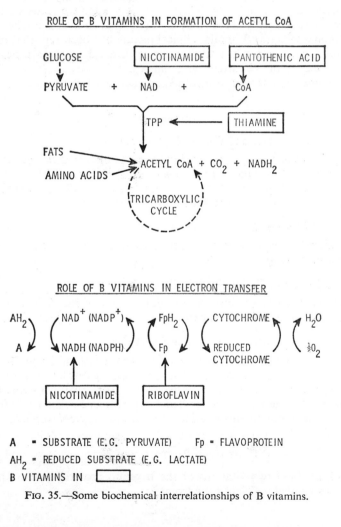

FIG. 35.—Some biochemical interrelationships of B vitamins.

the B vitamins discussed so far. Note that, as a general rule, deficiency of this group results in lesions of skin, mucous membranes and the nervous system.

## FOLIC ACID AND VITAMIN $B_{12}$

These two vitamins are included in the B group and are essential for the normal maturation of the erythrocyte; deficiency of either causes *megaloblastic anaemia*. Their effects are so closely interrelated that they are usually discussed together. A fuller discussion of diagnosis and treatment will be found in haematology textbooks.

**Folate** is present in green vegetables and some meats. It is easily destroyed in cooking and *dietary deficiency* may rarely occur. It is absorbed throughout the small intestine and in contrast to most of the other B vitamins (except $B_{12}$) clinical deficiency occurs relatively commonly in intestinal *malabsorption syndromes*. During *pregnancy* and lactation low blood folate levels are common and may be associated with megaloblastic anaemia. The active form of the vitamin is tetrahydrofolic acid, and this is essential for transfer of "one carbon" units: it is particularly important in *purine and pyrimidine* (and therefore DNA and RNA) synthesis. Competitive *folic acid antagonists* such as aminopterin have been used as cytotoxic drugs in the treatment of malignancy because they inhibit DNA synthesis.

**Vitamin $B_{12}$** is usually known as cyanocobalamin. When the cyanide group is replaced by adenine nucleotide this vitamin, like folate, has coenzyme activity in nucleic acid synthesis, and this conversion takes place in the body. Its biochemical functions are less clearly understood than those of folate, but it is probably required for normal metabolism of that vitamin. It is absorbed mainly in the terminal ileum and in *malabsorption syndromes* affecting this region deficiency can occur: it also occurs in the *blind loop syndrome* because of bacterial competition for the vitamin.

Absorption of the vitamin depends on its combination with intrinsic factor, secreted by the stomach, and in true *pernicious anaemia* malabsorption of $B_{12}$ is the result of absence of this factor.

Dietary deficiency of vitamin $B_{12}$ is very rare.

Deficiency of vitamin $B_{12}$, like that of folic acid, causes megaloblastic anaemia: unlike that of folate it can result in *subacute combined degeneration* of the cord. Although the megaloblastic anaemia of $B_{12}$ deficiency can be reversed by folate, this therapy should never be given in pernicious anaemia because it does not improve the neurological lesions and may even aggravate them.

Table XXXII summarises the synonyms of the B vitamins, their known biochemical actions, and the clinical syndromes associated with deficiency.

## ASCORBIC ACID

### Source of Ascorbic Acid

Ascorbic acid is found in fruit and vegetables and is especially plentiful in citrus fruits. It cannot be synthesised by man and other primates, nor by guinea-pigs and hamsters.

TABLE XXX

THE "B COMPLEX" VITAMINS

| Name | Synonyms | Biochemical function | Clinical deficiency syndrome |
|---|---|---|---|
| Thiamine | Aneurin Vitamin $B_1$ | Cocarboxylase (as thiamine pyro-phosphate) | Beri-beri (Neuropathy) Wernicke's Encephalo-pathy |
| Riboflavin | Vitamin $B_2$ | In flavoproteins (electron carriers) (as FAD and FMN) | Ariboflavinosis (affecting skin and eyes) |
| Nicotinamide | PP factor Niacin | In NAD and NADP (electron carriers) | Pellagra (dermatitis, diarrhoea, dementia) |
| Pyridoxine | Adermin Vitamin $B_6$ | Coenzyme in decarboxy-lation and deamination (as phosphate) | ? Pyridoxine responsive anaemia ? Dermatitis |
| Biotin | | Carboxylation co-enzyme | Probably unimportant clinically |
| Pantothenic acid | | In Coenzyme A | |
| Folic acid | Pteroyl glutamic acid | Metabolism of purines and pyrimidines | Megaloblastic anaemia |
| Vitamin $B_{12}$ group | Cobalamins | Coenzyme in synthesis of nucleic acid | Megaloblastic anaemia Subacute combined degeneration of the cord |

### Functions of Ascorbic Acid

Ascorbic acid can be reversibly oxidised in biological systems to dehydroascorbic acid and, although its functions in man are not well worked out, it probably acts as a hydrogen carrier. It seems to be required for normal collagen formation.

## Causes of Ascorbic Acid Deficiency

Deficiency of ascorbic acid causes scurvy and was commonly seen on the long sea voyages of exploration in the 16th, 17th and 18th centuries. Although Hawkins, as early as 1593, realised that oranges and lemons could cure the disease, and although in 1601 Sir James Lancaster introduced the regular use of oranges and lemons in the ships of the East India Company, a good description of scurvy can be found in Anson's account of his voyage around the world in 1740–44 (*Anson's Voyage Round the World*, edited by S. W. C. Pack in Penguin Books, Chapter 10); this demonstrates the importance of reading the literature.

Dehydroascorbic acid is easily oxidised further and irreversibly in the presence of oxygen, thus losing its biological activity; this reaction is catalysed by heat. Scurvy was at one time fairly common in bottle-fed infants, as the ascorbic acid was often destroyed in the preparation of the feeds. With our present knowledge of the aetiology of the disease, and with dietary supplementation, it is now rarely seen in this age group. It is most commonly seen in old people (especially men) living on their own on poor incomes, who do not eat fresh fruit and vegetables and who tend to cook in frying pans, where the combination of heat and the large area of food in contact with air irreversibly oxidises the vitamin.

## Clinical Effects of Ascorbic Acid Deficiency

Anson described "large discoloured spots", "putrid gums" and "lassitude" as characteristic of scurvy.

Many of the signs and symptoms of scurvy can be related to poor collagen formation.

1. Deficiency in *vascular walls* leads to a bleeding tendency with a positive Hess test, petechiae and ecchymoses ("large discoloured spots"), swollen, tender, spongy bleeding gums ("putrid gums") and, occasionally, haematuria and epistaxis. In infants subperiosteal bleeding and haemarthroses are extremely painful and may lead to permanent joint deformities.

2. There is *poor healing* of all wounds.

3. Deficiency of *bone matrix* causes osteoporosis and poor healing of fractures. In children bone formation ceases at the epidiaphyseal junctions, which look "frayed" radiologically.

4. There may be *anaemia*, possibly partly due to impairment of erythropoiesis. This anaemia may sometimes be cured by ascorbic acid alone. Bleeding aggravates the anaemia.

This florid form of the disease is rarely seen nowadays and the patient most commonly presents complaining that bruising occurs with only minor trauma.

All the signs and symptoms are dramatically cured by the administration of ascorbic acid.

### Diagnosis of Scurvy

The laboratory diagnosis of scurvy can only be made *before* therapy has started. Once ascorbic acid has been given it is difficult to prove that deficiency was previously present.

The *ascorbic acid saturation test* is based on the fact that if the tissues are "saturated" with ascorbic acid any further vitamin administered will be lost in the urine (see Appendix).

Ascorbic acid may be estimated in plasma, but levels vary widely in normal subjects. Leucocyte ascorbic acid is said to give a better indication of scurvy than that of plasma or whole blood. For practical purposes the saturation test is still probably the best one in the clinical diagnosis of scurvy.

## SUMMARY

1. Vitamins have important biochemical functions, most of which are now well understood. Unfortunately the relationship of these to clinical syndromes is not always obvious.

2. The fat soluble vitamins, especially "vitamin D", may be deficient in steatorrhoea. Both vitamin A and "vitamin D" are stored in the liver and deficiency takes some time to develop.

3. **Vitamin A** is necessary for the formation of visual purple and for normal mucopolysaccharide synthesis. Deficiency is associated with poor vision in dim light and with drying and metaplasia of epithelial surfaces, especially those of the conjunctiva and cornea.

4. **Vitamin D** as 1 : 25 dihydroxycholecalciferol is necessary for normal calcium metabolism and deficiency causes rickets in children and osteomalacia in adults.

5. Both vitamin A and vitamin D are toxic in excess.

6. **Vitamin K** is necessary for prothrombin formation and deficiency is associated with a bleeding tendency.

7. **Thiamine** deficiency causes beri-beri.

8. **Riboflavin** deficiency causes ariboflavinosis.

9. **Nicotinamide** can be manufactured from tryptophan in the body, but dietary deficiency causes a pellagra-like syndrome, which may also be seen in Hartnup disease, when tryptophan absorption is deficient, and in the carcinoid syndrome, when tryptophan is used in excess for 5-hydroxytryptamine synthesis.

10. **Pyridoxine** responsive anaemia may occur.

11. **Folic acid and vitamin $B_{12}$** deficiency produce megaloblastic anaemia, and deficiency of vitamin $B_{12}$ can also cause subacute combined degeneration of the cord. Compared with the other B vitamins, deficiency of these is relatively common in malabsorption syndromes, and vitamin $B_{12}$ deficiency can be a feature of the "blind-loop syndrome". Classical pernicious anaemia is due to intrinsic factor deficiency with consequent malabsorption of vitamin $B_{12}$.

12. **Ascorbic acid** deficiency causes scurvy.

## FURTHER READING

MARKS, J. (1968). *The Vitamins in Health and Disease.* London: J. &. A. Churchill.

GRAY, S. P. (1971). Laboratory studies in vitaminology. *Lab. Equipm. Digest*, **9**, 59.

# APPENDIX

One gram of ascorbic acid is administered orally at 8.00 hours.
The bladder is emptied at 11.00 hours and the urine discarded. Any urine passed between 11.00 and 14.00 hours is put in a collection bottle. At 14.00 hours the bladder is again emptied and the urine added to the collection.

Because of the ease with which ascorbic acid can be destroyed the specimen must be sent to the laboratory immediately for estimation. If "saturation" is complete (see below) the test can be terminated; if it is not the procedure is repeated on the following day.

## Interpretation

"Saturation" is said to be complete when at least 50 mg of the 1 g dose of ascorbic acid is excreted in the 3-hour urine collection.

In normal subjects "saturation" is complete on the first or second day of the test. In scurvy it takes much longer, but not usually more than a week.

**Warning.**—It should be noted that in malabsorption syndromes "saturation" may appear not to occur, even after several weeks. This is not diagnostic of scurvy, but is probably due to the fact that the absorption of the ascorbic acid is slower than usual and that it does not reach the kidneys in normal amounts during the 3 hours. Scurvy is very rare in such syndromes.

# Chapter XX

# PREGNANCY AND
# ORAL CONTRACEPTIVE THERAPY

## ASSESSMENT OF FOETO-PLACENTAL FUNCTION

UNTIL recently an indication for induction of labour was clinical evidence of foetal distress. Tests are now available for monitoring foetal well-being: they must only be used in conjunction with careful clinical assessment.

### URINARY "OESTRIOL"

#### Production of Oestriol by the Foetus and Placenta

Soon after fertilisation the ovum is implanted in the uterine wall: secretion of chorionic gonadotrophin by the developing placenta maintains the corpus luteum of the luteal phase and the continued secretion of oestrogen and progesterone from the corpus luteum prevents the onset of menstruation. Oestriol secretion at this stage is very low and it only rises very slowly during the first weeks of pregnancy.

Chorionic gonadotrophin secretion reaches a peak at about 13 weeks of pregnancy, and then falls. At this stage the foeto-placental unit takes over hormone production, and secretion of both oestrogen and progesterone rises rapidly.

The foetus and the placenta are now an integrated endocrine unit and both are required for the production of oestriol. The hormone passes into the maternal circulation and ultimately into the urine.

#### Value of Urinary Oestriol Estimation

The 24-hourly urinary oestriol secretion reflects both *placental and foetal function*. Accurate estimation of low concentrations of oestrogen fractions is a lengthy procedure. Results of foeto-placental monitoring, if they are to be useful, are required on the same day as the completion of collection of the 24-hour urine specimen. Several rapid analytical methods for urinary total oestrogen estimation have been devised but, because of their relative insensitivity, they are only of value after about the 28th week of pregnancy, when oestriol excretion is high. Most of the total oestrogen at this time is oestriol.

Urinary oestriol excretion continues to rise throughout the later weeks of pregnancy, but unfortunately the "normal range" of excretion at any particular stage is so wide that a single reading, unless very low, is of little diagnostic help: for instance, the range at 36 weeks is 35–115 $\mu$mol/24 h (10–33 mg/24 h). A sudden drop of level in a series of estimations is of more ominous significance than a single low reading and may be of help in deciding whether to terminate pregnancy.

Estimation of oestriol in blood is becoming available.

### SERUM HUMAN PLACENTAL LACTOGEN (HPL)

HPL is a peptide hormone, of unknown physiological significance, synthesised by the placenta. It is detectable by rapid radioimmunoassay methods after about the eighth week of gestation, and has been used in the assessment of threatened abortion. Levels rise until the 36th week, after which they fall slightly. The concentration is a measure of *placental* function only, unlike urinary oestriol which indicates foetal function as well: however, comparative tests show that the two types of estimation have similar predictive value. Estimation of HPL is quicker and simpler than that of urinary oestriol, and because it is estimated on serum the necessity for a 24-hour collection of urine, with the consequent delay and possibility of inaccuracy, is obviated. The "normal range", like that of oestriol, is wide, and serial readings are more useful than a single one: a sudden drop is more likely to indicate placental dysfunction than an isolated low value.

Some conditions predisposing to foetal risk, such as diabetes mellitus, rhesus incompatibility and pre-eclampsia, may have above "normal" levels, probably because of the associated large placenta. In such situations a falling level is still significant.

### HEAT-STABLE ALKALINE PHOSPHATASE

The placenta produces one of the alkaline phosphatase isoenzymes and the level of this rises in late pregnancy. This enzyme is stable at a temperature at which alkaline phosphatase from other sources is inactivated. Hopes that estimation of this heat-stable enzyme might be of value as a test of placental function have been disappointed, as results are too variable to be useful.

## BIOCHEMICAL TESTS ON AMNIOTIC FLUID

Amniotic fluid may be sampled through a needle inserted through the abdominal wall after 14 to 18 weeks of gestation (*amniocentesis*). The fluid may be examined for three purposes:

1. **Assessment of foetal maturity.**—The predictable rise in concentration of some constituents of the amniotic fluid may be used to assess

*foetal age. Creatinine* is most commonly used, levels greater than 180 $\mu$mol/l (2 mg/dl) usually indicating a foetal age of 36 weeks or more.

The assessment of *pulmonary maturity* is even more important. The concentrations of lecithin and sphingomyelin are approximately equal until about the 35th week of gestation, when lecithin concentrations rise rapidly: this probably coincides with the development of lung maturity. A lecithin/sphingomyelin (L/S) ratio of more than 2 indicates a very low probability of postnatal "respiratory distress syndrome".

2. **Measurement of bilirubin levels.**—This estimation is used, in conjunction with maternal antibody titres, to assess the effects on the foetus of rhesus (or other blood group) incompatibility. Normally amniotic fluid bilirubin levels decrease during the last half of pregnancy. The level at any stage may be correlated with the severity of haemolysis: used with tests of maturity the optimum time for termination of pregnancy, or the need for intra-uterine transfusion, may be assessed.

3. In early pregnancy some *congenital abnormalities* may be detected by chemical or enzymatic assay on cultures of cells in the fluid. These tests are performed only in special centres, and only on subjects with a genetic history of the condition.

All samples of amniotic fluid must be *fresh, free of blood,* and *protected from light.*

## METABOLIC EFFECTS OF PREGNANCY AND ORAL CONTRACEPTIVE THERAPY

The level of many plasma constituents is affected by steroid hormones: for example, the "normal" range of plasma urate and iron differs in males and females after puberty. It is therefore not surprising to find that pregnancy affects plasma concentrations of many substances, because of the very high circulating levels of oestrogens and progestogens. Oral contraceptive therapy is said to prevent ovulation by mimicking pregnancy, as the tablets contain synthetic oestrogens and "progestogens", and many of the metabolic changes during pregnancy are also found in some women taking the "pill". Most of these changes have been attributed to the oestrogen rather than the progestogen fraction, but it may be very difficult to be sure exactly which hormone is producing the change: sometimes it may be due to interaction between hormones. The mechanism of the changes is poorly understood, but steroids are known to affect protein synthesis.

### EFFECT ON CARBOHYDRATE METABOLISM

Steroids are known to affect carbohydrate metabolism. Cushing's syndrome is associated with diabetes and patients on large doses of corticosteroids may become diabetic. Cortisone has been given to subjects suspected of being "prediabetic" in spite of a normal glucose tolerance curve to uncover "latent diabetes" (p. 198).

During pregnancy many women develop a *diabetic type of glucose tolerance curve*, which usually reverts to normal after parturition. Similarly, some subjects taking oral contraceptive preparations develop reversible "diabetes", very rarely, if ever, severe enough to be accompanied by a raised fasting blood glucose level. There is no good evidence that such tablets cause true diabetes in normal subjects, but in those with latent diabetes the disease may be precipitated.

There may also be *renal glycosuria* both in pregnancy and in subjects taking oral contraceptives.

### EFFECT ON SPECIFIC CARRIER PROTEINS AND ON SUBSTANCES BOUND TO THEM

The plasma level of many specific carrier proteins is increased in pregnant subjects and in those taking oral contraceptive preparations. If this fact is not recognised an erroneous diagnosis may be made. In most cases the rise in carrier protein is accompanied by a proportional increase of the substance bound to it, without any change in the unbound fraction. As the protein bound fraction is a transport form and because, in most cases, it is the free substance that is physiologically active, this rise in concentration is only of clinical importance in interpretation of tests.

### Thyroxine-Binding Globulin (TBG)

The level of TBG is increased both in pregnant subjects and in those taking oral contraceptive preparations. This causes a rise in *plasma thyroxine* ($T_4$) and in *protein bound iodine* (PBI) concentrations, with a fall in *resin uptake*. The true cause of the rise in $T_4$ (or PBI) will be recognised if the resin uptake test is also performed (p. 166); if either is estimated alone the $T_4$ (or PBI) may lead to a false diagnosis of hyperthyroidism, and the resin uptake to one of hypothyroidism. Since the free thyroxine level is unaffected thyroid function is normal, a fact reflected in a normal neck uptake of iodine.

### Cortisol-Binding Globulin (CBG)

The level of cortisol-binding globulin, and therefore of plasma cortisol, increases as a result of pregnancy and of oral contraceptives. The free cortisol level, and therefore adrenal function, is normal in subjects on the "pill", although slightly raised in pregnancy. If these facts are not recognised a false diagnosis of Cushing's syndrome may be made.

**Transferrin**

The plasma concentration of the iron binding protein, transferrin, also increases within a day or two of starting oral contraceptive preparations. This is reflected in a rise in total iron binding capacity (TIBC). The rise in pregnancy occurs after about the 20th week. *Plasma iron* concentrations also increase but, unlike the substances already discussed, the rise is probably independent of that of the binding protein.

**Caeruloplasmin**

The copper binding protein, caeruloplasmin, is increased in concentration in subjects taking oral contraceptives and in pregnancy. It also rises during oestrogen therapy. There is an associated rise in *plasma copper* levels, but this is unlikely to lead to misdiagnosis. The increase in this bluish protein may impart a green colour to the plasma.

**Lipoproteins**

The plasma level of some lipoproteins is increased in subjects taking oral contraceptive preparations and is associated with an increased plasma triglyceride, and sometimes cholesterol, concentration. Plasma triglyceride level appears to be related to oestrogen content, while cholesterol levels are higher when progestational preparations are taken. In late pregnancy triglycerides are also increased but, compared with subjects taking oral contraceptives, there is a more marked rise in plasma cholesterol.

**Other Proteins**

The reduction of serum protein concentration in pregnancy has been thought to be due to dilution following fluid retention. Albumin and total protein levels have been reported to fall slightly during oral contraceptive therapy, but this change is unlikely to cause diagnostic confusion. It seems that only binding proteins are *increased* in these subjects.

PLASMA FOLATE AND VITAMIN $B_{12}$ LEVELS

Plasma folate levels have been reported to be reduced during oral contraceptive therapy. The relationship to the low folate levels and anaemia of pregnancy is not certain.

Total plasma vitamin $B_{12}$ levels are also slightly reduced during pregnancy.

LIVER FUNCTION TESTS

Some subjects develop a cholestatic jaundice during pregnancy and the same people are prone to a similar syndrome when taking oral contraceptive preparations. Although changes in flocculation tests and in transaminase levels have been reported in subjects taking oral contraceptives this is probably an extremely rare complication in normal people. We have not seen it in a series comprising several hundreds of subjects.

## HORMONE SECRETION

The effect of oral contraceptives on cortisol binding globulin and therefore on plasma cortisol has already been mentioned. No effect has been demonstrated on 17-oxo and 17-hydroxysteroid excretion. *Gonadotrophin levels* are very low in subjects taking oral contraceptives because of feed-back suppression. These revert to normal when the tablets are stopped.

The changes which may occur during pregnancy and oral contraceptive therapy are summarised in Table XXXIII.

### TABLE XXXIII

SOME METABOLIC EFFECTS OF PREGNANCY AND ORAL CONTRACEPTIVE THERAPY
WHICH MAY CONFUSE DIAGNOSIS

| Test | Effect | Comments |
|---|---|---|
| *Hormone Levels:* | | |
| *Adrenal hormones* | | |
| Plasma cortisol | Raised | Secondary to rise in binding protein. Adrenal function normal |
| *Thyroid hormones* | | |
| Plasma T$_4$ ⎫ | Raised ⎫ | Secondary to rise in binding protein. |
| PBI ⎬ | ⎬ | |
| Resin uptake | Lowered ⎭ | Thyroid function normal |
| *Pituitary hormones* | | |
| Urinary gonadotrophin | Lowered on oral contraceptives | ⎫ |
| Plasma gonadotrophin | Low normal | ⎬ Feed-back suppression ⎭ |
| *Carbohydrate metabolism* | | |
| Glucose tolerance test | "Diabetic" ⎫ | Reversible abnormalities |
| "Renal threshold" | Reduced ⎬ | |
| *Iron metabolism* | | |
| TIBC | Raised on oral contraceptives and in late pregnancy | |
| Plasma iron | Relatively raised | |

## SUMMARY

### Assessment of Foeto-Placental Function

1. After the 28th week of pregnancy estimations of daily oestriol excretion, in conjunction with the clinical findings, may help in assessing placental function and the health of the foetus.

2. Plasma levels of human placental lactogen (HPL) can be used as an index of placental function after the 8th week of gestation.

3. Amniotic fluid may be examined

   (a) to determine the maturity, especially pulmonary, of the foetus;
   (b) to assess the effects on the foetus of rhesus or other blood group incompatibility;
   (c) to detect inborn errors.

### Metabolic Effects of Pregnancy and Oral Contraceptive Therapy

Pregnancy and oral contraceptive therapy produce similar metabolic effects which, if not recognised, may lead to misdiagnosis. The more important of these are listed in Table XXXIII.

## FURTHER READING

ABDUL-KARIM, R. W., and BEYDOUN, S. N. (1972). Amniotic fluid: the value of prenatal analysis. *Postgrad. Med.*, **52**, 147 and 236.

CHARD, T. (1975). *Human Placental Lactogen Levels as a Guide to Foetal Well-being during Pregnancy* (Medical Monograph 8). Amersham: Radiochemical Centre.

*Lancet* (1973). Leading Article: Fetoplacental function tests. **2**, 947.

WILDE, C. E., and OAKEY, R. E. (1975). Biochemical tests for the assessment of fetoplacental function. *Ann. clin. Biochem.*, **12**, 83.

# Chapter XXI

# BIOCHEMICAL EFFECTS OF TUMOURS

TUMOURS of endocrine tissue, whether benign or malignant, often produce an excess of the hormone normally elaborated by that organ. Most of these syndromes (for instance, hyperparathyroidism due to a parathyroid adenoma) have been discussed in the relevant chapters. Two rare syndromes are associated with neoplasia of cells which, although normally producing hormones are, because of their scattered nature, less commonly thought of as endocrine organs. These are the carcinoid syndrome, due to an excessive production of 5-hydroxytryptamine by malignant tumours of argentaffin cells, and the syndrome which is due to oversecretion of the catecholamines, adrenaline (epinephrine) and noradrenaline (norepinephrine), by the cells of the sympathetic nervous system (phaeochromocytoma and neuroblastoma).

Many malignant tumours of non-endocrine cells can elaborate hormones and other peptides normally foreign to them. These interesting syndromes are more common than had previously been realised, and the second part of the chapter will be devoted to a brief discussion of such abnormalities.

## THE CARCINOID SYNDROME

### NORMAL METABOLISM OF 5-HYDROXYTRYPTAMINE

Small numbers of *argentaffin cells* (that is, cells which will reduce, and therefore stain, with silver salts) are normally found in tissues derived from embryonic gut. The commonest sites for these cells are the ileum and appendix, derived from embryonic midgut, but some are also found in such foregut derivatives as the pancreas and stomach, and in the rectum, which originates from the hindgut. They synthesise *5-hydroxytryptamine* (5-HT: serotonin) from tryptophan, the intermediate product being *5-hydroxytryptophan* (5-HTP). 5-HT is deaminated and oxidised to *5-hydroxyindole acetic acid* (5-HIAA) by monoamine oxidases (Fig. 36): the latter enzymes, as well as the aromatic amino acid decarboxylase, are found in other tissues as well as in argentaffin cells. 5-HIAA is normally the main urinary excretion product of argentaffin cells.

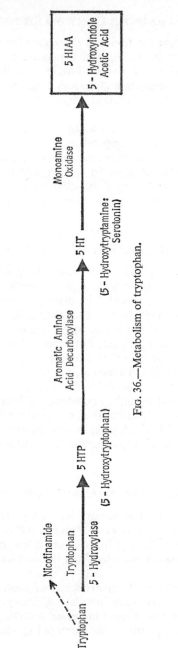

Fig. 36.—Metabolism of tryptophan.

## Causes of the Carcinoid Syndrome

The most usual site for tumours of argentaffin cells, which may be benign or malignant, is the ileum or appendix: neoplasms at the latter site rarely metastasise and may be histochemically different from the other small intestinal argentaffin tumours. Less commonly the neoplasm is bronchial, pancreatic or gastric in origin, any other site being extremely rare.

The carcinoid syndrome is usually associated with an excess of circulating 5-HT. Ileal and appendiceal tumours do not produce the clinical syndrome until they have metastasised, usually to the liver, but primary tumours at other sites do cause symptoms. It is thought that at least some of the secretions of intestinal neoplasms are inactivated in the liver, whereas at other sites they are released in an active form into the systemic circulation.

## Clinical Picture of the Carcinoid Syndrome

The clinical syndrome includes *flushing, diarrhoea* and *bronchospasm.* There may be *fibrotic lesions of the heart,* typically *right-sided,* except in the case of bronchial carcinoid. The diarrhoea may be so severe as to cause a malabsorption syndrome. The relationship of 5-HT to the signs and symptoms is not clear; it can increase gut motility and therefore may contribute to the diarrhoea, but other substances secreted by the tumour probably cause the other symptoms. One of the possibilities is prostaglandins; another is kallikrein, a proteolytic enzyme acting on a plasma protein to release bradykinin. Bradykinin has been shown to cause vasodilation and bronchial constriction. Histamine has also been suggested as a cause for the flushing, especially in tumours elaborating a preponderance of 5-HTP. A *pellagra* type syndrome can rarely develop because of diversion of tryptophan from nicotinamide to 5-HT synthesis (Fig. 36).

## Diagnosis of the Carcinoid Syndrome

In the carcinoid syndrome urinary 5-HIAA secretion, estimated on a 24-hour specimen of urine, is usually greatly increased. An excretion of more than 130 $\mu$mol (25 mg) in 24 hours is diagnostic, provided that walnuts and bananas are excluded from the diet for 24 hours before the collection is made.

In very rare cases, usually of bronchial or gastric tumours, the cells lack aromatic amino acid decarboxylase. Urinary 5-HIAA may not be obviously increased, and there is increased secretion of 5-HTP. If there is a strong clinical suspicion of the carcinoid syndrome and the urinary

5-HIAA excretion is normal, it may be more useful to estimate total 5-hydroxyindole excretion (which includes 5-HTP, 5-HT and 5-HIAA). This is very rarely necessary in practice.

## CATECHOLAMINE SECRETING TUMOURS

The adrenal medulla and the sympathetic ganglia consist of sympathetic nervous tissue: this is derived from the embryonic neural crest and is composed of two types of cells—the *chromaffin cells* and the

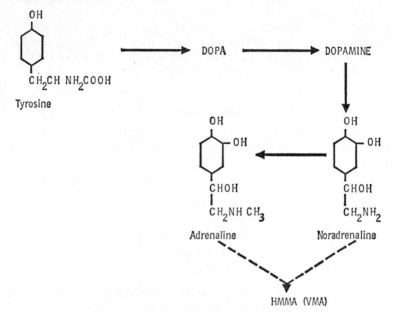

Fig. 37.—Synthesis and metabolism of the catecholamines.

*nerve cells*—both of which can elaborate the active catecholamines. Adrenaline (epinephrine) is almost exclusively a product of the adrenal medulla, while most noradrenaline (norepinephrine) is formed at sympathetic nerve endings.

### METABOLISM OF THE CATECHOLAMINES

Adrenaline and noradrenaline are formed from tyrosine via dihydroxyphenylalanine (DOPA), and dihydroxyphenylethylamine (DOPamine). DOPA, DOPamine, adrenaline and noradrenaline are all catecholamines (dihydroxylated phenolic compounds). The most important breakdown product of adrenaline and noradrenaline is 4-hydroxy, 3-methoxy-mandelic acid (HMMA; vanillylmandelic acid, VMA) (Fig. 37): this is excreted in the urine.

## ACTION OF THE CATECHOLAMINES

Both adrenaline and noradrenaline act on the cardiovascular system. Noradrenaline produces generalised vasoconstriction, and therefore hypertension and pallor, while adrenaline may cause dilatation of muscular vessels, with variable effects on blood pressure.

Adrenaline increases the rate of glycogenolysis (p. 181) and this, together with its other anti-insulin effects, may cause hyperglycaemia.

For a more detailed discussion of the action of these two hormones the student is referred to textbooks of pharmacology.

## CATECHOLAMINE SECRETING TUMOURS

Tumours of the sympathetic nervous system, whether adrenal or extra-adrenal, can produce catecholamines. Such tumours are of two main types:

1. Tumours of chromaffin tissue. *Phaeochromocytoma.*
   In adrenal medulla 90 per cent.
   Extra-adrenal 10 per cent.

2. Tumours of nerve cells. *Neuroblastoma.*
   In adrenal medulla 40 per cent.
   Extra-adrenal 60 per cent.

Although both these types of neoplasm secrete catecholamines their age incidence and the clinical pictures are quite different.

**Phaeochromocytoma** occurs mainly in *adult* life and is usually *benign.* The symptoms and signs can be related to excessive secretion of catecholamines and include *paroxysmal hypertension* accompanied by *anxiety, sweating,* a throbbing *headache* and either *facial pallor* or *flushing* during the attack. The attack may also be accompanied by hyperglycaemia and glycosuria. Occasionally hypertension may be persistent. This tumour accounts for only about half a per cent of all cases of hypertension, but if a young adult presents with high blood pressure for no obvious reason its presence should be sought, as the condition may be cured by surgery. Table XXXIV summarises some relatively rare causes of hypertension which should be excluded in such patients before "essential" hypertension is diagnosed, and in which biochemical investigations aid the diagnosis.

## TABLE XXXIV

### METABOLIC CAUSES OF HYPERTENSION

*Investigations*

| | |
|---|---|
| Renal disease | Plasma urea or creatinine.<br>Examination of urine for casts and protein.<br>Clearance tests if necessary (p. 20). |
| Primary aldosteronism | Plasma potassium and bicarbonate (p. 51). |
| Phaeochromocytoma | Urinary HMMA (VMA). |

A familial syndrome in which phaeochromocytoma, medullary carcinoma of the thyroid and neurofibromatosis coexist has been described.

**Neuroblastoma** is a very *malignant* tumour of sympathetic nervous tissue occurring in *children*. Secretion of catecholamines in these cases is often as high as, or higher than, that of a phaeochromocytoma, but the clinical syndrome described above is rare: the reason for this is not clear, but may be due to release of hormones into the blood stream in an inactive form. Some of the tumours secrete DOPamine in excess.

### DIAGNOSIS OF CATECHOLAMINE SECRETING TUMOURS

Because HMMA (VMA) is the major catabolic product of catecholamines, chemical diagnosis of the tumours can usually be made by estimating its excretion in a 24-hour specimen of urine. An excretion of more than twice "normal" is diagnostic, provided that all vanilla products, bananas and coffee are excluded from the diet for 24 hours before the test. Slightly raised excretion may be found in cases of essential hypertension.

Rarely HMMA excretion is normal, but that of total catecholamines is increased. Estimation of the latter should be reserved for cases clinically highly suggestive of phaeochromocytoma, but with a normal excretion of HMMA.

If hypertension is paroxysmal, urine should be collected during and immediately after the attack: this may be the only time at which increased excretion can be demonstrated.

## HORMONAL EFFECTS OF TUMOURS OF NON-ENDOCRINE TISSUE

This section will best be understood after the student has read the chapters concerned with the hormones discussed.

Hormone secretion at sites other than that of the tissue normally producing it is known as *"ectopic"*; *inappropriate* secretion (p. 50) may or may not be ectopic in origin (see ADH), but ectopic secretion is always inappropriate since it is not under normal feed-back control.

Many tumours of non-endocrine tissue may secrete hormonal substances very similar to, and probably identical with, the natural hormone: some such tumours have been shown to secrete two or more hormones.

## MECHANISM OF ECTOPIC HORMONE PRODUCTION

It is not yet certain why ectopic hormone secretion should occur, but although many theories have been propounded we feel that the one below is by far the most likely. For further discussion the interested student should consult the references at the end of the chapter.

DNA, through RNA formation, controls peptide and protein synthesis, the sequence of bases in the DNA molecule determining the structure of the peptide synthesised. All cells in the body are derived from one cell, the fertilised ovum. The DNA complement of this cell is determined by that of the unfertilised ovum and that of the sperm, and during the subsequent repeated cell division the DNA is replicated so that every daughter cell has a genetic complement identical with that of the fertilised ovum. Unless mutation were to occur during differentiation of tissues (and there is evidence that it does not), every cell in the body must have the potential to produce any peptide coded for in the fertilised ovum.

It is thought that during functional cell differentiation (which, of course, proceeds at the same time as histological changes) various parts of the DNA molecule are consecutively "repressed" (stopped from functioning), and "derepressed": for instance in the early stages of blastula formation no RNA is produced and DNA replication is dominant, while at other times different proteins are manufactured. In the fully differentiated cell much of the genetic information is probably repressed and only the peptides essential for cell metabolism and those concerned with the special function of the organ are made. If the cell undergoes neoplastic change, the altered histological picture reflects altered chemical function probably resulting from the changed pattern of repression of the DNA molecule. It is now easy to see how a cell could revert to synthesising a peptide or protein which is foreign to it in its fully differentiated state.

If this were the correct explanation, the ectopic hormone should be chemically identical with that normally produced by endocrine tissue. Hormonal substances have been extracted from tumour tissue and although in no case has the complete structure been worked out, evidence based on biological activity and chemical, physical and immunological properties is strongly suggestive that many of these are identical with the naturally occurring hormones. As might also be expected if the above theory were correct, tumours have been described secreting two or more peptide hormones.

It seems to us, then, that derepression is by far the most likely explanation for this interesting set of syndromes. However, this derepression cannot be non-specific because some hormones are secreted much more commonly by one type of tumour than another. This may reflect a differing chemical background in different parts of the body.

The student should remember that this theory is not yet proven. We hope we may have interested him in the subject and that he will be stimulated to read further.

## HORMONAL SYNDROMES

Many of these syndromes have been discussed in the relevant chapters. The following may be due to hormone secretion by the tumour:

Hypercalcaemia (Parathyroid Hormone)
Hyponatraemia (Antidiuretic Hormone)
Hypokalaemia (Adrenocorticotrophic Hormone)
Polycythaemia (Erythropoietin)
Hypoglycaemia ("Insulin-like substance")
Gynaecomastia (Gonadotrophin)
Zollinger-Ellison Syndrome (Gastrin)
Hyperthyroidism (Thyrotrophin)
Carcinoid Syndrome (5HT and 5HTP)
Hypertrophic Pulmonary Osteoarthropathy (?Growth Hormone)

It should be noted that all the hormones mentioned above are peptides except 5-HT and 5-HTP: the syndrome associated with the latter may be the result of overproduction of a single enzyme (see below).

None of these hormones produced at ectopic sites is under normal feed-back control. Secretion therefore continues under conditions in which it should be absent, and is inappropriate.

While absolute proof of hormone secretion depends on its isolation from the tumour tissue, in many cases strong presumptive evidence enables the clinician to diagnose the syndrome with a high degree of confidence.

### Hypercalcaemia Due to Parathyroid Hormone (PTH) Secretion ("Pseudohyperparathyroidism")

**Tumour types and incidence.**—In our experience this is probably the most common of these syndromes. Parathyroid hormone (PTH), or a peptide very similar to it, has been extracted from a wide variety of tumour types, and presumptive evidence for its secretion is available in a great many cases.

**Biochemical syndrome.**—As discussed in Chapter X, inappropriate PTH secretion, whether ectopic or parathyroid in origin, is associated with high plasma calcium and low plasma phosphate concentrations, unless accompanied by renal failure. If both these findings are present, and especially if there is no obvious evidence of bony metastases, it is probable that the hypercalcaemia is due to ectopic PTH secretion.

The hypercalcaemia of some cases of malignancy is said to be due to lysis of bone by secondary deposits, with release of calcium and phosphate into the blood stream. Some mammary tumours have been shown to produce osteolytic sterols which are normally only found in significant amounts in plants: the relationship between these and the hypercalcaemia of malignancy is not proven. As discussed on p. 247 the authors feel that these are very rare causes and that most cases (including many of those of myeloma) are due to ectopic PTH secretion. The student should keep an open mind and study the cases he meets, noticing particularly the level of phosphate *in relation to that of plasma urea* (p. 247).

The hypercalcaemia of malignancy, unlike that of primary hyperparathyroidism, may often be suppressed by large doses of corticosteroids (p. 251).

With the development of improved therapy of malignant disease it is now important to control hypercalcaemia in these patients, who should have plasma calcium estimated at frequent intervals. Severe hypercalcaemia may prove lethal before therapy of the primary lesion can be started: if this abnormality is treated the patient may have several years of useful life to come. The treatment is discussed on p. 255.

Hypokalaemia is a common accompaniment of hypercalcaemia (p. 261); it should always be sought. Its presence does not necessarily indicate simultaneous production of ACTH.

## Hyponatraemia Due to Inappropriate Antidiuretic Hormone (ADH) Secretion

**Tumour types and incidence.**—Ectopic ADH production is most commonly associated with the relatively rare oat-cell carcinoma of the bronchus, although the syndrome has been reported in a wide variety of tumours. Mild hyponatraemia may be overlooked in a severely ill patient, in whom such a finding is common; it is our experience that if further evidence is sought the syndrome is found to be relatively common.

Inappropriate ADH secretion can occur with various non-malignant syndromes, but in many of these cases the hormone is thought to originate in the hypothalamic-posterior pituitary region and therefore not to be ectopic in origin.

**Biochemical and clinical syndrome.**—ADH causes water retention. Water retention without sodium retention causes hyponatraemia, and therefore a reduction in plasma osmolality (p. 50), which normally cuts off ADH production by feed-back control. If the low plasma osmolality is to be corrected the urinary osmolality must be even lower, so that the excess of water is excreted: this is the "appropriate" response. If, however, ADH is not under feed-back control, water retention continues in spite of low plasma osmolality, and the urine passed is relatively concentrated. Thus the essential points in the diagnosis of inappropriate ADH secretion are:

1. A low plasma osmolality.

2. In spite of this low plasma osmolality, the urinary osmolality is relatively high.

It should be noted that both findings are essential. A low plasma osmolality occurs, for instance, in Addison's disease, but the urine is dilute. A high urinary osmolality is present in dehydration but the plasma osmolality is also high.

Continued expansion of the plasma volume reduces aldosterone secretion (p. 51), and the urine is of relatively high sodium concentration (note that plasma osmolality can be attributed almost entirely to sodium salts: urinary osmolality depends on a great many other substances). Retention of fluid dilutes other plasma constituents and these patients tend to have low plasma urea and protein concentrations.

The fall of plasma sodium concentration is usually gradual and allows time for equilibrium between cells and extracellular fluid: symptoms are therefore slight.

If the apparatus for measuring osmolality (an osmometer) is not available a presumptive diagnosis can be made if the following findings are present:

1. A well hydrated, normotensive, relatively fit patient in spite of a low plasma sodium concentration.

2. A normal or low plasma urea concentration.

3. A urine of relatively high sodium concentration and a relatively high specific gravity and urea concentration.

In the hyponatraemia of Addison's disease the patient is usually dehydrated, hypotensive and severely ill by the time there is a marked fall in plasma sodium concentration. The plasma urea concentration is usually raised. If the patient has been overloaded with fluid of low sodium concentration in the presence of a normal feed-back mechanism the urine, though containing sodium, will be of low specific gravity and urea concentration.

**Treatment.**—In most cases, which are relatively asymptomatic, restriction of fluids is adequate. However, if the sodium concentration has fallen rapidly, and if the symptoms of cerebral oedema are present, intravenous mannitol may relieve the latter: the osmotic diuresis causes predominant water loss by the kidneys (p. 9), and helps to increase plasma sodium concentration.

### Hypokalaemia Due to Adrenocorticotrophic Hormone (ACTH) Secretion

**Tumour types and incidence.**—Inappropriate ACTH secretion is probably rarer than the two syndromes already discussed. It occurs most commonly with the relatively rare oat-cell carcinoma of the bronchus, but has been described less frequently with a variety of other tumours, especially those of thymic and pancreatic origin.

**Clinical and biochemical syndrome.**—The ACTH stimulates secretion of all adrenal hormones except aldosterone (p. 138). In spite of this the picture of ectopic ACTH secretion is more like that of primary aldosteronism than of Cushing's syndrome: the patient usually presents with hypokalaemic alkalosis and relatively rarely shows the Cushingoid clinical picture. The reason for this is not clear.

**Treatment.**—Potassium replacement is only effective if accompanied by administration of aldosterone antagonists or triamterene (p. 70).

### Polycythaemia Due to Erythropoietin Secretion

The association of polycythaemia and renal carcinoma (hypernephroma) has been recognised for some years. It is now thought that the erythroid hyperplasia is often due to excessive production of an erythropoietin-like molecule by the tumour; erythropoietin stimulates marrow erythropoiesis. As this hormone is a normal product of the kidney this is not an example of ectopic hormone secretion: however, the syndrome has been reported rarely in association with other tumours, especially hepatocellular carcinoma (primary hepatoma).

### Hypoglycaemia Due to an "Insulin-like" Hormone

Severe hypoglycaemia has been reported in association with many tumours: these are usually large neoplasms of retroperitoneal mesenchymal tissues resembling fibrosarcomata, or hepatic tumours. Various theories have been suggested for the aetiology of the hypoglycaemia, but there is evidence that at least some of the neoplasms produce a substance with insulin-like biological properties. This is said usually to be immunologically distinct from insulin, although immuno-reactive insulin has rarely been demonstrated. In our experience the syndrome is very rare.

### Gynaecomastia Due to Gonadotrophin Production

Various types of tumour, including carcinoma of the bronchus, breast and liver, have been reported to cause gynaecomastia and be associated with high circulating levels of chorionic gonadotrophin or luteinising hormone; follicle stimulating hormone activity is rare. In children with hepatoblastoma there is precocious puberty. Although this syndrome is rarely recognised it is possible that mild degrees of gynaecomastia are overlooked.

### Zollinger-Ellison Syndrome Due to Gastrin Production

This is discussed on p. 281.

### Hyperthyroidism Due to "Thyrotrophin" Production

Some tumours of trophoblastic cells (choriocarcinoma, hydatidiform mole and one case with a testicular teratoma) have been shown to elaborate a thyrotrophin-like substance. In spite of markedly raised plasma levels of hormone, clinical signs of thyrotoxicosis are rare: tachycardia is the commonest finding. Laboratory findings include unequivocally raised plasma $T_4$ and protein bound iodine levels and the neck uptake of radio-iodine is also high.

### Growth Hormone (GH) Secretion

GH has been demonstrated in cells cultured from bronchial carcinoma. Although some of the patients have hypertrophic pulmonary osteoarthropathy, this clinical syndrome is not invariable: its relationship to the presence of GH is not proven.

### Carcinoid Syndrome Due to 5-Hydroxytryptamine (5-HT) and 5-Hydroxytryptophan (5-HTP) Production

All the syndromes so far discussed have been due to peptide hormones and, as we have seen, derepression of DNA could lead to their production. In the carcinoid syndrome, reported as a rare accompaniment of oat-cell tumours of the bronchus, there is usually excessive urinary excretion of 5-HTP and 5-HT, out of proportion to that of 5-HIAA. How can we explain this overproduction of a non-peptide hormone?

It should be stressed that the following theory is speculative and is meant only as a working hypothesis.

The student should study Fig. 36, p. 439: the production of 5-HTP from tryptophan requires only one enzyme, tryptophan-5-hydroxylase, while the decarboxylase and monoamine oxidase are normally found in non-argentaffin tissue. If derepression resulted in excessive production of this one enzyme there would be an excess of 5-HTP. 5-HT and 5-HIAA could be produced in other tissues in relatively smaller amounts than are usual in the carcinoid syndrome. It would, of course, be more difficult to explain overproduction of a substance synthesised by a pathway requiring several enzymes.

## NON-HORMONAL PEPTIDES AS INDICATORS OF MALIGNANCY

The production of non-hormonal peptides by derepression of DNA will be less clinically obvious than the production of hormones. Recently such circulating peptides have been identified by immunological techniques. Two such proteins are discussed briefly here: both are normally present in foetal life, but production appears to be largely repressed in normal adults.

α **Fetoprotein** may be found in the serum of many patients with *hepatocellular carcinoma* (primary hepatoma) and *teratoma* (p. 301). False positives may occur in hepatitis.

**Carcinoembryonic antigen (CEA)** appears to be produced by *malignant tumours of the gastro-intestinal tract*. As it may also be detected in non-malignant disease of the gastro-intestinal tract, its value remains to be established.

## SUMMARY

### The Carcinoid Syndrome

1. Argentaffin cells manufacture 5-hydroxytryptamine (5-HT) which is converted to 5-hydroxyindole acetic acid (5-HIAA) and excreted in the urine.

2. Tumours of argentaffin tissue are usually found in the intestine, and these do not produce typical symptoms of the carcinoid syndrome until they have metastasised.

3. The carcinoid syndrome is usually associated with an increased secretion of 5-HIAA in the urine.

### Catecholamine Secreting Tumours

1. Tumours of sympathetic nervous tissue are associated with increased urinary excretion of the catecholamines, adrenaline and noradrenaline, and their breakdown product, hydroxymethoxymandelic acid (HMMA: VMA).

2. Phaeochromocytoma is a rare tumour, usually of adult life, occurring most commonly in the adrenal medulla. It is associated with hypertension and other symptoms of increased catecholamine secretion.

3. Neuroblastoma is a tumour of childhood, occurring in the adrenal medulla or in extra-adrenal sympathetic nervous tissue. Catecholamine secretion is increased, but symptoms are rarely referable to this.

### Hormonal Effects of Tumours of Non-Endocrine Tissue

1. Many malignant tumours produce hormonal substances normally foreign to them, which may be identical with the hormones produced in endocrine glands.

2. The commonest of these is parathyroid hormone. Hypercalcaemia should be sought and treated in cases in which the malignancy is thought to be responsive to therapy.

### Non-Hormonal Peptides as Indicators of Malignancy

Some non-hormonal peptides may be detected by immunological techniques and have been used to diagnose malignancy of specific tissues.

## FURTHER READING

GRAHAME-SMITH, D. G. (1968). The carcinoid syndrome. *Hosp. Med.*, **2**, 558.

WOLF, R. L. (1974). Phaeochromocytoma. *Clinics in Endocr. & Metab.* **3**, 609.

ANDERSON, G. (1973). Paramalignant Syndromes. In *Recent Advances in Medicine*, 16th edit., Chap. 1, p. 1. Eds. Baron, D. N., Compston, N. and Dawson, A. M. London: J. & A. Churchill.

RATCLIFFE, J. G., and REES, L. H. (1974). *Brit. J. hosp. Med.*, **11**, 685.

*Lancet* (1967). Editorial: Hormones and Histones? **1**, 86.

# Chapter XXII

# THE CEREBROSPINAL FLUID

CEREBROSPINAL fluid (CSF) is formed from plasma by the filtering and secretory activity of the choroid plexus, and is reabsorbed into the blood stream by the arachnoid villi. The mechanism of its production is not fully understood, but it seems in some ways to be an ultrafiltrate of plasma, since it contains very little protein. Some active secretion of, for example, chloride may occur.

Circulation of CSF is very slow, allowing long contact with cerebral cells: their uptake of glucose may account for the relatively low concentration of this sugar in the CSF.

Concentrations of substances in the CSF should always be compared with those of plasma, as alterations in the latter are reflected in the CSF even when cerebral metabolism is normal.

## EXAMINATION OF THE CSF

Biochemical investigation of the CSF is usually of relatively little value compared with simple inspection and bacteriological and cytological examination of the fluid. Textbooks of bacteriology should be consulted for further details.

### TAKING THE SAMPLE

CSF should be taken into sterile containers and sent for *bacteriological examination first*. Any remaining specimen can be used for relevant chemical investigations, but if the tests are performed in the opposite order bacteriological contamination may occur. If possible, a few millilitres of CSF should be taken into two or three separate containers (see below—Appearance of CSF).

Specimens for glucose estimation, like those for blood glucose, should be taken into a tube containing fluoride. This minimises glycolysis, by any cells present, after the specimen has been taken.

### APPEARANCE

Normal CSF is completely clear and colourless and should be compared with water: slight turbidity is most easily detected by this method.

## Colour

**Bright red blood** may be due to:

1. A recent haemorrhage involving the subarachnoid space.
2. Damage to a blood vessel during puncture.

If CSF is taken into three tubes, all three will be equally blood-stained in the first case and progressively less so in the second.

*Xanthochromia* (yellow coloration).—This may be due to:

1. The presence of *altered haemoglobin* several days after a cerebral haemorrhage.

2. The presence of large amounts of *pus*. The cause will be obvious from the gross turbidity of the fluid, and from the presence of pus cells on microscopy.

3. The presence of *cerebral tumours* near the surface of the brain or spinal cord, impairing circulation of the CSF. Specimens from these cases have a very high protein content and tend to clot spontaneously after withdrawal, due to the presence of fibrinogen.

4. *Jaundice* due to a rise in unconjugated bilirubin may impart a yellow colour to the CSF.

## Turbidity

This is due to an *excess of white cells* (*pus*). Slight turbidity will, of course, occur after haemorrhage, but the cause of this will be apparent from the colour of the specimen.

## Spontaneous Clotting

This occurs when there is an excess of fibrinogen in the specimen, usually associated with a high total protein content.

The following are the most frequently requested chemical estimations.

### PROTEIN CONTENT

Normal CSF is mainly plasma ultrafiltrate and contains very little protein. The normal figures are usually quoted as 0·2 to 0·4 g/l. Quantitative estimation at this level is relatively inaccurate and little significance can be attached to concentrations up to 0·6 g/l.

## Total Protein

CSF will have a high protein content under the following conditions:

1. In the presence of *blood* (due to haemoglobin and plasma proteins).

2. In the presence of *pus* (due to cell protein and exudation from inflamed surfaces).

These two causes will be obvious from inspection and microscopic examination of the specimen, and nothing further is to be gained by estimating protein.

3. In non-purulent inflammation of the cerebral tissues, when there is a moderate rise of protein concentration. It is when such conditions are suspected that the estimation is most useful:

Tuberculous meningitis
Syphilitic meningitis
Multiple sclerosis
Encephalitis
Polyneuritis.

4. In blockage of the spinal canal, when stasis results in fluid reabsorption. There is xanthochromia and protein concentrations are very high (usually 5 g/l or more).

### Tests for Abnormal Ratios of Proteins in the CSF

These tests are not useful in the presence of blood or pus. They may, however, detect abnormalities when the total protein is *only equivocally increased*.

In the inflammatory conditions mentioned above CSF immuno-globulin, especially IgG, concentration is increased. Various methods have been used to detect this increase; it has yet to be proved that any of them is more sensitive or more diagnostically specific than the others. These tests are most valuable if multiple sclerosis is suspected—in this condition they are positive in about 50–60 per cent of cases.

### 1. Qualitative Tests

Excess of immunoglobulin may be detected by tests similar to the flocculation tests used on serum (p. 316).

(*a*) *Pandy reaction.*—The abnormal CSF becomes turbid when added to phenol.

(*b*) *Nonne-Appelt reaction.*—The abnormal CSF becomes turbid when added to ammonium sulphate.

(*c*) *Lange colloidal gold reaction.*—Gold sols are precipitated when the ratio of protein fractions is abnormal. The CSF is serially diluted. No reliance should be placed on Lange "curves", in which precipitation at different dilutions was said to indicate specific clinical conditions.

## 2. Demonstration of Increased γ Globulin Levels

(*a*) *Electrophoresis* has been used, but because it is necessary to concentrate the CSF, the test is of little routine value.

(*b*) *Zinc sulphate* precipitates γ globulins. The resultant turbidity can be compared with that of standards.

## 3. Immunological Techniques

The level of immunoglobulins can be quantitated. IgG concentration is increased in, amongst other conditions, multiple sclerosis. There is little evidence that such tests give more precise information than the relatively simple and cheap qualitative ones.

### GLUCOSE CONTENT

Provided that CSF for glucose estimation has been mixed with fluoride a low glucose concentration occurs in:

1. Infection.
2. Hypoglycaemia.

1. CSF glucose is normally metabolised only by cerebral cells. If many leucocytes and bacteria are present these also utilise the sugar and abnormally low levels are obtained. If obvious pus is present the estimation of CSF glucose adds nothing to diagnostic precision. It is most useful when the CSF is clear and *tuberculous meningitis* is suspected, although levels are not as low in this condition as in pyogenic meningitis.

2. CSF glucose concentration parallels that of blood, although there is a lag before changes in blood glucose are reflected in the CSF. In the presence of hypoglycaemia (which may cause coma) CSF glucose levels may be low although there is no primary cerebral abnormality. In case of doubt both blood and CSF concentrations should be measured.

In hyperglycaemia CSF glucose levels will be high.

### CHLORIDE CONTENT

The chloride concentration in the CSF is higher than that in plasma by about 20 mmol/l. This is mostly accounted for by the Donnan effect, the additional difference probably being due to secretion of the ion by the choroid plexus.

Changes in *CSF chloride concentration parallel those in plasma.* The level has been said to be lowered specifically in tuberculous meningitis, but there is evidence that this is merely a reflection of plasma chloride levels: tuberculous meningitis is of insidious onset and the patient often presents after some weeks of vomiting with chloride loss. CSF chloride estimation has no value as a diagnostic aid.

## PROCEDURE FOR EXAMINATION OF CSF

**CSF very cloudy.**—Send for bacteriological examination. Chemical estimations unnecessary.

**CSF heavily bloodstained** in three consecutive specimens. Cerebral haemorrhage. Chemical estimation unnecessary.

**CSF clear or only slightly turbid.**—Send for bacteriological examination and for estimation of *glucose* and *protein* concentration. If there is any possibility that coma is due to hypoglycaemia send blood for glucose estimation at the same time. If multiple sclerosis or neurological syphilis is suspected, tests indicating increased CSF immunoglobulin levels may be of value.

**CSF xanthochromic.**—Send specimen for protein estimation. Examine under microscope for erythrocytes.

## SUMMARY

1. Biochemical analysis of the CSF is less important than simple inspection and bacteriological examination.

2. Estimation of CSF protein concentration is most useful if non-purulent inflammation of cerebral tissues is suspected.

3. Tests indicating increased CSF immunoglobulin concentration are of most value when disseminated sclerosis or neurological syphilis are suspected.

4. Estimation of CSF glucose concentration is most useful in cases of suspected tuberculous meningitis.

5. CSF chloride estimation has no value as a diagnostic aid.

Chapter **XXIII**

# CHEMICAL PATHOLOGY AND THE CLINICIAN

CHEMICAL pathology is the study of physiology and biochemistry applied to medical practice. In essence, like haematology, it is a clinical subject, and as such it overlaps clinical medicine and surgery in particular. An understanding of the basis of chemical pathology, as we have seen, is essential for much diagnosis and treatment.

The medical student or the clinician need not know technical details of laboratory estimations unless they are ones which he is likely to be called upon to perform himself. However, correct interpretation of laboratory results requires some understanding of both the acceptable analytical reproducibility and of physiological variations in normal individuals, and this subject will be discussed in the next chapter. It is also desirable that the clinician should be aware of the speed with which tests can be completed, and of laboratory organisation in so far as it affects this; above all he should realise that the technique of collecting specimens can affect results drastically, and should co-operate with the laboratory in its attempt to produce rapid, accurate answers, quickly identifiable with the relevant patient. To this end he should understand the importance of accurately completed forms, correctly labelled specimens, taken at the right time by the right technique, and of speedy delivery to the laboratory. In an emergency *therapy based on correctly estimated results from a wrongly labelled or collected specimen may be as lethal as faulty surgical technique*: moreover, even if the error is recognised, *precious time could have been saved by a few minutes' thought and care in the first place. An emergency situation warrants more, not less, than the usual accuracy in collection and identification of specimens.*

## REQUEST FORMS

### CLINICAL INFORMATION

Control of accuracy of results is largely the concern of the laboratory and most departments take stringent precautions to this end. However, when very large numbers of estimatons are being handled it is impossible to be sure that every result is correct; fortunately the error rate is very low. The clinician can play his part by co-operating with the pathologist in minimising the chance of error, not only by taking suitable specimens, but also by giving *relevant* clinical information. "Unlikely" results are checked in most laboratories and, for instance, if the blood urea concentration of a patient had fallen by 33 mmol/l (200 mg/dl) in 24 hours, both estimations would be repeated before the latest was reported: on the other hand, if it were known that haemodialysis had been performed in the interval the result would have been an expected one and time, money and worry would have been saved. In this example "post-haemodialysis" would be more informative, and take no longer to write in the "Clinical Details" space than "chronic renal failure".

## PATIENT IDENTIFICATION

Accurate information about the patient, including *surname* and *first names* correctly and consistently spelt, and legibly written, *age or date of birth* and *hospital case number* are essential for comparing current with previous results on the same patient. If the laboratory uses "cumulative" reporting, the results on each patient on successive days are entered on one form (Fig. 38): this type of form enables the clinician to see the progress of the patient more easily than he can when looking at a single result and, in addition, the laboratory can detect sudden changes, so that the cause can be sought. For the system to work successfully, accurate patient identification is essential: a surprising number of patients have the same names, even when these are apparently uncommon; it is less likely that they will also have the same age in years, and even less probable, the same date of birth; they will not have the same hospital number. Any of these items may be written inaccurately on the form, and, unless there is complete agreement with previous details, results can be entered on the wrong patient's record: this can lead to confusion, and even danger to the patient. Some large departments use computers to report results: computers cannot think, telephone the ward, or make intelligent guesses, and if the information fed into them is inaccurate it may either be rejected or, worse still, results may be reported as belonging to another patient.

## LOCATION OF THE PATIENT AND CLINICIAN

It is obvious that if the *ward or department* is not stated it may take time and trouble to find out where results should be sent. The consultant's *name*, and the *signature of the doctor requesting the test* are desirable if urgent or alarming results are to be notified rapidly, and advice given about treatment.

The forms designed by pathology departments ask only for information essential to ensure the most efficient possible service to the

**DEPT. OF CHEMICAL PATHOLOGY, W.M.S. Tel. 2440**

Clinical details:—

Thyrotoxicosis with vomiting.

Case No. 12345   Consultant Dʳ X
Surname B:—   Ward A.2
First Names E.—   Sex F
Date of Birth 10.10.44

| SERUM | H = Haemolysed | | I = Insufficient | | U = Unsuitable | | | | * Duplicated Result | | |
|---|---|---|---|---|---|---|---|---|---|---|---|
| Lab. No. | ⊠ | D8 | ⊠ | D85 | ⊠ | D87 | D87 | D87 | V3L | X26 | 352 |
| Date | 4/2/69 | 5/2/69 | 21/2/69 | 24/2/69 | 28/2/69 | 3/3/69 | 6/3/69 | 7/3/69 | 14/3/69 | 12/4/69 | 12/7/69 |
| Ca (mg/100ml) | 13.8⁺ | 13.2 | 9.2⁺ | 8.8 | 9.4 | 10.6 | 12.8⁺ | 12.6 | 11.4 | 9.0 | 9.8 |
| PO₄ (as P) (mg/100ml) | 3.6 | 3.4 | 2.9 | 1.8 | 4.0 | 4.7 | 3.3 | 2.3 | 4.0 | 3.2 | 3.3 |
| Alk.phos. (K.A. units) | 1 | 19 | 15 | 14 | 16 | 14 | 15 | 1 | 1 | 20 | 22 |
| T. Protein (g/100ml) | 7.7 | 7.9 | 6.5 | 6.3 | 7.2 | 7.2 | 8.3 | 7.9 | 8.1 | 7.5 | 7.5 |
| Urea (mg/100ml) | 50 | 47 | 26 | 19 | 20 | 27 | 59 | 59 | 26 | 27 | 27 |
| 'Corrected Calcium' mg/100l | 13.3 | 12.7 | 9.7 | 9.6 | 9.4 | 10.6 | 11.8 | 12.1 | 10.7 | 8.7 | 9.5 |
| Signature | ⊘ | ⊘ | ⊘ | ⊘ | ⊘ | ⊘ | ⊘ | ⊘ | ⊘ | ⊘ | ⊘ |

Annotations within the chart: STEROID SUPPRESSION TEST STARTED 17.2.60 — STEROIDS STOPPED 27/2/69 — STEROIDS — VOMITING AGAIN — CARBIMAZOLE STARTED

FIG. 38.—A sample of cumulative reporting of the calcium group of tests on a patient with hypercalcaemia, showing the effect of steroids and carbimazole.

In some hospitals all types of result are reported on the same sheet.

clinician and the patient. All pathologists have met the form containing as the only information "Smith"; sometimes not even the investigations requested are stated. Unless the pathologist is endowed with psychical powers it is difficult for him to help the clinician under these circumstances.

## ADDRESSOGRAPH SYSTEMS

Some hospitals use "Addressograph" labels. All the required information (except clinical details) can be printed on a set of these labels when the patient is admitted, and one label can be attached to each request form and another to its accompanying specimen. With this system the possibility of error is much reduced.

# COLLECTION OF SPECIMENS

## COLLECTION OF BLOOD

If the laboratory obtains a clinically improbable result on a specimen, it will usually check this on that specimen. If, as is usual, the second result agrees well with the first, a fresh specimen should be obtained. Before doing this it is essential to try to find out why the first one gave a false answer (if it was false). Contamination of the syringe, needle or tube into which the specimen was collected, although an obvious possibility, is relatively rare in these days of disposable apparatus. It should not be accepted as the cause until other more common ones have been excluded.

The errors to be discussed below, and which arise outside the laboratory are, in our experience, relatively common. Any examples given are genuine ones.

### Effect on Results of Procedures Prior to Venepuncture

(*a*) **Oral medication.**—Specimens should not be taken to measure a substance just after a large oral dose of the same substance has been given. For example, in the presence of iron deficiency, plasma iron concentration may rise to normal, or even high, levels for a few hours after a large dose of oral iron.

(*b*) **Interfering substances.**—Previous administration of a substance may affect plasma levels for some time. For instance, administration of iodide in cough medicines, ointment or, even more seriously, in radiological contrast media, will result in apparently raised protein bound iodine levels for some days, weeks or even months. Iodine, of course, not only affects blood levels, but will cause a falsely low thyroid uptake of radio-iodine (p. 167). The effect of interfering substances on analytical methods may be more widespread than generally recognised.

(c) **Palpation of the prostate.**—The prostate contains tartrate labile acid phosphatase and the serum concentration of this enzyme is used as an index of spread of carcinoma of the gland (p. 351). However, palpation of a non-malignant prostate may release relatively large amounts of this enzyme into the blood stream. These false levels may persist for several days after rectal examination, passage of a catheter, or even after straining at stool. For example, a specimen was received by the department from a patient who had had a rectal examination a few hours previously. The tartrate labile acid phosphatase level was 2·4 K.A. units (the upper normal for this laboratory is 0·9 K.A. units). Three days later a further specimen gave a level of 1·5 K.A. units— still raised. Eight days after the examination, however, the concentration was only 0·2 K.A. units.

Any marginally raised tartrate labile acid phosphatase level should be checked on a specimen taken a few days later. If possible, blood for this estimation should be withdrawn before performing a rectal examination. If, as is often the case, the result of such an examination suggests the need for the estimation it is best, if possible, to wait a week before taking blood. To save time, the specimen may be taken immediately and *a note made on the request form to the effect that rectal examination has been performed.* If the result is normal no further action is required and time has been saved: if it is not, another specimen will be requested.

### Effect on Results of Technique of Venesection

(a) **Venous stasis.**—When blood is taken a tourniquet is usually applied proximally to the site of puncture to ensure that the vein "stands out", and is easier to enter with the needle. If this occlusion is maintained for more than a short time the combined effect of raised intravenous pressure and anoxia of the vein wall results in passage of water and small molecules from the lumen into the surrounding extracellular fluid. Large molecules, such as protein, and erythrocytes and other cells, cannot pass through the vein wall: their concentration therefore rises. It should be remembered that there will not only be a rise in total protein levels but also in all protein fractions, including immunoglobulins, and day-to-day variations in these can often be attributed to this factor, as well as to changes in posture (p. 321).

Many plasma constituents are, at least partially, bound to protein in the blood stream. Prolonged stasis can falsely raise calcium concentrations (Fig. 39), possibly to high or equivocal levels. If such levels are found it is important to take another specimen, without stasis, for analysis. Other important protein bound substances include lipids, and $T_4$ or PBI.

Examples of the effects of stasis in a normal subject are given below.

| Stasis for | 0 | 2 | 4 | 6 minutes |
|---|---|---|---|---|
| Calcium (mmol/l) | 2·38 | 2·45 | 2·52 | 2·58 |
| Total protein (g/l) | 72 | 74 | 77 | 80 |
| Albumin (g/l) | 39 | 40 | 42 | 43 |
| Haemoglobin (g/dl) | 14·7 | 14·8 | 15·1 | 15·5 |

Prolonged stasis, with associated local anoxia, may also cause such intracellular constituents as potassium to leak from the local cells into the plasma, causing falsely high potassium results.

Many patients have "bad veins", difficult to enter without stasis. Under such circumstances a tourniquet may be applied until the needle is in the vein lumen; if it is now released, and a few seconds allowed before removal of blood, a suitable specimen will be obtained.

(b) **Site of venepuncture.**—Many patients requiring chemical pathological investigations are receiving intravenous infusions. In veins in the same limb, whether proximal or distal to the infusion site, the drip fluid has not mixed with the whole of the blood volume; local concentrations will therefore be unrepresentative of that circulating in the rest of the patient. Blood taken from the opposite arm will, however, give valid results.

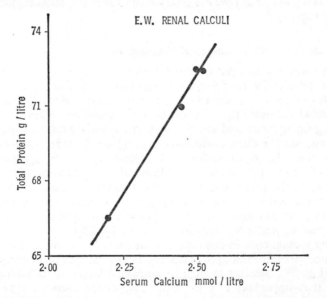

FIG. 39.—Relationship between plasma total protein and calcium concentrations.

The following example illustrates the point well. The clinical details were given as "post-op". On the previous day the electrolytes had been normal, the plasma urea 16·7 mmol/l (100 mg/dl), and the plasma total protein 67 g/l. The relevant results on the day in question were as follows:

| | |
|---|---|
| Plasma sodium | 65 mmol/l |
| Plasma potassium | 2·2 mmol/l |
| Plasma bicarbonate | 11 mmol/l |
| Plasma protein | 52 g/l |
| Plasma urea | 8·7 mmol/l (52 mg/dl) |

The patient was "doing well". As all measured constituents were diluted it was assumed that dextrose was being infused. A plasma sugar of 50 mmol/l (900 mg/dl) supported this view. A call to the ward confirmed that the specimen had been taken from the arm into which a dextrose infusion was flowing. Analysis of blood taken from the opposite arm gave results almost identical with those of the previous day. (It should be noted that during dextrose infusion there will be apparent hyperglycaemia, even in blood from the opposite arm.)

Such an extreme example as this is easily detected and, although time has been wasted and the patient subjected to two venepunctures, no serious harm is done. If, for example, physiologically normal saline is being infused, electrolyte results might *appear* to be correct. Under such circumstances the wrong therapy might be applied.

If no veins are available in another limb a suitable sample can be obtained by stopping the infusion, disconnecting the tubing from the needle, aspirating 20–30 ml of blood through this needle and discarding before taking the specimen for analysis.

### Containers for Blood

Many hospital laboratories issue a list of the types of container required for each specimen, and this will vary slightly from hospital to hospital. For instance, most departments require that fluoride be added to blood taken for glucose estimation: this inhibits glycolysis, which would otherwise continue in the presence of erythrocytes.

To ensure accuracy of results laboratories will only accept blood in the correct containers. However, errors can arise if blood is decanted from one container to another. Oxalate and sequestrene (ethylenediamine tetracetate, EDTA) act as anticoagulants by removing or chelating calcium. Estimation of the latter is therefore invalidated by the presence of these chemicals. The potassium salt of sequestrene is usually used and this will invalidate potassium estimation: sodium oxalate (and sodium heparin instead of lithium heparin) would, of course, upset sodium estimations.

As an example blood was received from the out-patient department apparently in the correct tube. Clinical details were "uretero-sigmoidostomy". A calcium value of 0·4 mmol/l (1·6 mg/dl) was obtained. The plasma potassium was 7·5 mmol/l in spite of the fact that the patient felt very well. Enquiry confirmed that blood had been taken, at the same time, into a sequestrene bottle (for haematological investigations), and that to bring the blood level in the tube "to the mark" some had been tipped into that for chemical pathology.

### Effects of Storage and Haemolysis of Blood

Erythrocytes contain very different concentrations of many substances from those of the surrounding plasma (for instance the potassium concentrations are about 25 times as high). If haemolysis occurs the contents will "leak out" and false answers will be obtained on plasma. Plasma from haemolysed blood is red and this will be detected by the laboratory. To minimise the chance of haemolysis blood should be treated gently. The plunger of the syringe should not be drawn back too fast, and there should be an easy flow of blood. The needle should be removed from the syringe before the specimen is expelled *gently* into the correct container.

The maintenance of differential concentrations across the red cell wall requires energy, which is supplied by glycolysis. In whole blood outside the body the erythrocytes will soon use up available glucose (hence the need for fluoride in blood glucose specimens), after which no energy source remains: concentrations in erythrocytes and plasma will tend to equalise by passive diffusion across cell membranes. If blood is left unseparated for more than an hour or two the effect on plasma levels will, therefore, be the same as that of haemolysis, with the important difference that, to the naked eye, the plasma looks normal. If the container is undated, or wrongly dated, the error may not be detected. It is important to separate plasma from red cells before storing (even in the refrigerator) overnight.

An example of the effect of allowing whole blood to stand at room temperature on plasma potassium and blood glucose levels is given below.

| Blood separated after | 0 | 4 | 8 | 24 hours |
|---|---|---|---|---|
| Potassium (mmol/l) | 4·0 | 4·3 | 4·8 | 6·4 |
| Glucose (mmol/l) | 4·84 | 3·94 | 3·00 | 1·94 |

### COLLECTION OF URINE

Many urine estimations are carried out on timed specimens. Because of large circadian variations in excretion only qualitative tests, or those of tubular concentrating ability, are performed on random collections. Results are expressed as units/time (for example, mmol/24 h), and to calculate this figure the concentration (for example, mmol/l) is multiplied by the total volume collected. Clearance estimations, too, depend on comparing the amount of, for example, urea excreted per 24 hours with its concentration in the blood (p. 21). In both these examples the accuracy of the final answer depends largely on that of the urine collection: this latter is surprisingly difficult to ensure. In some cases the difficulty is insurmountable unless a catheter is inserted, and this is undesirable because of the risk of urinary infection: for instance, the patient may be incontinent or, because of prostatic hypertrophy or neurological lesions, be incapable of complete bladder emptying. However, more often errors arise because of a misunderstanding on the part of the nurse, doctor or patient collecting the specimen.

Let us suppose that a 24-hour collection is required between 8 a.m. on Monday and 8 a.m. on Tuesday. The volume *secreted by the kidneys* during this time is the crucial one: urine already in the bladder at the start of the test and secreted some time before should not be included; that in the bladder at the end of the test and secreted between the relevant times *should* be included. The procedure is therefore as follows:

8 a.m. on Monday—Empty bladder completely. Discard specimen. Collect all urine passed until:
8 a.m. on Tuesday—Empty bladder completely. *Add this to the collection.*

The error of not carrying out this procedure is very great for short (for example, hourly) collections of urine.

A preservative must usually be added to the urine to prevent bacterial growth and destruction of the substance being estimated. Before starting the collection the bottle containing the correct preservative should be obtained from the laboratory.

## COLLECTION OF FAECES

Rectal emptying is much more erratic than bladder emptying, and cannot usually be performed to order. Estimations of 24-hourly faecal content of, say, fat may vary by several hundred per cent from day to day. If the collection were continued for long enough the *mean* 24-hourly output would be very close to the true daily loss from the body (which includes that in faeces in the rectum at any time). There must be a reasonable compromise on time, and most laboratories collect for between 3 and 5 days. Provided that the patient is not constipated the mean daily loss is usually reasonably representative: if no stool is passed during this period, excess faecal loss of any kind is most un-likely, and the test probably unnecessary. To render the collection more accurate many departments use coloured "markers" (see Appendix to Chapter XI).

Faecal estimations and collections are time-consuming and un-pleasant for all concerned. It is important that *every* specimen passed during the time of collection is sent to the laboratory if a worthwhile answer is to be obtained. Administration of purgatives or enemas during the test alters conditions and invalidates the answer.

## LABELLING SPECIMENS

It is important to label a specimen accurately to correspond with the accompanying form in all particulars. The date, and sometimes the time of taking the specimen should be included, and should be written *at the time* of collection. If it is done in advance the clinician may change his mind, and the information will be incorrect; there is also the danger of using a container with one patient's name on it for another patient's specimen.

### Blood Specimens

Specimens in wrongly labelled tubes may cause danger to one or more patients. The date of the specimen is important, both from the clinical point of view, and as discussed on p. 464, to assess the suit-ability of the specimen for the estimation requested. It is important to state the time when a specimen was taken, particularly if the con-centration of the substance being measured varies during the day: for instance blood glucose varies according to the time since the last meal. If more than one specimen is sent for the same estimation on one day *each must be timed* so that it is known in which order they were taken.

### Urine and Faecal Specimens

Timed urine specimens should be labelled with the date and time of starting and completing the collection, so that the volume per unit time is known. Faecal collections are best labelled, not only with date and time, but with the specimen number in the series of 5-day collections, so that the absence of a specimen is immediately obvious.

## SENDING THE SPECIMEN TO THE LABORATORY

If, in an emergency, a result is needed quickly, many types of esti-mations can be carried out in a short time. However, it is more econ-omical in staff, reagents and time, as well as easier to organise, if esti-mations are batched as far as possible. For this reason most laboratories

## TABLE XXXV
### SOME EXTRA-LABORATORY FACTORS LEADING TO FALSE RESULTS

| Cause of error | Consequence |
|---|---|
| Keeping blood overnight before sending to laboratory. | High plasma K, total acid phosphatase, LD, HBD, AST. |
| Haemolysis of blood. | As above. |
| Prolonged venous stasis during venesection. | High plasma Ca, total protein and all protein fractions, lipids, $T_4$ and PBI. |
| Taking blood from arm with infusion running into it. | Electrolyte and glucose results approaching composition of drip fluid. Dilution of everything else. |
| Putting blood into "wrong" bottle or tipping it from this into Chemical Pathology tube. | e.g. EDTA or oxalate cause low Ca, with high Na or K. |
| Blood for glucose not put into fluoride tube. | Low glucose (fluoride inhibits glycolysis by erythrocytes). |
| Palpation of prostate by PR, passage of catheter, enema, etc., in last few days. | High tartrate labile acid phosphatase. |
| Inaccurately timed urine collection. | False timed urinary excretion values (e.g. per 24 hours). False and erratic renal clearance values. |
| Loss of stools during faecal fat collection. Failure to collect for long period between markers. | False faecal fat result. |
| Therapy with iodine, IVP, angiogram, etc. | Falsely low radio-iodine result. Falsely high PBI result. |
| Therapy with antithyroid drugs. | False radio-iodine result. |

like to receive non-urgent specimens early in the morning. A constant "trickle" of specimens may cause delay in reporting the whole batch, and possibly in noticing a result requiring urgent treatment.

If a patient is seen for the first time later in the day and the results are not required urgently, the specimen should be sent to the laboratory with a note to that effect. Plasma can then be separated from cells and stored overnight.

In cases of true clinical emergency, the department should be notified. The clinical details warranting urgency may be given on the form, but preferably the laboratory should be warned before the specimen is taken, so that they may be prepared to deal with it quickly. Usually a specimen not known to require urgent attention, and certainly one accompanied by no information about clinical details, will be assumed to be non-urgent. *It is the clinician's responsibility to indicate the degree of urgency.*

Table XXXV summarises some of the errors which have been discussed in this chapter.

## SUMMARY

The clinician's responsibility for maintaining accuracy and speed of reporting of results includes:

Taking a suitable specimen of blood

(*a*) At a time when a previous procedure will not interfere with the result.

(*b*) From a suitable vein.

(*c*) With as little stasis as possible.

(*d*) With precautions to avoid haemolysis.

Putting the specimen in the correct container.

Labelling the specimen accurately.

Completing the form accurately, including *relevant* clinical details.

Ensuring that the specimen reaches the laboratory without delay, and that plasma is separated from cells before storing the specimen.

In the case of urine and faeces, collecting accurate and complete timed specimens.

## FURTHER READING

CHRISTIAN, D. G. (1970). Drug interference with laboratory blood chemistry determinations. *Amer. J. clin. Path.*, **54**, 118.

PANNALL, P. (1971). Pitfalls in the interpretation of blood chemistry results. *S. Afr. med. J.*, **45**, 1184.

ZILVA, J. F. (1970). Collection and preservation of specimens for chemical pathology. *Brit. J. hosp. Med.*, **4**, 845.

# Chapter XXIV

# REQUESTING TESTS
# AND INTERPRETING RESULTS

## REQUESTING TESTS

### WHY INVESTIGATE?

THE doctor now has at his disposal a great many tests. These can often provide helpful information if used critically: if used without thought the results are at best useless, and at worst misleading and dangerous. Writing a request form should not be considered as casting a magic spell, which will benefit the patient merely by doing it.

Investigation should be used to improve the management of the patient: it should not be used to show how "clever" the doctor is. This may seem obvious, but is often forgotten. It is more truly intelligent and satisfying to know what one hopes to gain from investigation and to achieve this aim as economically as possible in time and money. One test is not necessarily better than another because it is newer, more expensive or more difficult to perform: if it *is* better it should be used instead of (not as well as) the other one.

Far from helping the patient, over-investigation may harm him by delaying treatment, causing him unnecessary discomfort or danger, or more insidiously, by using money that may be more usefully spent on other aspects of his care. Of course, under-investigation is just as undesirable as over-investigation: the cost to the patient of omitting a necessary test is just as high as carrying out unnecessary ones.

Before requesting an investigation the doctor should ask himself:

1. Will the answer, whether it be high, low, or normal, affect my diagnosis?

2. Will the answer affect the treatment?

3. Will the answer affect my estimate of the patient's prognosis?

4. Can the abnormality I am seeking exist without clinical evidence of it? If so, is such an abnormality dangerous, and can it be treated?

If, after careful thought, the answer to *all* these questions is a clear "no", there is no need for the test. If it is "yes" to *any* of them, it should be performed.

Sometimes, even if the clinical diagnosis is obvious, the test may still be necessary. For example, a patient may have overt myxoedema. Once treatment is started both the clinical and biochemical features are obscured. Later, another doctor seeing the patient for the first time may not be sure that the patient was ever hypothyroid, and may need to stop the therapy to verify the diagnosis. *One* unequivocally abnormal result before treatment is started (such as a very low free thyroxine index) provides objective documentation: more than one in these circumstances is unnecessary. To perform *all* available tests routinely, such as estimation of neck uptake with TSH stimulation, *and* serum $T_4$ *and* TSH is unnecessary; such a "battery" should be reserved for genuine diagnostic problems.

## Why Not to Investigate?

The student is warned against the following unqualified statements.

1. "It would be nice to know." Ask yourself if it will help the patient.

2. "We would like to document it fully." Will this extra documentation make any difference to your management of the patient?

3. "Everyone else does it." They may be right, but do you know their reasons? Perhaps they too do it because everyone else does it. Do not accept anything from anyone (not even this book) uncritically. Re-assess dogma continually. If you lack experience, at least look at the reasoning behind what you read or are taught; use the statements of "experts" as working hypotheses until you are in a position to make up your own mind.

## How Often Should I Investigate?

This depends on:

1. How quickly numerically significant changes are likely to occur. For instance, serum protein fractions are most unlikely to change significantly in less than a week, if in that, and the plasma urea concentration will not become significantly abnormal after 12 hours "anuria".

2. Whether a change, even if numerically significant, will alter treatment. For instance, transaminase levels may alter over 24 hours during acute hepatitis. Once the diagnosis is made this is unlikely to affect treatment. On the other hand, potassium concentrations may alter rapidly in patients on large doses of diuretics, and these *may* indicate the need for treatment.

Unless the patient is receiving intensive therapy of some kind, investigations are very rarely required more than once every 24 hours.

## WHEN IS AN INVESTIGATION "URGENT"?

The only reason for asking for an investigation to be carried out urgently is that an earlier answer will alter treatment. This situation is very rare. For example, the doctor should ask himself how often treatment would really be different within the next 12 hours if the urea was 10 mmol/l (60 mg/dl) or 50 mmol/l (300 mg/dl).

# INTERPRETING RESULTS

Before considering diagnosis or therapy based on a result received from the laboratory the clinician should ask himself three questions:

1. If it is the first estimation performed on this patient, is it normal or abnormal?

2. If it is abnormal, is the abnormality of diagnostic value or is it a non-specific finding?

3. If it is one of a series of results, has there been a change, and if so is this change clinically significant?

## IS THE RESULT NORMAL?

### Normal Ranges

The normal range of, for example, blood urea is often quoted as between 3·3 and 6·7 mmol/l (20 and 40 mg/dl). It is clearly ridiculous to assume that a result of 6·5 mmol/l (39 mg/dl) is normal, while one of 6·9 mmol/l (41 mg/dl) is not. Just as there is no clear-cut demarcation between "normal" and "abnormal" for body weight and height, the same applies to any other measurement which may be made.

The majority of a normal population will have a value for any constituent near the mean value for the population as a whole, and all the values will be distributed around this mean, the frequency with which any one occurs decreasing as the distance from the mean increases. There will be a range of values where "normals" and "abnormals" overlap (Fig. 40): all that can be said with certainty is that the *probability* that a value is abnormal increases the further it is from the mean until, eventually, this probability approaches 100 per cent. For instance, there is no reasonable doubt that a plasma urea value of 50 mmol/l (300 mg/dl) is abnormal, whereas with one of 7·5 mmol/l (45 mg/dl) it is possible that it is normal for the individual concerned. It should also be noted that a "normal" result does not necessarily exclude the disease sought: because of intersubject variation, a value within the "normal" range may, nevertheless, be abnormal for the individual.

To stress this uncertainty on the borders of the normal values it is better to quote limits between which values for 90 and 95 per cent of the

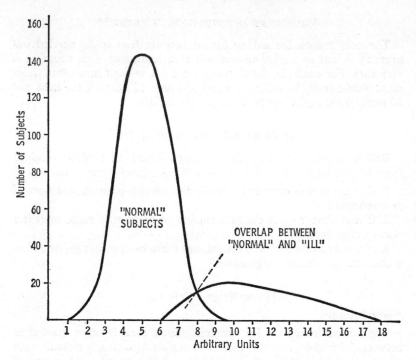

FIG. 40.—Theoretical distributions of values for "normal" and "ill" subjects, showing overlap at upper end of "normal range".

"normal" population fall, than to give a "normal range". Statistically the 95 per cent limits are two standard deviations from the mean. Although the probability is high that a value outside these limits is abnormal, obviously 2·5 per cent of the "normal" population may have such a value at either end. The range of variation for a single subject is usually less than that for the population as a whole.

In assessing a result one can only take all factors, including the clinical picture, into account, and reach some estimate of the probability of its being normal.

## Physiological Differences

Certain physiological factors affect interpretation of results. For instance, the "normal" levels of plasma urate or iron vary with *sex*, being higher in males than in females (pp. 383 and 392); the plasma urea concentration tends to rise with *age* especially in male subjects, and normal values in children are often different from those in adults;

in different parts of the world mean values of many parameters are different, either because of *racial* or *environmental* factors. It is clear, therefore, that we are not so much expressing "normal" values as the most usual ones for a given population. A blood urea level remaining at 7·5 mmol/l (45 mg/dl) at the age of 20 suggests mild renal impairment, which may progress to clinically severe damage in later life: the same value at the age of 70 suggests the same degree of renal impairment, but usually the subject will die of some other disease before this becomes severe. In other words, this rising mean value of urea with age is not strictly normal and probably does reflect disease. Note that we are talking of *mean* values for a population of a certain age: in an individual there may be no change with advancing years.

## Differences Between Laboratories

From the above discussion will be seen that, even if the same method is used in the same laboratory, it is difficult to define a normal range clearly. With some constituents interpretation becomes even more difficult if results obtained in different laboratories are compared, because different analytical methods may be used. For most estimations agreement between reliable laboratories is close. However, with certain constituents, such as serum proteins and especially albumin, different methods, even in the best hands, give different results: this reflects the fact that different techniques measure different properties of protein, and that different methods of fractionation do not necessarily separate exactly the same fractions. It should be noted that the definition of "international units", in which the results of enzyme assays are expressed, does not include the temperature at which the assay is performed. Different laboratories, for various technical reasons, may use temperatures varying from 25°C to 37°C: results at these temperatures will be very different, although apparently expressed in the same units. If reproducibility is acceptable, one method is often no better than another for clinical purposes, provided that the results are compared with the "normals" for the laboratory in which the estimation was performed and provided that serial estimations are carried out by the same method.

### IS THE ABNORMALITY OF DIAGNOSTIC VALUE?

Serum or plasma values express extracellular concentrations. Moreover, some abnormalities are non-specific and of no diagnostic or therapeutic import.

## Relationship Between Plasma and Cellular Levels

Intracellular constituents are not easily estimated, and plasma levels do not always reflect the situation in the body as a whole; this is particularly true for such constituents as potassium, which have very high intracellular concentrations compared with those in the surrounding fluid. A normal, or even high, plasma potassium concentration may be associated with cellular depletion, if conditions are such that the equilibrium across cell membranes is disturbed (for instance, in acidosis and dehydration).

### Relationship Between Extracellular Concentrations and Total Body Content

The numerical value of a concentration depends not only on what we are measuring, but also on the amount of water in which it is dissolved (for instance, mmol/l). A low plasma sodium is not necessarily (or even usually) due to sodium depletion: it is more often due to water excess. Under such circumstances there may even be excess of sodium in the body (p. 56). Conversely, hypernatraemia is more often due to water deficit than sodium excess (p. 46). It is very important to recognise this fact and adapt therapy accordingly. It has already been noted that protein concentrations can be affected by stasis during venesection, but if this factor is eliminated, significant day-to-day variations of protein concentration over a short period of time can be used to assess changes of hydration of the patient (in other words, the amount of protein is not, but that of water is, changing significantly).

Another cause of falsely low plasma sodium concentration is gross lipaemia or hyperproteinaemia (p. 36).

### Non-specific Abnormalities

Circulating levels of, for example, albumin, calcium and iron, vary considerably in diseases unrelated to the primary defect in metabolism.

The concentration of albumin, as well as of all other protein fractions (including immunoglobulins) and of protein-bound substances, may fall by as much as 15 per cent after as little as 30 minutes recumbency, possibly due to fluid redistribution in the body. This effect may, partly at least, account for the non-specific low albumin concentration found in quite minor illnesses. In-patients usually have blood taken early in the morning, while recumbent, and tend to have lower values for these parameters than out-patients.

Routine laboratory methods for estimating calcium measure the total protein bound + ionised concentrations: changes in albumin levels are associated with changes in those of the calcium bound to it, without an alteration of the physiologically important ionised fraction, and this can occur either artefactually, as discussed on p. 461, or with true changes in albumin. It is most important not to attempt to raise the total calcium level to normal in the presence of significant hypoalbuminaemia.

Plasma iron is very labile, and levels can fall in the presence of anaemia other than that of iron deficiency: giving iron, especially by parenteral routes, to patients with anaemia and a low plasma iron can be dangerous unless other, more reliable, evidence of iron deficiency is present (p. 397).

Note that different methods may give different answers for albumin concentration on the *same specimen*. Even when the same method is used on different specimens from the *same patient* the value obtained depends upon the amount of stasis used during venesection, and on whether the patient is ambulant or not. Moreover, a very low albumin concentration may be the *cause* of oedema, or may be the *result* of overhydration. A decision on management based primarily on whether the result is above or below an arbitrary figure of, say, 20 g/l, is to misunderstand the difficulties of interpreting such a figure.

## HAS THERE BEEN A CLINICALLY SIGNIFICANT CHANGE?

To interpret day-to-day changes in results, and to decide whether the patient's biochemical state has altered, one must know the degree of variation to be expected in results from a normal population.

### Reproducibility of Laboratory Estimations

In reliable laboratories most estimations should give results reproducible to well within 5 per cent: some (such as calcium) should be even more reproducible. Changes of less than the reproducibility of the method are probably clinically insignificant.

### Physiological Variations

Physiological variations occur in both plasma levels and urinary excretion rates of many substances and false impressions may be gained from results of several types of investigation if this fact is not taken into account.

Physiological variations may be regular or random.

1. **Regular variations.**—Regular changes occur throughout the 24-hour period (circadian or diurnal rhythms, like that of body temperature), or the month: there may be seasonal variations. The causes of these are often obscure. There is a marked diurnal variation in the urinary excretion of, for example, electrolytes, steroids, phosphate and water. For this reason estimates of excretion carried out on accurate 24-hour collections are more valuable than measurements of concentration of random specimens. In interpreting such results the effect of diet and of fluid intake should be remembered: because of these effects it is difficult to give "normal" values for many urinary constituents (for example, of electrolytes).

Blood glucose concentration varies with the time after a meal, and the concentration of plasma protein and of protein bound substances varies with posture. Plasma iron shows very marked circadian variation, apparently unrelated to meals or other activity: it may fall by 50 per cent between morning and evening. The circadian variation of plasma cortisol is of diagnostic importance (p. 146), and it should be remembered that superimposed on this regular variation "stress" will cause acute rises. To eliminate the unwanted effect of circadian variation, blood should, ideally, always be taken at the same time of day (preferably in the early morning, with the patient fasting). This is not always possible, so that these variations should be borne in mind when interpreting results. Correct interpretation of blood glucose levels requires an estimation on blood taken with the patient fasting, or at a set time after a known dose of glucose (p. 194).

Some constituents show monthly cycles, especially in women (again, compare body temperature). These can be very marked in the case of plasma iron, which may fall to very low levels just before the onset of menstruation (p. 393). There are also, probably, seasonal variations in some constituents: study of these is in the very early stages.

Although some of these changes, such as those of blood glucose related to meals, have obvious causes, many of them appear to be regulated by a so-called "biological clock", which may be, but often is not, affected by the alternation of light and dark.

2. **Random variations.**—Day-to-day variations in, for instance, plasma iron levels are very large, and may swamp regular changes (p. 393). The causes of these are not clear, but they should be allowed for when interpreting serial results. The effect of "stress" on plasma cortisol levels and the many factors affecting serum protein concentration have already been mentioned.

### CONSULTATION WITH THE LABORATORY STAFF

The object of citing the examples given in this and in the preceding chapter, is not to confuse the clinician, but to stress the pitfalls of interpretation of a figure taken in isolation. On most occasions, if care has been exercised while taking the specimen, a diagnosis can be made and therapy instituted *by relating the result to the clinical state of the patient.* However, if there is any doubt about the correct type of specimen required, or about the interpretation of a result, consultation between the clinician and chemical pathologist or biochemist can be helpful to both sides.

Laboratory errors do, inevitably, occur, even in the best-regulated departments. However, a discrepant result should not be assumed to be due to this. Consultation may help to find the cause. The estimation may already have been checked, and if it has not the laboratory is usually willing to do so in case of doubt. If it has already been checked, a fresh specimen should be sent to the laboratory after consultation to determine why the first specimen was unsuitable. If the result is still the same every effort must be made to find the cause. Laboratory staff of all grades often take an active interest in patients whom they are investigating. On their side they often take trouble to keep the clinician informed of changes requiring urgent action, and they may suggest further useful tests: the clinician should reciprocate by giving the pathologist information relevant to the interpretation of a result, and of the clinical outcome of an "interesting" problem. Such exchange of ideas and information is in the best interests of the patient.

## SUMMARY

The clinician should use the laboratory intelligently and selectively, in the best interests of the patient. In interpreting results the following facts should be borne in mind:

1. The "normal range" only indicates the *probability* of a result being normal or abnormal.

2. There are physiological differences in normal ranges and physiological variations from day to day.

3. There are small day-to-day variations in results due to technical factors and "normal ranges" may vary with the laboratory technique employed.

4. Using plasma or serum, extracellular concentrations are being measured. These depend on the amount of water in the extracellular compartment, as well as that of the constituent measured, and may be a very poor reflection of intracellular levels.

5. Changes in a given constituent may be non-specific, and unrelated to a primary defect in the metabolism of that constituent.

Finally, when in doubt, two heads are better than one. Pathologists and clinicians tend to see things from slightly different angles and full consultation between the two is in the patient's best interest.

## FURTHER READING

RUSSE, H. P. (1969). The use and abuse of laboratory tests. *Med. Clin. N. Amer.* **53**, 223.

BOLD, A. M., and WILDING, P. (1975). *Clinical Chemistry Conversion Scales for S.I. Units. with Adult Normal (Reference) Values.* Oxford: Blackwell Scientific Publications.

A very salutary collection of essays on the general subject of common sense in medicine is:—

ASHER, R. (1972). *Richard Asher Talking Sense.* London: Pitman Medical. (The short section entitled "Logic and the Laboratory" on pp. 166–167 is excellent).

# INDEX

# INDEX

The main page references are in **bold** type

Abetalipoproteinaemia, **222**
Acetazolamide and acidosis, 70, **99**
Acetest, **215**
Acetoacetate, and acetone, 93, **189**, 214, 215
Acetyl cholinesterase, **355**
Acetyl CoA, 93, **181**
  fatty acid metabolism and, 183, 912
  ketone production and, 189, 192
  pantothenic acid and, 424
  thiamine and, 421
Achlorhydria, **281**, 282, 287
Acid, gastric, **280**, 287
Acidosis, 76, **92**
  bicarbonate depletion and, 99
  calcium, ionized and, 24, 95, **235**
  causes, **92**
  diabetes and, 92, 190, **200**
  effects, **93**
  glomerular filtration rate and, 12, 18, **89**, 96
  hyperchloraemic, **90**, 95, 99
  "lactic", 93, **190**, 203
  "mixed", **98**
  potassium excess and, **65**
  potassium, plasma level, effect on, **64**, 99
  renal failure and, 12, 15, 18, 89, **95**
  renal tubular, 90, **95**
  respiratory, **96**
  tubular damage and, **14**, 95
Acromegaly, **116**
  diabetes and, 116, 187, 191, **199**
  diagnosis, **117**
ACTH (*see* Adrenocorticotrophic hormone) 112, 136, **138**, 141
ACTH stimulation test (*see also* Tetracosactrin), **143**, 148, **150**
Acute oliguric renal failure, **18**
  treatment, **22**
Acute tubular necrosis, **18**
  treatment, **22**
Addison's disease, **148**
  ACTH in, 141, 150
  pseudo, 48
  sodium depletion and, **49**, 149
Adenohypophysis, **112**
ADH (*see also* Antidiuretic hormone), **7**, 39, 45, 50, 114, 446
Adipose tissue, metabolism of, **223**

Adrenal hypofunction, primary, 49, **148**
  secondary, **150**
Adrenaline, 134, **441**
  glycogen and, 181
  hyperglycaemia and, 187
  neuroblastoma and, **442**
  phaeochromocytoma and, **442**
  stress and, 187
Adrenocorticotrophic hormone, 112, 136, **138**, 141
  Addison's disease and, 149, **150**
  adrenal steroid synthesis and, 138
  congenital adrenal hyperplasia and, 153
  Cushing's syndrome and, 144, **148**
  ectopic production, 144, 145, **448**
  hypopituitarism and, 121, 150, **151**
  measurement of, **141**
  secondary adrenal hypofunction and, 118, 121, 150, **151**
Agammaglobulinaemia, **330**
ALA (*see* δ Aminolaevulinic acid), 405
Alanine transaminase, **347**
  liver disease and, **297**, 302, 304, 305, 354
ALA synthetase, **405**, 413
Albinism, **370**
Albumin, 35, 53, 299, 305, **319**, 338, 435
  bilirubin, plasma and, **290**, 320
  calcium, plasma and, **235**, 249, 253, 320, 338
  flocculation tests and, 316
  liver disease and, 299, 305
  nephrotic syndrome and, 320, 325
  oedema and, **53**, 320, 338
  thyroxine binding and, 162, 338
  water distribution and, 35
Albustix, **341**
Alcohol, hypoglycaemia and, 206
Aldactone, 70
Aldolase, **352**
Aldosterone, 3, **37**, 136
  congenital adrenal hyperplasia and, 153
  control of, 37
  hypopituitarism and, 119, 150
  potassium in plasma and, 52, **67**, 68, 70
  renin-angiotensin and, 37
  secondary adrenal hypofunction and, 119, 150

**Aldosterone,** sodium, urinary and, **56**
  synthesis of, **140**
**Aldosteronism, primary, 51**
  **secondary, 52**
**Alkalosis, 100**
  aldosteronism, primary and, 52
  calcium, ionized and, 70, 100, **235**
  causes, **100**
  Cushing's syndrome and, 145, 448
  glomerular filtration rate and, 90
  "milk-alkali syndrome" and, 248
  potassium depletion and, **64,** 70, 71,
    90
  pyloric stenosis and, 90, **100**
  respiratory, **102**
  tetany and, 70, 100, **235**
**Alkaptonuria, 212, 369**
**Allopurinol, 384**
**ALS, 352**
**ALT, 297, 347**
**Amenorrhoea, 125**
**Ametazole hydrochloride,** 282
**Amino acids,** absorption, 268
  gluconeogenesis and, **181,** 182
  insulin secretion and, 185, **206**
  liver disease and, 304
  reabsorption, in kidney, 2, 364
    in tubular damage, 14, 365
**Aminoaciduria, 304, 364**
  tubular damage and, 14, 365
**δ Aminolaevulinic acid,** 405, 407, 409,
    411
  acute intermittent porphyria and, 409
  hereditary coproporphyria and, 410
  lead poisoning and, 413
  porphyria variegata and, 409
  symptomatic cutaneous hepatic
    porphyria and, 411
**Ammonia,** 77, 87
  intestinal production, 289
  uretic transplantation and, 99
  urinary, 87
**Ammonium chloride loading test,** 109
**Amniotic fluid analysis, 360, 432**
**α Amylase, 349**
  acute pancreatitis and, **276,** 349
  carbohydrate digestion and, 267
  pancreatic function and, 276
  renal failure and, 277
  salivary, 267, 349
**Amyloidosis,** myelomatosis, 334
  nephrotic syndrome and, 338
**Anaemia,** investigation of, **396**
  iron absorption and, 390
  "iron deficiency without", 397
  iron overload and, 398
  iron, plasma and, 393
  malabsorption and, 275
  megaloblastic, 275, 425
  myelomatosis and, 333

**Anaemia,** pyridoxine-responsive, 393
  renal failure and, 22, 393
  scurvy and, 427
  sideroblastic, 393
  thalassaemia and, 393
**Analbuminaemia,** 319
**Androgens,** 124, **136**
  Addison's disease and, 149
  congenital adrenal hyperplasia and,153
  Cushing's syndrome and, 145
  ovary and, 124, 136
  synthesis in adrenal, **140**
  testis and, 124, 136
**Aneurine, 420**
**Angiotensin,** 37
**Anoxia** and acidosis, **93,** 98, 190
**Antibiotics,** jaundice and, 307
  macrocytic anaemia and, 277
  steatorrhoea and, 277
**Antidiuretic hormone,** 5, 7, **39**
  control of, 39
  "inappropriate" secretion of, 50, **446**
  osmolality, plasma and, 9, 39, 447
    urine and, 9, 39, 447
  sodium plasma concentration and, 39,
    447
  specific gravity, urinary and, 39, 46
  water depletion and, 46
**Antinuclear factor** in hepatitis, 301
$α_1$ **Antitrypsin,** 321
**Apoferritin,** 390
**Argentaffinoma, 440**
**Arginine,** cystinuria and, 365
  insulin secretion and, 185
**Ariboflavinosis,** 422
**Ascorbic acid, 426**
  iron overload and, 402
**Ascorbic acid** saturation test, 428, **430**
**Aspartate transaminase, 297, 346**
  in liver disease, **297,** 302, 304, 305, 354
**AST, 297, 346**
**Australia antigen, 300,** 305

**Bantu siderosis,** 402
**Bence Jones protein,** 251, **332**
  macroglobulinaemia and, 335
  myelomatosis and, 334
  tests for, **342**
**Benedict's test,** 210, **214**
  interpretation, 211, **214**
**Benign paraproteinaemia,** 336
**Bicarbonate, depletion** and acidosis, 99
  duodenal and pancreatic function, 276
  kidney and, 4, **85**
**Bicarbonate, blood,** actual, 92
  buffering and, 82, **88**
  erythrocyte and, 83
  hydrogen ion homeostasis and, 82, 88,
    91, 94, 98
  ketosis and, **93, 200**

Bicarbonate, blood, potassium and, 65, 100
  primary aldosteronism and, 52, 443
  pyloric stenosis and, 100
  renal glomerular insufficiency and, 12, 95
  renal tubular disease and, 15, 95
  standard, 92, 97
Bile acids, 308
Bile formation, 308
Bile pigments, 290, 297
Bile salts, 266, 267, 277, 291, 308
  cholesterol and, 267, 291
  enterohepatic circulation, 267
  lipid absorption and, 266, 277
  metabolism, 308
  micelle formation and, 266
Biliary calculi, 308
Biliary cirrhosis and jaundice, 303
Biliary obstruction, 302
  liver cell damage and, 295
  steatorrhoea and, 277
  urinary urobilinogen in, 302
Biliary secretion, potassium content, 63
  sodium content, 44
Bilirubin, 290
  albumin binding, 290, 320
  amniotic fluid and, 433
  urinary, 297, 301, 302, 305, 313
Biotin, 420, 424
Bisalbuminaemia, 319
"Blind loop syndrome", 277
Blood gases, 105
  specimen collection, 109
Blood glucose, diabetes mellitus and, 191, 192, 193
  growth hormone and, 116, 186, 191
  insulin and, 185
  maintenance of, 186
  measurement of, 188
  oral contraceptives and, 434
  post-prandial (2 hours), 194
  pregnancy and, 434
Blood urea nitrogen (see also Urea), 19
Bone marrow, investigation of anaemia and, 393, 395, 397
  iron overload and, 393, 395, 399, 401
  myelomatosis and, 335
Bovril test, 122, 131
Bradshaw's test, 342
Bradykinin, 440
Bromsulphthalein retention test, 298, 312
Buffering, ammonia and, 87
  bicarbonate and, 82
  definition, 77
  haemoglobin and, 83
  phosphate and, 87
  proteins and, 81
BUN (see also Urea), 19

C₃, 323
Caeruloplasmin, 322, 372
  oral contraceptives and, 373, 435
  pregnancy and, 373, 435
  Wilson's disease and, 372
Calciferol (see Vitamin D), 237
Calcitonin, 239
Calcium, deficiency, 241
  intake, 234
  intestinal absorption of, 235, 269
  loss, 235
  metabolism, 235
  parathyroid hormone and, 236
  renal failure and, 22, 245
  thyroid hormone and, 239
  vitamin D and, 237, 269
Calcium, ionised, 235
  excess, effects of, 239
  hyperparathyroidism, primary and, 242
    secondary and, 243
  lowered, 240
  parathyroid hormone secretion and, 236
  pH and, 24, 95, 235
  raised, 239
  reduced, effects of, 240
Calcium, plasma, accuracy of estimation of, 235
  acidosis and, 24, 95, 235
  acute pancreatitis and, 245, 276
  alkalosis and, 70, 100, 235
  control, 236
  ectopic parathyroid hormone and, 243, 446
  EDTA, cause of misleading results, 463
  excess, effects of, 239
  familial hypophosphataemia and, 249
  hyperparathyroidism, primary and, 242
    secondary and, 244
    tertiary and, 243
  hypoparathyroidism and, 246
  malignant disease and, 243, 247, 446
  "milk-alkali syndrome" and, 248
  nephrotic syndrome and, 338
  osteoporosis and, 248
  Paget's disease and, 248
  pseudohypoparathyroidism and, 245
  renal calculi and, 24, 239
  renal failure and, 22, 239
  "resistant rickets" and, 249
  sarcoidosis and, 247
  steatorrhoea and, 245, 273
  venous stasis and, 250
  vitamin D and, 237, 244, 261
Calcium preparations, 261, 262
Calcium, protein bound, 235
  raised, 249

Calcium, protein bound, reduced, 253
Calcium, pyrophosphate and pseudo-
    gout, 386
Calcium, urinary, factors affecting, 3, 251
    hypercalcaemia, differential diagnosis
        of and, 251
    renal calculi and, 24, 239
    renal function and, 22, 239
Calculi, biliary, 309
Calculi, renal, 23, 239
    composition of, 24
    cystinuria and, 25, 365
    formation of, 23
    hypercalcaemia and, 24, 239
    hyperparathyroidism and, 243
    investigation of, 29
    treatment of, 25
    xanthine and, 26, 386
Carbohydrate, absorption of, 267
    chemistry of, 179
    digestion of, 267
    oral contraceptives and, 434
    pregnancy and, 434
Carbon dioxide, bicarbonate and, 4, 82
    erythrocytes and, 83
    kidney and, 4, 84
    lungs and, 82
    total plasma, 92
Carbonate dehydratase (see Carbonic
    anhydrase)
Carbonic anhydrase, erythrocyte and, 83
    kidney and, 85
Carcinoembryonic antigen, 450
Carcinoid syndrome, 438
    carcinoma of the bronchus and, 449
    malabsorption and, 272, 440
    nicotinamide deficiency in, 423, 440
Cardiac failure, enzymes in, 354
    oedema in, 53
    renal circulation in, 53
Carotenes, 417
Casts, renal, 20
Catecholamines, 441
CEA, 450
Cerebrospinal fluid, 452
    chloride, 455
    examination, 452
    formation, 452
    glucose, 455
    protein, 453
Chloride, absorption of, 269
    CSF, 455
    depletion, 90
    hydrogen ion homeostasis and, 90
    kidney and, 3
    pyloric stenosis and, 101
    uretic transplantation and, 91, 99
Chlorpromazine, and jaundice, 303, 307
    hypothalamus and, 114
Cholecalciferol, 237

Cholestasis, 295
    diagnosis, 303
Cholesterol, 222, 225
    absorption, 223, 266
    bile salt circulation and, 225, 267
    metabolism, 225
Cholesterol, plasma, diet and, 225
    malabsorption and, 273
    nephrotic syndrome and, 228, 339
    oral contraceptives and, 435
    pregnancy and, 435
    raised, 228, 231
    thyroid disease and, 173, 228
Cholestyramine, 225
Cholinesterase, 355
    liver disease and, 356
Chorionic gonadotrophin, 112, 127, 133,
    431
Chromaffin cells, 441
Chylomicrons, 220, 223
    lipid absorption and, 223, 267
    plasma turbidity and, 223, 231
Circadian rhythms, cortisol, 138, 475
    iron, 393
Cirrhosis, 304
    electrophoretic pattern in, 325
    iron, plasma and, 394, 398, 401
    Wilson's disease and, 373
Cirrhosis, biliary, and jaundice, 296
CK, 348, 353, 354
Clearance, renal, 21
    differential protein, 338
Clinistix, 211, 214
Clinitest, 211, 214
Clomiphene test, 120, 123, 132
Coagulation factors in liver disease, 289,
    299
    vitamin K and, 273, 419
Coeliac disease, 271
Colchicine, 385
Collecting ducts, ADH and, 9
Colloid osmotic pressure, 35
Complement, serum, 322
Congenital adrenal hyperplasia, 153
Congenital hyperbilirubinaemia, 306
Conn's syndrome, 51, 67
Copper in Wilson's disease, 372
Corticosterone, 135
Corticotrophin releasing factor, 138
Cortisol, 135
    physiology of, 135
    stress and, 138, 145
    synthesis of, 138
Cortisol binding globulin, 135
    nephrotic syndrome and, 338
    oral contraceptives and, 434
    pregnancy and, 434
"Cortisol", plasma, 135, 140
    collection of blood for, 157

"Cortisol", plasma, congenital adrenal hyperplasia and, 153
corticosteroid releasing factor and, 138
corticosteroid therapy and, 151
Cushing's syndrome and, 145
hypopituitarism and, 119, 151
nephrotic syndrome and, 338
oral contraceptives and, 434
pregnancy and, 434
Cortisol secretion rate, 146
Cortisol, urinary, 140
Cushing's syndrome and, 145
hypopituitarism and, 121
Cortisone, 136
congenital adrenal hyperplasia, treatment of and, 154
glucose tolerance test, 198
suppression test for hypercalcaemia, 251, 260
Countercurrent, 5
exchange, 7
multiplication. 5
tubular damage and, 14
Creatine kinase, 348, 353, 354
muscle disease and, 354
myocardial infarction and, 353
thyroid disease and, 173
Creatinine, 20
amniotic fluid and, 433
clearance, 21
excretion, 21
kidney and, 4
metabolism of, 20
renal disease and, 16
CRF, 138
Crigler-Najjar syndrome, 306
Cryoglobulinaemia, 336
CSF (see Cerebrospinal fluid), 452
Cushing's syndrome, 144
ACTH in, 144, 148
causes, 144
chemical abnormalities in, 67, 144
diagnosis, 145
tests in, 145
Cystic fibrosis, 272
Cystine and renal calculi, 25, 365
Cystinosis, 366
Cystinuria, 25, 365
Cytochromes, 388

11-Deoxycortisol, 139, 140
Desferrioxamine, 401
Dexamethasone suppression test, 146, 147, 158
Dextrose, cerebral oedema, treatment of, 35
1: 25-DHCC, 1, 234, 237
Diabetes insipidus, 46, 112
hereditary nephrogenic, 46

Diabetes insipidus, unconscious patient and, 47
Diabetes mellitus, 191
cholesterol in, 192, 230
hyperosmolar coma in, 201
insulin levels in, 192
ketosis and, 200
latent, 191
lipoproteins and, 192, 228
maturity onset, 192
pancreatic damage and, 191, 274
treatment, 199, 203
triglycerides in, 192
types, 191
Dialysis, haemo- and peritoneal, 23
"Diamox", 70, 99
Diarrhoea stool, potassium content, 63
sodium content, 44
Dibucaine number, 356
Differential protein clearance, 338
1:25-Dihydroxycholecalciferol, 1, 234, 237
Dihydroxyphenylalanine, 441
Dihydroxyphenylethylamine, 441
Di-iodotyrosine, 162
Disaccharidases, 268
deficiency, 279
DIT, 162
Diuretics, 69
action, 69
potassium loss and, 67, 69
urate retention and, 386
Diurnal rhythms, cortisol, 138, 475
iron, 393, 476
DOPA, 370, 441
DOPamine, 441
"Drip", intravenous, cause of misleading results, 462
Drugs and inherited disease, 373
Dubin-Johnson syndrome, 306
"Dumping" syndrome, 275
Duodenal aspirate, 276
Dwarfism, pituitary, 121
Dyshormonogenesis, thyroid, 171
"Dytac", 70

Ectopic hormone secretion, 50, 443
"Edecrin", 70
EDTA, calcium, plasma effect in vitro, 463
clearance, 21
potassium, plasma effect in vitro, 463
Effective thyroxine ratio, 166
Electrolytes, sweat, 44, 63
fibrocystic disease and, 272
Electrophoresis, immuno, 317
isoenzyme, 348
lipoprotein, 219, 228
patterns, 324
zone, 316

Electrophoresis, of CSF, 455
Embden-Meyerhof pathway, 181
Endocrine adenomatosis, multiple, 207, 281
Endogenous triglycerides, 222
  plasma turbidity and, 231
Enterohepatic circulation of bile salts, 267, 308
Enteropathy, protein losing, 280, 316
Enzymes, duodenal and pancreatic function, 276
Enzyme units, 345
Epinephrine (see also Adrenaline), 441
Erythrocyte sedimentation rate, 316
Erythropoietin, 1, 448
  renal carcinoma and, 448
ESR (see Erythrocyte sedimentation rate), 316
Essential fructosuria, 212
Essential paraproteinaemia, 336
Ethacrynic acid, 70
Ethylene diamine tetra-acetate, 463
  calcium, plasma, effect in vitro, 463
  clearance, 21
  potassium, plasma, effect in vitro, 463
Exophthalmos, 169
Extracellular fluid compartment, 32

Faecal fat, 263, 272
  collection, 285
  split and unsplit, 274
Familial hypophosphataemia, 249
Fanconi syndrome, 14, 67, 365
  cystinosis and, 366
  galactosaemia and, 371
  rickets and, 249
  Wilson's disease and, 372
Fat absorption, 223, 266
Fatty acids, 183, 218, 223
  calcium absorption and, 234, 269, 273
  free plasma, 218
    diabetes mellitus and, 192
    fasting and, 187
    glycogen storage disease and, 210
    hormones and, 183
    ketones and, 183
Ferric chloride test, 215
  phenylketonuria and, 369
Ferritin, 388, 394
  plasma levels, 394, 396
α Fetoprotein, 301
FFA (see Free fatty acids), 183, 218, 223
Fibrinogen, electrophoresis and, 317
  ESR and, 316
  liver disease and, 299
Fibrocystic disease of the pancreas, 272
Fistula fluid, electrolytes in, 48, 63
Flavoproteins and riboflavin, 421
Flocculation tests, 316

Fluid balance, assessment of, 41
  coma and, 47
  normal, 31
Fluoride number, 356
Folic acid, 425
  oral contraceptives and, 435
  pregnancy and, 435
Folic acid antagonists, 379, 425
Follicle stimulating hormone, 112, 122
  menstrual cycle and, 124
Free fatty acids, 183, 218, 223
  diabetes mellitus and, 192
  fasting and, 187
  glycogen storage disease and, 210
  growth hormone and, 115, 187
  insulin and, 185
Free thyroxine index, 166
Fructose, 179
  urinary, 212
  von Gierke's disease and, 210
Fructose intolerance, hereditary, 206, 208, 212
Fructosuria, essential, 212
Frusemide, 70
FSH (see also Follicle stimulating hormone), 112, 122, 124
Functional hypoglycaemia, 205
Furosemide, 70

Galactorrhoea, 117
Galactosaemia, 206, 208, 371
  hypoglycaemia and, 206, 371
Galactose, 179, 371
  urinary, 212
  von Gierke's disease and, 210
Galactose-1-phosphate uridyl transferase, 371
Gallstones, 309
Gastrectomy, "dumping syndrome" and, 275
  malabsorption and, 275
Gastric juice, 44, 63, 100
Gastrin, 280
  plasma levels, 283
  Zollinger-Ellison syndrome and, 281, 283
Gastro-intestinal secretions, potassium content, 63
  sodium content, 44
G-cells, 280, 281
Gerhardt's test, 215
GFR (see also Glomerular filtration rate), 11, 16, 21
Gigantism, 116
Gilbert's disease, 306
Globulins, α, 321
Globulins, β, 322
Globulins, γ, 323
  CSF, 455
  ESR and, 316

**Globulins,** flocculation tests and, 316
  raised, 299, 304, **323**
  reduced, 325, **330**
**Glomerular filtration rate,** Addison's
    disease and, 149
  hydrogen ion homeostasis and, **89**
  measurement of, 21
  reduced, effect of, **11**, 16, 68
**Glomerulonephritis, 14,** 18
**Glomerulus, renal,** disease of, 11, **13**
  nephrotic syndrome and, 14
**Glucagon, 186**
  glycogen storage disease and, 210
  insulinoma and, **207, 216**
**Glucocorticoids, 135**
  Addison's disease and, **149**
  Cushing's syndrome and, **144**
  glucose, blood and, 135, 144, 149, 183,
    187
  hypoadrenalism, secondary and, 150
**Gluconeogenesis, 181**
  Cushing's syndrome and, 144
**Glucose,** cell hydration and, 35, 201
  kidney and, 2
  metabolism, **183**
  osmotic diuresis and, 10, 47
  tubular damage and, 14
**Glucose, blood,** fasting and, 187, 194
  diabetes mellitus and, **194**
  fluoride, *in vitro* effect on, 464
  growth hormone and, 115, 186
  insulin and, **185**
  measurement of, **188**
  newborn and, 208
**Glucose, CSF, 455**
**Glucose oxidase,** 188
  Clinistix and, 211, 214
**Glucose-6-phosphatase,** 180
  von Gierke's disease and, 210
**Glucose-6-phosphate,** 180
  glycogenolysis and, 181
**Glucose-6-phosphate** dehydrogenase,
    182, 374
  haemolytic anaemia and, 182, 374
**Glucose tolerance test, 195, 215**
  acromegaly and, 116
  Addison's disease and, 149
  Cushing's syndrome and, 144
  disaccharidase deficiency and, 279
  indications for, 195
  intravenous, 198
  malabsorption and, 198, 274
  oral contraceptives and, 434
  pregnancy and, 434
  thyrotoxicosis and, 173, 198
  types of curves, 196
**Glucose, "true", 188**
**Glucuronates** in urine, 211
**Glutamate dehydrogenase,** 353

**Glutamate oxaloacetate transaminase,**
    346
  liver disease and, 297, 302, 304, 305,
    354
**Glutamate pyruvate transaminase, 347**
  liver disease and, **297,** 302, 304, 305,
    354
**Glutaminase,** 87
γ **Glutamyltransferase,** 300, **352**
γ **Glutamyltranspeptidase,** 352
**Glutathione reductase,** 422
**Gluten sensitive enteropathy, 271**
**Glycerokinase,** 183
**Glycine,** iminoaciduria and, 365, **367**
  purines and, 380
**Glycogen,** 179, 181
  liver and, 181
**Glycogen storage disease, 210**
**Glycogenesis,** 181
**Glycogenolysis,** 181
**Glycolysis,** 181
α₁ **Glycoprotein,** 321
**Glycosuria, 189,** 211
  Cushing's syndrome and, 144
  diabetes mellitus and, 193. 200
  glomerular filtration rate and, 189,
    193, 200
  phaeochromocytoma and, 442
  stress and, 187
  tests for, 214
  tubular damage and, 14
**Glycosuria, renal, 14,** 198
  glucose tolerance test and, 198
  pregnancy and, 434
**Goitre, euthyroid, 173**
  Graves' disease and, 169
  thyroid dyshormonogenesis and, 171
**Gonadotrophin, chorionic,** 112, 123, **127,**
    **133,** 431
  malignant disease and, 449
  stimulation test, **133**
  trophoblastomas and, 127
**Gonadotrophins, pituitary,** 112, **122**
  clomiphene and, 123, 126, **132**
  hypogonadism and, 125
  hypopituitarism and, 119
  menstrual cycle and, 124
  oral contraceptives and, 436
  ovulation and, 127
  releasing hormone and, 122, 126, 131,
    **132**
**Gout, 383**
  treatment, 384
**Graves' disease, 169**
**Growth hormone,** 112, **115**
  acromegaly and, 116
  bronchial carcinoma and, 449
  control, **115**
  deficiency, **121**
  effects, **115**

Growth hormone, excess, **116**
  fasting and, 187
  glucose, blood and, 115, 183
  glucose ingestion and, 117, **186**
  hypertrophic pulmonary osteo-
    arthropathy and, 449
  hypopituitarism and, 119
  measurement of, 116
  pituitary dwarfism and, 121
  stimulation tests, 130
Growth hormone releasing hormone, 115
GTT (*see* Glucose tolerance test), **195, 215**
Gynaecomastia in malignant disease, **449**

HAA, 300
Haemotocrit, haemoconcentration and
  dilution and, 41
Haemochromatosis, **399**
Haemoconcentration, 41, 315
  calcium, plasma and, 249, 461
Haemodialysis, 23
Haemodilution, 41, 315
  calcium, plasma and, 253
Haemoglobin, 337, 388
  breakdown of, 290
  haemoconcentration and dilution and,
    41
  iron and, 388
  oxygen saturation and, 105
Haemolysis, anaemia and, 394
  effects on results *in vitro*, 345, **464**
  enzymes, plasma and, 345, 355
  G-6-PD deficiency and, 182, 374
  gallstones and, 309
  iron, plasma and, 394
  jaundice and, **292, 304**
Haemosiderin, 388
  urinary, in haemochromatosis, 401
Haemosiderosis, **399**
Halothane and jaundice, 307
Haptoglobin, 322
Hartnup disease, **366**
  nicotamide and, 366, 423
Hashimoto's disease, **171**
HBD, 348, 353
25-HCC, 1, **237**
HCG, 112, 123, **127, 133,** 431
Heavy chain disease, 336
Henderson-Hasselbalch equation, **78**
Henle, loop of, 5, 10
Heparin, clearing effect of, 223
Hepatitis, acute, 294, **301,** 304
  chronic active (aggressive), 295, 302
  chronic persistent, 295, 302
  extrahepatic obstruction, diagnosis
    from, 303
  iron, plasma and, 394
  sample handling, 312
Hepatitis associated antigen, **300**
Hepatocellular failure, **304**

Hepatolenticular degeneration, **372**
Hereditary nephrogenic diabetes
  insipidus, 46
Hexose monophosphate shunt, **181**
Hexose 1-phosphate uridylyltransferase,
  371
HGPRT, 381, 385
5-HIAA, **438**
  oat cell carcinoma of bronchus and,
    449
Hirsutism, **127**
"Histalog", 282, **287**
Histamine, carcinoid syndrome and, 440
  gastric secretion and, 282
Histidinaemia, **371**
HMMA, **442**
Homogentisic acid, 370
  urinary, 211, 370
HPL, **432**
5-HT, **438**
  oat cell carcinoma of bronchus and,
    449
5-HTP, **438**
  oat cell carcinoma of bronchus and,
    449
Human chorionic gonadotrophin, 112,
  123, **127, 133,** 431
Human placental lactogen, **432**
Hydrocortisone suppression test, **251, 260**
Hydrogen ion, bicarbonate and, in
  kidney, 4, **85**
  buffering of, 77, 79, 82, 83
  chloride and, **90**
  erythrocytes and, 83
  fate in body, 79
  GFR and, 12, 18, **89**
  kidney and, 4, **84**
  lungs and, **82**
  metabolic production, 76
  potassium and, 4, **64,** 90
  sodium and, 3, 37, **89**
  tubular damage and, 14, 18, **95**
β Hydroxybutyrate, 93, **189,** 214, 215
Hydroxybutyrate dehydrogenase, 348,
  353
25-Hydroxycholecalciferol, 1, **237**
11-Hydroxycorticosteroids, 140, 145,
  149, **157**
17-Hydroxycorticosteroids (*see* 17-oxo-
  genic steroids, total), 119, **140**
5-Hydroxyindole acetic acid, **438**
  oat cell carcinoma of bronchus and,
    449
3-hydroxykynurenic acid, 423
4-Hydroxy-3-methoxymandelic acid, **442**
17-hydroxyprogesterone, 140, **153**
Hydroxyproline, 367
Hydroxypyrazolopyrimidine, 384
17-Hydroxysteroids (*see* 17-oxogenic
  steroids total), 119, **140**

5-Hydroxytryptamine, 438
    oat cell carcinoma of bronchus and,
        449
5-Hydroxytryptophan, 438
    oat cell carcinoma of bronchus and,
        449
Hyperbilirubinaemia, congenital, 306
Hypercalcaemia, 249
    acute pancreatitis and, 276
    differential diagnosis, 249
    ectopic parathyroid hormone and,
        243
    effects of, 239
    hyperthyroidism, primary and, 242
        tertiary and, 243
    idiopathic of infancy, 247
    malignant disease and, 243, 247, 250,
        445
    myelomatosis and, 246, 248, 334
    pancreatitis and, 276
    renal calculi and, 24, 239, 243
    sarcoidosis and, 247
    steroid suppression test and, 251, 260
    thyrotoxicosis and, 173
    treatment of, 255
    urinary calcium and, 24, 251
    vitamin D and, 247
Hypercalcaemia, idiopathic, of infancy,
    247
Hypercalcuria, idiopathic, 24, 251
    renal calculi and, 24
Hyperchloraemic, acidosis, 90, 99
Hyperchlorhydria, 281, 287
Hypercholesterolaemia, 226
Hyperglycaemia, 187, 192, 194
    hyponatraemia and, 43, 57
    phaechromocytoma and, 442
    stress and, 187
Hyperinsulinism, 205, 207
Hyperkalaemia, causes, 68
Hyperlipoproteinaemia, 227
    effects on plasma osmolality, 36
Hypermagnesaemia, 258
Hypernatraemia, 43, 46, 54, 56, 201
Hyperosmolal coma, 201
Hyperoxaluria, primary, 25, 29
Hyperparathyroidism, primary, 242
    diagnosis, 249
    steroid suppression test and, 251, 260
Hyperparathyroidism, secondary, 243
Hyperparathyroidism, tertiary, 243
    steroid suppression test and, 251, 260
Hyperproteinaemia, 315
    effects on plasma osmolarity, 36
Hyperpyrexia, malignant, 374
Hypertension, congenital adrenal
        hyperplasia and, 153
    Cushing's syndrome and, 144
    metabolic causes, 443
    phaeochromocytoma and, 442

Hyperthyroidism, 168
    calcium in, 173, 239, 247
    ectopic TSH production and, 169,
        449
Hypertrophic pulmonary osteo-
        arthropathy, 449
Hyperuricaemia, 22, 25, 381
Hyperviscosity syndrome, 333, 335
Hypervitaminosis A, 418
Hypoalbuminaemia, causes, 320
Hypocalcaemia, differential diagnosis,
    253
Hypogammaglobulinaemia, 330
Hypoglycaemia, 204
    Addison's disease and, 149, 205
    alcohol induced, 206
    brain and, 183
    causes, 204
    children and, 208
    CSF glucose and, 455
    diagnosis, 207, 209
    "dumping syndrome" and, 275
    fasting, 205
    functional, 205
    galactosaemia and, 206, 208, 371
    glycogen storage disease and, 205, 210
    hepatic disease and, 205, 304
    hereditary fructose intolerance and,
        206
    hyperinsulinism and, 205, 207
    leucine sensitivity and, 206
    malignant disease and, 205, 448
    reactive, 205
    secondary adrenal hypofunction and,
        119, 143
Hypogonadism, 125
Hypokalaemia, causes, 66
Hypomagnesaemia, 257, 262
Hyponatraemia, 43, 49, 54, 56, 409
Hypoparathyroidism, 246
Hypophosphataemia, familial, 249
Hypopituitarism, 118, 150
    diagnosis of, 119, 151
    investigation of, 119, 151
Hypoproteinaemia, protein-losing
        enteropathy and, 280
Hypothalamic-pituitary-adrenal axis,
    136, 141
    corticosteroid therapy and, 151
    tests of, 120, 143
Hypothalamus, ADH and, 39
    pituitary hormones and tests for, 113,
        121, 141
    thirst and, 39
Hypothyroidism, 170, 226
    secondary, 121, 171
Hypouricaemia, 386
Hypoxanthine, 381
Hypoxanthine-guanine phosphoribosyl
        transferase, 381, 385

ICSH, 122
Idiopathic haemochromatosis, 399
  diagnosis, 400
Idiopathic hypercalcaemia of infancy, 247
Idiopathic hypercalcuria, 24, 251
Idiopathic steatorrhoea, 271
IgA, 299, 327
  myelomatosis and, 334
IgD, 328
IgE, 328
IgG, 299, 327
  myelomatosis and, 334
IgM, 299, 328
  macroglobulinaemia and, 335
Ileal secretion, potassium content, 63
  sodium content, 44
Iminoglycinuria, 367
Immunoelectrophoresis, 317
Immunoglobulins, 299, 325
  cerebrospinal fluid and, 455
  deficiency, investigation of, 330
"Inappropriate" hormone secretion, 50, 443
Infertility, male, 126
"Insensible loss", 41
Insulin, 185
  acromegaly and, 116
  diabetes mellitus and, 191
  fasting and, 187
  free fatty acids and, 185, 223
  glucose, blood, and, 185
  glycogen and, 185
  hypoglycaemia and, 204, 205, 207, 209
  leucine and, 185, 206
  malignant disease and, 205, 448
  resistance, 199
Insulin hypoglycaemia test adrenal
    hypofunction and, 130, 143
  gastric secretion and, 282, 287
  growth hormone and, 122
  pituitary function and, 121, 130
Insulinoma, 207
  tests for, 207, 216
Interstitial cell stimulating hormone, 122
Intestinal secretion, acidosis and, 99
  potassium content, 63
  sodium content, 44,
Intravenous infusions, cause of
    misleading results, 462
Intrinsic factor, 268, 425
  pernicious anaemia and, 278, 425
  Schilling test and, 278
Inulin clearance, 21
Iodine, deficiency, and goitre, 173
  and neck uptake, 167
  excess and neck uptake, 168
  and protein-bound iodine, 163, 177
Iodine metabolism, 160
Iron, absorption, 270, 390
  anaemia and, 393, 396

Iron, distribution in body, 388
  in cirrhosis, 394, 401
  in haemochromatosis, 399
  in haemosiderosis, 399
  excretion, 390
  haemochromatosis and, 399
  storage, 388
  transport, 392
Iron binding capacity, 394
  anaemia and, 394, 396, 397
  congenital transferrin deficiency and, 399, 403
  haemochromatosis and, 400
  iron deficiency and, 394
  nephrotic syndrome and, 338
  oral contraceptives and, 394
  percentage saturation, 396
  pregnancy and, 394
Iron overload, 398
  porphyria and, 411
Iron, plasma, anaemia and, 393, 396
  factors affecting, 392, 460
  haemochromatosis and, 400
  nephrotic syndrome and, 338
Iron therapy, 397, 398
  iron overload and, 398, 403
Islet cell hyperplasia, 207
Isocitrate dehydrogenase, 353
  liver disease and, 297, 354
Isoenzymes, 344
  alkaline phosphatase, 350
  lactate dehydrogenase, 348
Isoleucine and maple syrup urine disease, 370
Isomaltase, 268, 279
Isoniazid and pyridoxine deficiency, 423

Jaundice, classification, 292
  drug induced, 306, 307
  haemolytic, 292, 297, 304
  neonatal, 292, 305, 307
  "physiological", 292, 305
  pregnancy, 295

Kallikrein, 440
Kernicterus, 305
Ketoacidosis, 93, 200
  plasma potassium in, 68, 201
Ketonaemia, 189, 201
  tests for, 215
Ketones, 189
  tests for, 214, 215
Ketonuria, 189, 201
  tests for, 214
Ketosis, 93, 189, 200
  acidosis and, 93, 190, 200
  causes, 189
  diabetic, 200
    treatment, 203

Ketosis, fasting, 189
  glycogen storage disease and, 210
  salicylate poisoning and, 103
17-Ketosteroids (see 17-Oxosteroids), 141
Ketostix, 215
Krebs cycle, 181

Lactase, 268
  deficiency, 212, 279
Lactate dehydrogenase, 347
  isoenzymes of, 348
  liver disease and, 297, 354
  myocardial infarction and, 353
Lactic acidosis, 93, 190
Lactose, 179
  urinary, in pregnancy and lactation, 212
"Lag storage curve", 197, 206, 275
Lange colloidal gold reaction, 454
"Lasix", 70
LATS in hyperthyroidism, 169
LCAT, 222
LD, 347
  liver disease and, 297, 354
L Dopa, hypothalamus and, 115
Lead poisoning, porphyrin metabolism and, 413
Lecithin, 218, 226, 433
Lecithin cholesterol acyl transferase, 222
Lecithin-sphingomyelin ratio, 433
Lesch-Nyan syndrome, 385
Leucine, hypoglycaemia and, 206, 209
  insulin secretion and, 185
  maple syrup urine disease and, 370
  sensitivity, 206, 209
  test, 216
LH, 112, 122
  menstrual cycle and, 124
LH-FSH releasing hormone test, 121, 122, 131, 132
Light chain, 326
  Bence Jones protein and, 332
Lipaemia, 227
  effect on plasma osmolarity, 36
Lipase, 266
Lipase, lipoprotein, 223
Lipase, plasma, pancreatic function and, 276
Lipase, triglyceride, 223
Lipid, absorption, 223, 266
  metabolism in liver, 223
Lipids, plasma, analysis, 219
  classification, 219
  glycogen storage disease and, 210
  nephrotic syndrome and, 228
  oral contraceptives and pregnancy and, 435
Lipolysis, 223
Lipoprotein, 219
Lipoprotein lipase, 223

Liver disease, 294
  bile pigments and, 292, 297
  BSP retention and, 298
  enzymes, plasma and, 297, 299, 354
  hypoalbuminaemia and, 299
  immunoglobulins and, 299, 330
  iron, plasma and, 394, 400
  porphyrins, urinary and, 413
  proteins, plasma and, 299, 304, 325
  prothrombin time and, 299, 419
  symptomatic cutaneous hepatic porphyria and, 410
  vitamin K and, 299, 419
Liver function tests, 297
  oral contraceptives and pregnancy and, 435
Liver functions, 289
Long-acting thyroid stimulator in hyperthyroidism, 169
Loop of Henle, 5, 10
  countercurrent multiplication and, 5
Lundh test, 276
Luteinizing hormone, 112, 122, 124
Lysine in cystinuria, 365

Macroamylasaemia, 350
$\alpha_2$ Macroglobulin, 322
Macroglobulinaemia, 335
Magnesium, intestinal absorption, 269
  plasma, 257
  urine, 3
Malabsorption, anaemia in, 274, 275
  calcium and, 245, 253
  causes, 270
  differential diagnosis, 274
  results, 272
Malignancy, ectopic hormone production and, 443
  enzymes, plasma, in, 355
Malignant hyperpyrexia, 374
Maltase, 268
  deficiency, 279
Maltose, 179
Mannitol, osmotic diuresis and, 10
Maple syrup urine disease, 370
Megaloblastic anaemia, 425
  enzymes, plasma in, 355
  iron, plasma in, 393
  malabsorption and, 275, 277
  pernicious anaemia and, 278, 425
Melanin, albinism and, 370
Melanocyte stimulating hormone, 112
  Addison's disease and, 149
Menstrual cycle, 124
  iron, plasma and, 393
Methyltestosterone, jaundice and, 307
Metyrapone (metopyrone) test, 158
  Cushing's syndrome and, 148
  hypopituitarism and, 151
  17-oxogenic steroids and, 148, 158

Micelles, 266
"Midamor", 70
"Milk-alkali syndrome", 248
Mineralocorticoids, 136
  Addison's disease and, 149
  Cushing's syndrome and, 144
MIT, 162
Mitochondrial antibodies in hepatitis, 301
Monoamine oxidases, 438
Monoclonal gammopathy, 331
Monoiodotyrosine, 162
MSH in Addison's disease, 149
Mucoprotein, 321
"Mucosal block" and iron absorption, 390
Mucoviscidosis, 272
Multiple endocrine adenomatosis, 207, 281
Multiple myeloma (see Myelomatosis), 333
Muscle disease, 354
Myelomatosis, 333
  alkaline phosphatase and, 248, 250, 334
  IgA and, 334
Myocardial infarction, plasma enzymes in, 353
Myoglobin, 337, 388
Myopathies, 354
Myxoedema, 171

Natriuretic hormone, 38
Neck uptake tests, 167
  factors affecting, 167, 177
  interpretation, 167, 170, 172
Neomycin and malabsorption, 277
Neonatal period, enzyme levels in, 345
  hypoglycaemia in, 208
  jaundice in, 292, 305, 306, 307
Nephrogenic diabetes insipidus, 46
Nephrotic syndrome, 14, 338
  electrophoretic pattern in, 325
  hypoalbuminaemia and, 320
  iron and, 338
  thyroid function tests and, 166
Neuroblastoma, 443
Neurohypophysis, 112
Neutral fat, 266
Nicotinamide, 422, 440
  in Hartnup disease, 366, 423
Non-esterified fatty acids (see Free fatty acids), 183, 218, 223
Nonne-Apelt reaction, 454
Noradrenaline, 441
Norepinephrine, 441
"Normalised" thyroxine, 166
Novobiocin and jaundice, 308
Nucleic acid and uric acid, 379
5'-Nucleotidase, 350
  liver disease and, 300, 354

Obesity, Cushing's syndrome and, 144, 145
  urinary steroids in, 145
Ochronosis, 370
Oedema aldosteronism, secondary and, 52
  hypoalbuminaemia and, 53
  water excess and, 53
Oestradiol, 123
Oestriol, 123, 431
  pregnancy and, 431
Oestrogens, 123, 124
  total, 123, 431
Oestrone, 123
17-OHCS (see 17-oxogenic steroids, total), 119, 140
Oliguria, acute tubular necrosis and, 18
  chronic renal failure and, 19
  diagnosis, 19
  GFR and, 13
  water depletion and, 13
Oliguric renal failure, acute, 18
  chronic, 19
  treatment, 22
Oncotic pressure, 35
Opsin, 417
Oral contraceptives, carbohydrate metabolism and, 433
  carrier proteins and, 166, 434
  folate and, 435
  gonadotrophin levels and, 436
  iron, plasma and, 393, 435
  jaundice and, 307, 435
  metabolic effects, 435
  TIBC and, 394, 435
Ornithine in cystinuria, 365
Osmolality, 33, 36
  ADH, inappropriate and urinary, 51
  distal tubule and, 9
  kidney and, 4
  loop of Henle and, 5
  oliguria and, 12, 19
  osmotic diuresis and, 9
  plasma, 33
  proximal tubule and, 4
  sodium and plasma, 34, 39
  urinary, 12, 15, 19, 36
  water depletion and plasma, 44, 46
Osmolarity, 35, 36
Osmotic diuresis, 9, 47
  chronic renal failure and, 18
  glucose and, 10, 192
  mannitol and, 10
  sodium reabsorption and, 10
  urea and, 10
Osmotic pressure, 33
Osteoblasts and alkaline phosphatase, 241, 344, 350
Osteomalacia, 241, 244
  anticonvulsant drugs and, 237, 245

**Osteomalacia,** malabsorption and, 245
  renal tubular disorders and, **248**
**Osteoporosis, 248,** 273
  Cushing's syndrome and, 144
  myelomatosis and, 333
  scurvy and, 427
**Ovarian hormones, 123**
**Ovulation,** 124
  detection of, 127
  induction of, 127
**Oxalate** and renal calculi, 25
**Oxidative phosphorylation** and hydrogen
  ion homeostasis, 76, 190
**17-Oxogenic steroids,** total, 119, **140**
  congenital adrenal hyperplasia and,
    154
  Cushing's syndrome and, 146
  hypopituitarism and, 119
  hypothyroidism and, 173
**17-Oxosteroids,** 126, 128, **141**
  congenital adrenal hyperplasia and,
    154
  Cushing's syndrome and, 146
  hirsutism and, 128
  hypogonadism and, 126
  hypopituitarism and, 119
  hypothyroidism and, 173
  virilism and, 128
**Oxygen,** alveolar exchange, 105
  haemoglobin, saturation with, 105
  water solubility, 105
**11-Oxygenation index,** 154
**Oxytocin,** 112

**Paget's disease** of bone, 248, 351
**Palmitic acid,** 218
**Pancreas,** cystic fibrosis of, **272**
  gastrin secretion and, 281
**Pancreatic function,** tests of, **275**
**Pancreatic secretion,** 276
  potassium content of, 63
  sodium content of, 44
**Pancreatitis, acute,** 276, 349
  hypercalcaemia and, 276
  hypocalcaemia and, 245, **277**
  triglycerides and, 230
**Pancreatitis, chronic,** 272, 276
**Pancreozymin,** 276
**Pandy reaction,** 454
**Panhypopituitarism, 118**
  diagnosis, 119
**Pantothenic acid, 424**
**Paraproteinaemia, 331**
  assessment, 332
  benign, **336**
  essential, **336**
**Parathyroid hormone,** 236, 242, 246
  assay, 251, 253
  ectopic production, 243, **445**
  excess, effects of, **240**

**Parathyroid hormone,** "inappropriate"
  secretion, **242**
  osteomalacia and, **241**
  rickets and, **241**
  Vitamin D and, **237**
**P$_{CO_2}$,** hydrogen ion homeostasis
  assessment and, 91, 96, 98
  lung disease and, 106
**Pellagra, 422**
**Penicillamine,** cystinuria and, 26, 366
  Wilson's disease and, 373
**Pentagastrin, 282**
  stimulation test of gastric secretion,
    282, **287**
**Pentose shunt,** 181
**Pentoses,** 179
  urinary, **212**
**Pentosuria,** alimentary, 212
  essential, 212
**Pepsin,** 280
**Peptide** absorption, 268
**Perchlorate,** thyroid iodine uptake and,
    162, 177
**Peritoneal dialysis,** 23
**Pernicious anaemia,** 278, 347, **425**
  gastric secretion and, 278, 282
  vitamin B$_{12}$ and, 278, **425**
**pH, definition,** 77
  estimation in blood, 92
**pH, urinary,** ammonium chloride
  loading and, 95, 109
  renal calculi and, 24
  tubular damage and, 14, 95
**Phaeochromocytoma, 442**
**Phenistix,** 369
**Phenylalanine, blood,** 367, 369
**Phenylalanine hydroxylase,** 367
**Phenylketonuria,** 367
**Phenylpyruvic acid,** 369
**Phosphatase, acid,** 351
  prostatic palpation and, 352
**Phosphatase, alkaline,** 241, 299, **350**
  bony secondaries and, 248, 355
  cirrhosis and, 351
  childhood and, 345
  familial hypophosphataemia and, 249
  heat stable, 350, **432**
  hypercalcaemia and, 250
  hyperparathyroidism and, 243, 351
  hypocalcaemia, diagnosis of and, 253
  hypocalcaemia, treatment of and, 256,
    261
  liver disease and, **299, 354**
  malabsorption and, 253, 273
  myelomatosis and, 248, 250, 334
  5'-nucleotidase and, **350**
  osteomalacia and, **241,** 351
  Paget's disease and, 248, 351
  parathyroid hormone excess and, **241**
  pregnancy and, 350, **432**

Phosphatase, alkaline, renal failure and, 22, 245
  "resistant rickets" and, 249
  rickets and, 241
Phosphate, acromegaly and, 117
  buffering and, 3, 87
  calcium absorption and, 234, 255, 269
  Fanconi syndrome and, 14, 249, 365
  GFR and, 12
  hypercalcaemia, treatment of and, 255, 260
  kidney and, 3, 87, 249
  osteoporosis and, 248
  parathyroid hormone and, 236, 241, 244
  renal failure and, 16, 237, 244
  "resistant rickets" and, 249
  tubular damage and, 14, 249
  vitamin D and, 3, 237
Phosphaturia, "resistant rickets" and, 248
  tubular damage and, 14, 248
Phospholipids, 218, 226
  fat absorption and, 266
  metabolism, 226
Phosphoribosyl pyrophosphate in purine metabolism, 379
Phytate and calcium absorption, 234, 245, 269
PIF, 117
Pigmentation, ACTH and, 138, 149
  Addison's disease and, 149
  haemochromatosis and, 399
  hypopituitarism and, 119
Pitressin test, 28
Placenta, alkaline phosphatase and, 350, 432
Plasma cell myeloma, 333
  alkaline phosphatase and, 248, 250, 334
Plasmacytoma, 333
P$_{O_2}$, lung disease and, 105
Polyclonal response, 331
Polycythaemia, and renal carcinoma, 448
Polyuria, chronic renal failure and, 18
  hypercalcaemia and, 239
  osmotic diuresis and, 47, 192
  tubular damage and, 14
Polyvinyl pyrrolidone, 280
Porphobilinogen, 406, 409, 410
  test for, 407, 415
Porphyria, 408
Posture, effect on proteins, 321, 461
Potassium, acidosis and, 65
  Addison's disease and, 149
  aldosteronism, primary and, 52
    secondary and, 56
  alkalosis, extracellular and, 52, 65
  body content, 62

Potassium, body fluid, concentration, 32, 62
  cellular, 32, 62
  Cushing's syndrome and, 67, 145
  depletion, causes, 66
  diabetes mellitus and, 68, 201
  distribution of, 32
  EDTA, in vitro effect, 463
  effect of keeping blood on, 464
  GFR and, 12, 63
  haemolysis in vitro and, 464
  hydrogen ion and, 4, 64, 90
  hydrogen ion homeostasis and, 65, 90
  hypercalcaemia and, 261
  intestinal absorption of, 63, 269
  kidney and, 3, 63
  preparations containing, 73
  renal failure and, 16, 68
  sodium ion and, 3, 63
  treatment of disturbances, 70
  tubular damage and, 14
Potassium ferrocyanide, iron and, 388, 397
Prednisolone test, 303, 312
Pregnancy, carbohydrate metabolism and, 433
  carrier proteins and, 166, 434
  cholestasis and, 295, 435
  enzymes, plasma and, 345
  folate, plasma and, 435
  iron, plasma and, 393, 435
  monitoring of, 431
  reducing substances, urinary and, 212
  TIBC and, 394, 435
Pregnanediol, 124
Pregnanetriol in congenital adrenal hyperplasia, 154
Pregnenolone, 138
Probenecid, 384
Progesterone, 124, 127
  menstrual cycle and, 124, 127
Proinsulin, 185
Prolactin, 117
Prolactin-release inhibiting factor, 117
Proline, 367
Prostaglandins, 440
Protein, Bence Jones, 332, 334, 335, 337, 342
Protein-bound iodine, 163
  interfering substances and, 163, 177, 460
  oral contraceptives and pregnancy, 434
Protein, CSF, 453
Protein-losing enteropathy, 280, 316
  hypoalbuminaemia and, 280, 320
Proteins, adrenocortical hormones and, 135
  buffering and, 81, 82
  Cushing's syndrome and, 144

**Proteins,** growth hormone and, 115
  insulin and, 185
  liver and, 289, 299
  X and Y, 290
**Proteins, plasma,** aldosteronism,
  secondary and, 53
  cirrhosis and, 304, 325
  fractionation of, **316**
  functions of, 314
  hydration and, 41, 314
  liver disease and, 299, 304, 325
  nephrotic syndrome and, **325,** 338
  total, value of, 315
  venous stasis and, 315, **461**
  water, distribution of, 35, 314
**Proteinuria,** 20, 337
  Bence Jones, **332,** 334, 335
  nephrotic syndrome and, **338**
  orthostatic (postural), 337
  tests for, **342**
  tubular disease and, 15, 337
**Prothrombin time,** liver disease and,
  299, 419
  malabsorption and, 273, 419
**Protoporphyria, erythrohepatic, 411**
**Prussian blue** reaction, 388
**Pseudo Addison's disease,** 48
**Pseudocholinesterase, 356**
**Pseudogout, 386**
**Pseudohermaphroditism,** 153
**"Pseudohyperparathyroidism",** 242
**Pseudohypoparathyroidism,** 245
**Pulmonary embolism,** enzyme changes
  in, 354
**Purine,** fate of, **381**
  folic acid and, 379
  gout and, 384
  metabolism of, 379
  synthesis of, 379
**PVP,** 280
**Pyelonephritis,** 16
**Pyloric** stenosis, 90, **100**
**Pyrexia** and water loss, 47
**Pyridoxine, 423**
  iron, plasma and, 393
  responsive anaemia, 393, 397, 423
**Pyruvate,** 181, 421
  thiamine and, 421

**Radioiodine uptake tests,** 167
  factors affecting, 167, 177, 460
  interpretation, 167, 170, 172
**Reducing substances in urine,** 210
  alkaptonuria and, 370
  galactosaemia and, 371
  tests for, **214**
**Releasing hormones,** 113
**Renal blood flow** and sodium excretion,
  38
**Renal calculi, 23,** 239, 243

**Renal circulatory insufficiency, 13**
  Addison's disease and, 49
  tubular damage and, 15
**Renal concentration tests,** 20, **28**
**Renal failure,** acidosis and, 12, 14, 18,
  **95**
  acute, 18
  anaemia and, 22
  calcium and, 22, **245**
  causes of, 18
  chronic, 18
  enzymes, plasma and, 277, 348, 349
  hyperuricaemia and, 22, **386**
  iron, plasma and, 393
  phosphate and, 22, 245
  treatment of, 22
**Renal function, investigation of, 19**
  "Renal threshold", 3, 189, 193
**Renal tubular acidosis,** 95, 235
  potassium levels and, 67
**Renal tubule,** acidosis and, 14, **95**
  ADH and, 7
  amino acids and, 2, 364
  bicarbonate and, 4
  casts and, 15
  chloride and, 3
  creatinine and, 4, 21
  damage, effects of, 14
    causes of, **15**
  familial hypophosphataemia and, 248
  GFR and, 15
  glucose and, 2, 198
  hypokalaemia, 14, **67**
  hypophosphataemia and, 14, 248
  pH urinary and, 4, 95
  phosphate and, 3, 248
  polyuria and, 14
  potassium and, **3, 63**
  proteinuria and, 15
  sodium and, 3
  sodium, urinary and, 56
  urea and, 4, 21
  urea, urinary and, 15
  water and, **4**
**Renin,** 37
  aldosteronism and, **52**
  renal blood flow and, 37
  stimulus to, 37
**Resin uptake tests, 164**
  nephrotic syndrome and, 166
  oral contraceptives and, 166
  pregnancy and, 166
**"Resistant rickets",** 249
**Respirators, fluid balance and,** 47
**Rhodopsin,** 417
**Riboflavin,** 421
**Rickets,** 241, **244**
  renal tubular disorders and, **248**
  "resistant", 249
**Rothera's test, 214**

S value, 327
Salicylates, Gerhardt's test and, 215
  hydrogen ion homeostasis and, 103
  uric acid excretion and, 384
"Salt losing" syndrome, 153
Sarcoidosis, calcium in, 247
  proteins in, 323
Schilling test, 278
Scurvy, 427, 430
Secretin, 276
Sequestrene, calcium plasma, effect on
    *in vitro*, 463
  clearance, 21
  potassium, plasma, effect on *in vitro*,
    463
Serotonin, 438
S$_f$ units, 220
SGOT, 346
  liver disease and, 297, 302, 304, 305,
    354
SGPT, 347
  liver disease and, 297, 302, 304, 305,
    354
SHBD, 348
Sideroblastic anaemia, 393
Siderophilin (*see also* TIBC), 392
Siderosis, Bantu, 402
Small intestinal secretion, potassium
    content, 63
  sodium content, 44
SML classification of lipids, 220
Smooth muscle antibodies in hepatitis,
    301
Sodium, absorption of, 31, 269
  Addison's disease and, 49, 149
  ADH and, 39
  aldosterone and, 3, 37
  aldosteronism, primary and, 51
    secondary and, 52
  balance, assessment of, 42
  blood volume and, 38
  body content, 30
  cell hydration and, 34
  concentration changes, 43, 56
  concentration in body fluids, 44
  control of, 36
  Cushing's syndrome and, 144
  depletion, 48
  distribution of, 32
  effect of change in concentration, 34,
    43
  excess, 51
  excretion, 31
  faecal, 31, 44
  GFR and urinary, 38
  hydrogen ion and, 3, 89
  hydrogen ion homeostasis and, 89
  hypopituitarism and, 150
  kidney and, 3, 31
  natriuretic hormone and, 38

Sodium, osmolality, plasma and, 34, 201
  osmotic diuresis and, 10
  plasma, 43, 44, 56
  potassium ion and, 3, 37, 63
  preparations containing, 59
  sweat, 44, 272
  treatment of disturbances, 34, 57
  tubular damage and, 14
  urinary, 12, 14, 56
  water metabolism and, 40
Somatomedin, 115
Sphingomyelin, 218, 433
Spironolactone, 70
Sprue, 271
Starch, 179
Stearic acid, 218
Steatorrhoea, 272
  abetalipoproteinaemia and, 222
  causes of, 270
  diagnosis, 274
  effects of, 272
Stercobilin, 290
Stercobilinogen, 290
Steroid suppression test, 251, 260
Steroid therapy, 151
  hypokalaemia and, 67
Stimulation tests, principles, 111
Stones, renal (*see also* Calculi, renal), 23,
    239, 243
Stress, cortisol and, 138, 143, 145
  glucose and, 187
Sucrase, 268
  deficiency, 279
Sucrose, 179
Suppression tests, principles, 111
Suxamethonium sensitivity, 355
Svedberg coefficient, 327
Svedbergh flotation units, 220
Sweat, potassium content, 63
  sodium content, 44, 272
  water loss and, 31, 47
Synacthen, 138
Synacthen stimulation test, 143, 157
  Addison's disease and, 150, 158

Tangier disease, 222
TCA cycle, 181
Technetium uptake test, 167
Testicular failure, diagnosis, 126
Testis, 124
Testosterone, 124
  hirsutism and, 127
  virilism and, 128
Tetracosactrin stimulation tests, 138,
    143, 150, 157, 158
Thalassaemia, plasma iron and, 393, 397
Thiamine, 420
Thiocyanate and thyroidal radioiodine
    uptake, 177
Thiouracil and iodine binding, 162, 168

"Third factor", 38
Thirst, control of, 39
  water depletion and, 44
Thyrocalcitonin, 160, 239
Thyroglobulin, 162
Thyroid adenoma and thyrotoxicosis, 169
Thyroid antibodies, 171, 173
Thyroid dyshormonogenesis, 171
  hypothyroidism and, 171
Thyroid function tests, 163
  interfering substances, 168, 177, 460
Thyroid hormones, 160
  bone, action on, 173, 239
  ingestion, and hyperthyroidism, 169
  synthesis, 160
Thyroiditis, 171
Thyroid stimulating hormone (see TSH), 166
Thyrotoxicosis, 168
  hypercalcaemia and, 173, 247
  triiodothyronine and, 170
Thyrotrophic hormone (see TSH), 166
Thyrotrophin releasing factor, 131, 162, 170, 172, 177
Thyroxine, 162
  circulating, 162
  synthesis, 162
Thyroxine binding globulin, 162
  congenital deficiency, 166
  drugs and, 166
  nephrotic syndrome and, 166, 338
  oral contraceptives and, 166, 434
  pregnancy and, 166, 434
  protein bound iodine and, 165
  resin uptake tests and, 164
Thyroxine, free, 162, 166
  control of TSH and, 163
  index, 166
Thyroxine, plasma, 162
  nephrotic syndrome and, 166, 338
  "normalised", 166
  oral contraceptive therapy and, 166, 434
  pregnancy and, 166, 434
TIBC, 394
  congenital transferrin deficiency and, 399, 403
  haemochromatosis and, 400
  iron deficiency and, 394
  nephrotic syndrome and, 338
  oral contraceptives and, 394, 435
  percentage saturation of, 396
  pregnancy and, 394, 435
$T_m$, 3
Tocopherols, 419
Tolbutamide test, 207, 217
Tophi, gouty, 383
Total iron binding capacity (see TIBC), 394
Transaminases, 297, 346

Transaminases, liver disease and, 297, 302, 304, 305, 354
  myocardial infarction and, 353
Transferrin (see TIBC), 394
Transketolase, 421
Transplantation of the ureters, acidosis and, 91, 99
TRF, 131, 162, 170, 172, 177
TRH, 131, 162, 170, 172
  stimulation test, 131, 177
Tricarboxylic acid cycle, 181
Triglycerides, 218, 222, 228
  oral contraceptives and pregnancy and, 435
Tri-iodothyronine, 163
  circulating, 163
  synthesis of, 162
  thyrotoxicosis and, 170
Tri-iodothyronine suppression test, 170
Trophoblastic tumours, 127, 169
Tropical sprue, 271
Trypsin, 268
  faecal, pancreatic function and, 276
Tryptophan, 366
  carcinoid syndrome and, 438
  Hartnup disease and, 366
  nicotinamide synthesis and, 366, 438
TSH, 112, 162, 166, 170
  ectopic production, 169, 449
  hypopituitarism and, 118
  thyroid hormone synthesis and, 162
TSH stimulation test, 172
  hypopituitarism and, 121
$T_3$ suppression test, 170
$T_3$ thyrotoxicosis, 170
Tyrosinase, 370
Tyrosine, 367
UDP glucuronyl transferase, 290
Ultracentrifugation, 220, 318
  lipoproteins and, 220
Uraemia, glomerular disease and, 11
  hyponatraemia and, 43, 57
  iron, plasma and, 393
  tubular disease, 15
  water deficiency and, 13
Urate, 379
  GFR and, 12, 386
  metabolism of, 379
  renal calculi and, 25
  renal failure and, 16, 386
Urea, Addison's disease and, 149
  ADH, inappropriate and, 447
  cell hydration and, 34
  cerebral oedema and, 35
  clearance, 21
  GFR and, 11
  kidney and, 4
  liver disease and, 304
  metabolism, 19
  osmotic diuresis and, 10, 47

Urea, renal disease and, 17
  tubular damage and, 15
Ureteric transplantation, acidosis and,
  91, 99
Uric Acid, 379
  GFR and, 12, 386
  metabolism of, 379
  renal calculi and, 25
  renal failure and, 16, 386
Uricolysis, 381
Uricosuric drugs, 384
Uridyl diphosphate glucuronyl
  transferase, 290
Urine acidification test, 95, 109
Urine concentration test, 20, 28
Urine, timed collection of, 465
Urobilin, 291
  urinary, 301, 302, 305, 313
Urobilinogen, 291
  urinary, 301, 302, 305, 313

Valine and maple syrup urine disease,
  370
Van den Bergh reaction, 291
Vanillyl mandelic acid, 442
Vasa recta, countercurrent exchange
  and, 7
Villi, intestinal, 264
  steatorrhoea and, 271
Virilisation, congenital adrenal
  hyperplasia and, 153
  Cushing's syndrome and, 145
Virilism, 127
Visual purple, 417
Vitamin A, 417
  steatorrhoea and, 273, 418
Vitamin B₁, 420
Vitamin B₂, 420
Vitamin B₆, 420
Vitamin B₁₂, 425
  absorption, 268
  clearance, 21
  intestinal disease and, 275, 277, 425
  liver and, 289
  oral contraceptives and, 435
  pernicious anaemia and, 278, 425
Vitamin C, 426
Vitamin D, 237, 419
  calcium absorption and, 234, 269
  deficiency, causes of, 244
    effects of, 241
  excess, effects of, 246
  hypocalcaemia, therapy of, 256, 261
  idiopathic hypercalcaemia of infancy
    and, 247
  kidney and, 1, 3, 237, 245
  liver and, 237
  parathyroid hormone and, 237
  renal failure and, 237, 245
  "resistant rickets" and, 249

Vitamin D, sarcoidosis and, 247
  steatorrhoea and, 245, 272
  therapy, 256, 261
Vitamin E, 419
Vitamin K, 289, 419
  liver disease and, 299
  malabsorption and, 273
Vitamins, fat soluble, 416
  absorption of, 267
Vitamins, water soluble, 420
  absorption of, 269
VMA, 442
Von Gierke's disease, 210

Waldenström's macroglobulinaemia, 335
Water, absorption of, 269
  Addison's disease and, 149
  ADH and, 9
  balance, 30
  body content of, 30
  Cushing's syndrome and, 144
  depletion, effects of, 46
  distribution of, 32
  disturbances of metabolism, 44, 50
    treatment, 57
  excess, 50
  expired air and, 31
  faecal, 31
  kidney and, 4
  osmotic diuresis and, 9, 200
  sodium and, 40
  sweat and, 31
  tubular damage and, 14
Water load, 9
Water restriction, 9
Weight, body, assessing hydration and,
  42
Wilson's disease, 372

Xanthine, 381
  renal calculi and, 26
  xanthinuria and, 386
Xanthine oxidase, 381
  allopurinol and, 384
  uric acid and, 381
  xanthinuria and, 386
Xanthinuria, 386
Xanthochromia, 453
Xanthomatosis, 228, 230
Xathurenic acid, 423
Xerophthalmia, 418
Xerosis conjunctivae, 418
X protein, 290
Xylose absorption, 274, 286

Y protein, 290

Zollinger-Ellison syndrome, 281
  acid secretion and, 281, 287
  gastrin and, 281, 283
Zone electrophoresis, 316